SOIL BIOCHEMISTRY

Volume 8

BOOKS IN SOILS, PLANTS, AND THE ENVIRONMENT

SOIL BIOCHEMISTRY

Volume 8

edited by

JEAN-MARC BOLLAG
Center for Bioremediation and Detoxification
Environmental Resources Research Institute
The Pennsylvania State University
University Park, Pennsylvania

G. STOTZKY
Department of Biology
New York University
New York, New York

CRC Press
Taylor & Francis Group
Boca Raton London New York

CRC Press is an imprint of the
Taylor & Francis Group, an **informa** business

CRC Press
Taylor & Francis Group
6000 Broken Sound Parkway NW, Suite 300
Boca Raton, FL 33487-2742

First issued in paperback 2019

ISBN-13: 978-0-8247-9044-8 (hbk)
ISBN-13: 978-0-367-40236-5 (pbk)

Library of Congress Cataloging-in-Publication Data

(Revised for vol. 8)

Soil Biochemistry

    (V. 2: Books in soil science) (v. 3-5: Books in soils and environment) (v. 6-   : Books in soils, plants, and the environment)
    Vol. 2-   edited by A. Douglas McLaren and others.
    Vol. 8-   edited by Jean-Marc Bollag, G. Stotzky.
    1. Soil Biochemistry. I. McLaren, Arthur Douglas.
II. Peterson, George H. III. Series.
IV. Series: Books in soil science. V. Series: Books in soils and the environment. VI. Series: Books in soils, plants, and the environment.
S592.7.S64   631.4'17   66-27705

Visit the Taylor & Francis Web site at
http://www.taylorandfrancis.com

and the CRC Press Web site at
http://www.crcpress.com

Preface

The quality of the soil environment is a continuing major concern, as pollution is widespread and a dominant problem affecting natural ecosystems and endangering efficient agricultural production. These problems can be solved only through a better understanding of the basic reactions, most of them biochemical, that occur in the environment. The current trend to reduce the effects of pollution and to improve soil quality is the application of bioremediation procedures. To be successful in these endeavors, more basic studies in soil biochemistry are necessary to enable the acquisition of a thorough understanding of the processes that occur in nature.

This volume of *Soil Biochemistry*, which again covers a wide variety of topics, as soil biochemistry is an interdisciplinary science, addresses certain aspects of bioremediation. One approach to bioremediation is the introduction to the environment of microorganisms with specific catabolic capabilities. However, it is difficult to predict and monitor the fate and effects of these organisms and, consequently, the success of the bioremediation process. Two chapters describe traditional as well as molecular and immunological techniques for detecting specific microorganisms in soil. The fate of introduced genetically modified organisms is particularly discussed, as this requires the use of newer research tools, such as nucleic acid probes and antibodies that recognize specific microbial antigens.

One chapter evaluates the use of a white-rot fungus, *Phanerochaete chrysosporium*, for bioremediative purposes in soil and discusses current results of this application. A major problem with this and other methods is competition by the indigenous microbial populations with the introduced organisms—a problem whose study and resolution in such a heterogeneous system as soil are complex and difficult.

The interaction of xenobiotics, such as pesticides, with soil organisms is discussed in several chapters. Some authors emphasize the interactions with specific organisms or enzymes, whereas other authors emphasize generic microbial metabolism and degradation pathways. Inhibition of the nitrification process by allelochemicals released by plants is discussed, and doubts are expressed that the various hypotheses proposed to explain this effect of allelochemicals have sufficient support in experimental studies. As in previous volumes, microbial mineralization of various compounds under anaerobic conditions is considered, and its importance in the global carbon cycle is stressed. One chapter discusses the formation of soil organic matter, particularly in forest soils, noting that the humification process is affected by the concentration of available microbial and plant products, as well as by environmental factors. This chapter also emphasizes that CPMAS ^{13}C-NMR spectroscopy can be a major analytical technique with which to determine the chemicals or chemical groups involved in the humification process.

We extend our sincere thanks to the authors who contributed to this volume. We encourage other experts in the field of soil biochemistry, especially younger investigators with new concepts and techniques, to contribute to future volumes.

Jean-Marc Bollag
G. Stotzky

Contents

Contributors

John M. Bremner Department of Agronomy, Iowa State University, Ames, Iowa

John A. Bumpus Department of Chemistry and Biochemistry, University of Notre Dame, Notre Dame, Indiana

Richard G. Burns Biological Laboratory, University of Kent, Canterbury, Kent, United Kingdom

Jonathan P. Carter School of Biological Sciences, University of East Anglia, Norwich, Norfolk, United Kingdom

Henry L. Ehrlich Department of Biology, Rensselaer Polytechnic Institute, Troy, New York

Dennis D. Focht Department of Soil and Environmental Sciences, University of California, Riverside, California

Ingrid Kögel-Knabner Soil Science Group, University of Bochum, Bochum, Germany

James M. Lynch Horticulture Research International, Littlehampton, West Sussex, United Kingdom

Gregory W. McCarty Environmental Chemistry Laboratory, U.S. Department of Agriculture, Beltsville, Maryland

Andreas Schäffer Departments of Plant Protection, Product Safety and Ecochemistry, CIBA-GEIGY Ltd., Basel, Switzerland

Kadiyala Venkateswarlu   Department of Microbiology, Sri Krishnadevaraya University, Anantapur, India

Caroline S. Young   Biological Laboratory, University of Kent,
Canterbury, Kent, United Kingdom

SOIL BIOCHEMISTRY

Volume 8

1

Detection, Survival, and Activity of Bacteria Added to Soil

CAROLINE S. YOUNG and RICHARD G. BURNS *University of Kent, Canterbury, Kent, United Kingdom*

I. INTRODUCTION

The microbial world appears to have an almost limitless metabolic potential, and as comprehension of this has increased so has the desire to use the enormous biochemical diversity to advantage. Advances in molecular biology, fermentation technology, and microbial physiology, coupled with the search for and discovery of microorganisms with novel metabolic capabilities, have presented numerous opportunities. Microbial biotechnology is already being exploited in the chemical and pharmaceutical industries, where commercial-scale bioreactors produce antibiotics, insulin, growth hormones, amino acids, proteins, enzymes, vitamins, hemoglobins, and a host of fragrances and flavorings. In addition, the large-scale generation of microbial biomass supplies food for human beings and domestic animals.

Despite many exciting commercial developments and no shortage of imaginative ideas, microorganisms have not yet been applied widely to the environment outside of the laboratory or away from the contained and carefully monitored industrial reactor. This is partly because soil is a heterogeneous environment that cannot be manipulated easily to ensure the success of a potential inoculant. As a consequence, the activity and persistence of soil microbial amendments are difficult to predict. In addition to these constraints, a major public debate is taking place concerning the

safe use of microbes in the environment. This debate is focused
on a number of issues, including perceived human health hazards
associated with release [1], the potential disruption to the indigenous flora and fauna as a result of the successful establishment of
an introduced species [2], and ethical concerns regarding the manipulation of the natural environment. The arguments become more
passionate when the deliberate release of genetically modified microorganisms is advocated rather than that of naturally occurring species [3].

Intentional or unintentional introduction of microorganisms to new
environments, via natural or anthropogenic dispersal, has been occurring since the beginning of time (e.g., sewage applications to
soil, transport of microorganisms over long distances by air and
water currents and by animals). This is an important point to
bear in mind when risk assessments are made concerning novel
inoculants. Unfortunately, there is a dearth of knowledge concerning the effects of allochthonous microorganisms on the microbial ecology of soil, and the best methods for measuring these effects have not been decided. In other words, soil microbiologists
are largely ignorant of the multispecies dynamics that occur in
natural soil microbial communities, and there are few standardized
techniques for counting introduced species or assessing their activities. As a consequence, the capacity to select and manipulate
microorganisms in vitro has far outstripped the ability to forecast
their behavior when added to soil. This gulf between the potential
of microorganisms to enhance soil processes and the information that
would allow confident introduction of bacteria, fungi, algae, and viruses to soil is well recognized [4]. In an attempt to bridge this
gulf, dozens of major research programs are under way throughout the world involving industry, academia, and government insti-
tutes, all aimed at the safe, predictable, and profitable exploitation
of soil microbial inoculants.

A. Examples of Soil Inoculants

As the 21st century approaches and the world's population contin-
ues to expand at an alarming rate, it will be necessary to consider
novel ways to increase food production while, at the same time,
preserving the environment. To achieve these objectives, the
widespread and routine introduction of beneficial microorganisms
into soil will probably have an important role. This is not a new
idea, but the potential applications of microbial inoculants have increased in recent years and many of these uses are concerned with
the maintenance and improvement of soil fertility and the promotion
of plant growth [5]. The benefits may result from the involvement
of inoculants in fundamental chemical and biological processes that

are essential to plant nutrition (e.g., release of nitrate, solubilization of phosphate), the enhancement of nodulation and nitrogen fixation (e.g., effective rhizobial strains, the production of plant growth-promoting agents (e.g., auxins, cytokinins), the initiation and retention of soil aggregate stability (e.g., polysaccharide production, filamentous growth), or the suppression of plant pathogens by antagonistic metabolites (e.g., enzymes, antibiotics).

Nitrogen Fixation

For the past 70 years, the most widely adopted use of microbial inoculants has involved rhizobia and legumes. The first patents for rhizobial inoculants were issued at the beginning of the century, and commercial exploitation followed soon after. Given the importance of legumes as edible seeds and sources of protein, oils, animal fodder, and green manure, many of these early bacterial formulations sold well. The demand existed then, as it does today, because nodulation does not occur in soils in the absence of appropriate *Rhizobium* species. Cultures were mixed with seed before sowing or, less efficiently, were poured onto the soil at planting. Very often these preparations contained very low numbers of viable rhizobia (because of losses during storage and preparation by the farmer) or did not survive long enough, either in soil or on the seed, to infect the plant. As a result, low success rates were common (as measured by nodulation of the host legume) and applications of nitrogen fertilizer were still required. Many of these problems still exist today, and much recent research is aimed at selecting or genetically manipulating *Rhizobium* to produce "improved" strains which will outcomplete indigenous rhizobia and form effective nodules in a high proportion of the crop [6]. In addition, application techniques are being modified to deliver a high viable inoculum at seed germination [7]. Many of the ecological constraints limiting the success of rhizobia added to soil are described in detail elsewhere [8] and are relevant to the understanding of other bacterial inoculants.

It is often assumed that enhancement of plant growth after inoculation is a direct response to the introduced bacteria. In the case of rhizobia, the appearance of nodules, coupled with sensitive measurements of nitrogen fixation using ^{15}N and acetylene reduction techniques, permits unequivocal correlation of plant response with the inoculant. However, there is a danger of confusing cause and effect when assessing the function of other microbial inoculants. An example is provided by the many experiments in which soil has been amended with a free-living nitrogen-fixing genus *Azotobacter*. Improvements in germination, plant growth, or yield after soil and seed treatments (bacterization) with *Azotobacter chromococcum* have

been attributed solely to N fixation [9], although there are many
alternative explanations [10]. These include the direct physical
protection from soil pathogens or plant growth inhibitors resulting
from a dense initial seed coating of the inoculant, an indirect disruption of the developing rhizosphere population away from "harmful" and toward "beneficial" bacteria, a "green manuring" effect
caused by the addition of large amounts of microbial biomass and
the subsequent release of plant nutrients as the cells are lysed
and degraded, and the production of plant growth regulators by
the inoculant species. The significance of microbial metabolites as
growth promoters and the caution that should be exercised when
interpreting a plant response are illustrated by the work of Frankenberger and colleagues [11–13]. They showed an increase in
growth of *Zea mays* and *Lycopersicon esculentum* after the addition
of L-methionine and L-ethionine to soil, possibly as a result of the
microbial formation of ethylene from the amino acid precursors.
Furthermore, radish yields were enhanced in glasshouse and field
experiments after application of the auxin precursor, L-tryptophan.
However, even these data are not conclusive, as the increase in
plant growth could have been a response to auxins produced by
the soil microbiota, a change in the rhizosphere population, or
even something nonmicrobiological, such as direct uptake of L-
tryptophan by the plant and its subsequent metabolism within plant
tissues.

A number of nonleguminous plant roots are colonized by diazotrophs and there have been numerous attempts to enhance this association. *Azospirillum* species, in particular, have received attention [14], because their intimate association with the rhizosphere of
many agriculturally important cereals suggests that fixed nitrogen
may be readily accessible to the plant. However, once again the
direct contribution of nitrogen fixation to increased crop yields has
been questioned and growth regulators have been implicated [15].

Inoculating rice soils with cyanobacteria has long been a standard practice in India, even though results have generally been inconsistent [16]. In contrast, there are comparatively few studies
of the colonization of cyanobacteria in temperate soils and, in those
that exist, the increased nitrogen fixation measured is not agron-
omically significant [17]. Extensive field experiments, with a large
number of cyanobacterial species [18], revealed that *Nostoc* and
Calothrix species become dominant but that their value in agricultural soils is likely to be limited to those that are irrigated. However, if in addition to nitrogen fixation cyanobacteria can be shown
to bring about an improvement in soil structure and promote an increase in total microbial biomass and soil organic matter [19], then
the overall contribution of phototrophic inocula in soil fertility may
be significant. Certainly microalgae, such as *Chlamydomonas* and

Scenedesmus species, have been touted as "soil conditioners" on many occasions [20–22].

Phosphate Solubilization

After nitrogen, phosphorus is the most important limiting nutrient in agriculture. It has been known since the 1920s that some microorganisms can solubilize phosphate from inorganic and organic phosphates using acidic metabolites or phosphatases. Field experiments in Russia and Eastern Europe were carried out in the 1950s with phosphatase-positive strains of *Bacillus megaterium* as seed inoculants, but the results were inconclusive because of inadequacies in experimental design and techniques. In any case, *B. megaterium* may not be an ideal choice of inoculant, because despite its phospholytic nature, it is likely that a high proportion of the inoculant will sporulate and, therefore, become dormant in response to the shock of being introduced into soil. The application of phosphate-solubilizing microbes has been studied with regard to their potential in tropical soils [23,24], in which available phosphate is often below the levels required by the crop. Establishment of the introduced species and their performance were severely reduced by high temperatures and desiccation. However, Gaind and Gaur [25] evaluated some thermophilic *Bacillus* species and, encouragingly, found that they could solubilize tricalcium phosphate at 30 to 45°C. Furthermore, mung bean seeds inoculated with *Bacillus* species showed improved nodulation, as well as higher straw and grain yields, compared to uninoculated controls. The beneficial effect of enhanced phosphate uptake on nodulation has also been demonstrated with co-inoculation of mycorrhizae and rhizobia [26], although improved phosphate capture rather than solubilization may account for this.

Bioremediation

A need to reclaim polluted soil for agricultural, recreational, and commercial exploitation has prompted fundamental and applied research and development in soil bioremediation. Some successes using microbes for the degradation of pesticides and pollutants in soil have been reported. Acclimation, directed mutation, or genetic manipulation of bacteria has resulted in a large number of strains capable of degrading recalcitrant xenobiotic chemicals. In general, these strains have been assessed in vitro [27–29], and comparatively little work has been conducted with regard to their survival and expression of activity in the natural environment [30–32]. In contrast to the conditions adopted in vitro, soil can be regarded as a suboptimal environment for microbial growth and clearly the potential use of natural or recombinant bacteria for bioremediation of soils (or any other purpose) depends on the survival of the inoculants and the stability of the relevant genetic information. In addition to the usual physical, chemical, and biological barriers to colonization by the inoculant, the concentration of the target contaminant may be inadequate to support microbial growth. Furthermore, even when its concentration is sufficient, sorption and the uneven distribution of the pollutant will reduce its bioavailability. These factors suggest that large-scale bioremediation of soil will usually require cultivation, aeration, irrigation, and nutrient amendments in addition to repeated inoculation with the appropriate microorganisms [33]. This in situ process is termed *landfarming*. In other situations, excavation of the soil followed by its biotreatment in a reactor (either on site or off site) may be necessary [34]. Frequently, soils contain a complex mixture of organic pollutants, and the biodegradation of these compounds is likely to be carried out by an interactive community of microbial species. Establishing a mixed culture of inoculants in soil is likely to be more difficult than integrating an individual species with the indigenous microbiota. One advantage of using genetically modified microorganisms is that all the relevant biochemical properties could be expressed in a single species.

Accelerated biodegradation of pesticides in soil after inoculation with appropriate microbial species has been demonstrated with phenylcarbamates [35], phenoxyacetic acids [36], DDT [37] and parathion [38]. More recently, laboratory studies have demonstrated successfully the biodegradation of chlorobenzenes [39], bromacil [40], phenoxyacetates [41,42], chlorophenols [43], and chloropropham [44], but progress to large-scale field bioremediation is slow. Golovleva et al. [45] inoculated field plots with a strain of *Pseudomonas aeruginosa* genetically modified to degrade the acaricide kelthane. The inoculants not only survived in soil but also were able to degrade kelthane efficiently: 32% was broken down after 1 week and 92% after 3 months. Petroleum hydrocarbons are important pollutants in terrestrial and aquatic environments, and the difficulties of microbial cleanup have been widely stated [46,47]. Landfarming methods have been used [48,49], but the composition of the mixed cultures of acclimated microbial species applied is not always defined. van den Berg et al. [50] investigated the biorestoration of contaminated subsoil (by bacteria isolated from this soil) at a petrol station in The Netherlands in which the concentration of oil components (not defined) was between 1000 and 5000 mg kg^{-1}. Their initial studies revealed that the major constraints to biodegradation of the oil by the indigenous microbiota were the absence of organic matter, low microbial numbers and diversity, low solubility of the oil, poor motility of introduced bacteria, and suboptimal pH, water, and oxygen.

Biological Control

Biological control of soilborne plant pathogens has long been the focus of research [51], and interest has grown with increasing concern about the impact of chemical pesticides on the environment. However, with the exception of insecticidal strains of *Bacillus thuringiensis* [52,53], which represents about 1.5% of the global insecticide market, biocontrol agents have had comparatively little commercial success. This is partly because of their unreliability in the field rather than an inability to isolate promising antagonists in vitro. One of the few successful soil-inhabiting bacterial biocontrol agents is *Agrobacterium rhizogenes* strain 84, which is used to control crown gall of fruit trees caused by biotypes of *Agrobacterium* [54,55]. Strain 84, which is nonpathogenic and is applied as a root dip, produces an antibiotic (agrocin 84) that prevents establishment of the pathogenic strain in the rhizosphere. Severity of other diseases, such as take-all (*Gaeumannomyces graminis* var. *tritici*), can be reduced by using naturally occurring antagonistic bacteria [56] often isolated from suppressive soils. Most of the promising species are fluorescent pseudomonads [57], and their antagonism may be due to production of antibiotics [58,59], inhibition of plant growth-suppressive cyanogenic bacteria [60,61], release of Fe-chelating siderophores [62], or reduction of manganese [63]. Once again, the relation between cause and effect may be misleading, and the second and third of these processes may enhance or suppress plant growth *directly* by changing the availability of iron [64] and manganese [65]. The antagonistic properties of pseudomonads and their potential in the control of soilborne plant pathogens have been reviewed by Défago and Haas [66]. Many of the other putative biocontrol agents belong to fungal genera and are evaluated elsewhere [67].

Soil Aggregation

The microbial contribution to soil aggregation has been well documented [68,69] and include enmeshing soil particles within filamentous structures and adhesion resulting from extracellular polysaccharide production. Efforts to reverse structural degeneration and to stabilize agricultural soils, sand dunes, and spoil heaps have concentrated on chemical additives [70–72], but microbial inoculation has also been attempted [73]. One of the problems with using allochthonous heterotrophs as soil inoculants is that many of the target soils have their nutritional and energy resources fully exploited by the indigenous microbiota, reducing the opportunity for colonization. In contrast, phototrophic inoculants, such as unicellular algae and cyanobacteria, do not compete for scarce C, N,

and energy and have been used successfully in laboratory, green-
house, and field experiments designed to improve soil structure
[74]. With appropriate temperature, light, and moisture levels,
phototrophs colonize soil rapidly and bring about significant improvements in aggregation as well as increasing soil biomass and
organic carbon (including polysaccharides) [75,19].

It is clear from the aforementioned examples that the potential
impact of microbial inoculation of soil is great. However, the realization of this potential is dependent on reliable tracking of bacteria after their application and a thorough knowledge of their effects on target and nontarget processes.

II. DETECTION AND ENUMERATION

The successful use of beneficial microorganisms in agriculture, and
in the environment in general, is dependent on a number of fundamental developments in methodology. Paramount among these is the
ability accurately and sensitively to detect bacterial cells and their
activity once they have been introduced into soil. Precise monitoring of inoculants in soil is important in predicting their survival,
proliferation, and movement from the site of introduction, as well
as any changes that they may cause in the indigenous microbiota.
Unequivocal detection of released bacteria may even suggest ways
in which they could be killed (e.g., by targeting with specific bac-
tericides) if they were discovered to produce undesirable (and un-
predicted) side effects. There is little doubt that public and scientific confidence in the use of soil inoculants will be greatly
increased if there is routine monitoring of released bacteria.

Dilution and culture methods for enumerating bacteria have been
used by soil microbiologists for decades. These traditional techniques still have great value, but there is a need for improved
sensitivity. Some reliable methods already exist for bacterial identification and enumeration in the diagnosis and treatment of disease
and for monitoring microbial contamination of food and water [76].
These methods are now being modified and evaluated for the detection of bacteria in soil.

A. Culture Methods and Direct Observation

Conventional plate count methods, whereby the bacteria are cultured on media and the resulting colonies counted by eye or with
a microscope, are widely used for enumerating microorganisms added
to soil. For culture methods to be applicable to the tracking of soil
inoculants, they must be not only sensitive but also specific, reproducible, and practical. Plate count techniques can be modified to
fulfill these requirements to some extent [77], although recovery

rates less than 100% are often recorded after inoculation. However, this may not be a constraint, provided that the efficiency is at least equal to the desired detection level and consistent enough to permit the use of a correction factor.

Plate count methods remain popular with researchers for reasons other than tradition. The advantages include the ease of processing a large number of soil samples, relatively rapid production of results (depending on the growth rate of the microorganism), and technical simplicity. However, it is well known that care is needed in interpretation because the numbers calculated from colony counts may be very different from the number of microorganisms that were present or functioning in the original sample. This is because plate counts do not distinguish between active and dormant bacteria [78]; for example, bacterial endospores germinate and produce colony-forming units on nutrient-rich agar. In addition, plate counts do not reveal the existence of viable but nonculturable bacteria [79,80], whose numbers (along with those of dormant bacteria) may be expected to increase with time. It may also be difficult to know whether a single colony observed on an agar medium arose from an individual cell or from a cluster of cells that were not dispersed during the dilution and agitation phases that are pretreatments in most methods. Another potential inaccuracy is that the recovery rates may not be constant throughout a period of prolonged incubation for reasons other than nonculturability. For example, if microbial inoculants proliferate, they may produce extracellular polysaccharides, develop biofilms, and show greater adhesion to soil particles and to each other [81]. Thus, what were efficient extractants and dispersants at the beginning of a survival experiment may prove to be inappropriate as time progresses. Changing extraction efficiencies with time are difficult to measure without the availability of a specific stain that permits direct counting of the target species and, therefore, a reliable comparison with methods that depend on dilution and growth. In general, adsorption of bacteria to soil particles results in the recovery of much less than 100% of the cells and plate counts may be as low as 1 to 4% of direct counts [77,82]. As a result of all these factors, an increasing discrepancy with time between viable counts and direct counts would be predicted [82,83].

Detection limits for plate counts (i.e., the sensitivity of the method for reliable counting of low numbers of cells) are frequently not determined. When the inoculated species either proliferates or declines very slowly in soil, ignorance of a detection limit may not be a problem. However, in studies in which bacterial numbers are reported as "declining to zero" it is important to know the sensitivity of the method adopted. Detection thresholds for culture methods vary and, along with those determined using other techniques, are reported in Table 1. This variability is a reflection

Table 1 Examples of Detection Limits for Different Methods of
Enumerating Bacteria in Soil

Method	Organisms and marker	Detection limit (g^{-1} soil)
Plate count	*Flavobacterium balustinum, Arthrobacter globiformis*: spontaneous resistant mutants	100
Plate count, most probable number	*Erwinia carotovora*: genetically modified to be kanamycin resistant	1–10
Plate count	*Pseudomonas aeruginosa*: chromosomal addition of lac operon and kanamycin resistant genes	10
Plate count	*Pseudomonas fluorescens*: with introduced plasmid containing lac operon and kanamycin resistant genes	<10
Most probable number, then DNA hybridization	*Rhizobium leguminosarum, Pseudomonas putida*: both carrying chromosomal Tn5 transposons	100 10
Total DNA extraction, then DNA probes	*Bradyrhizobium japonicum*: naturally occurring specific DNA sequence	4.3×10^4
Total DNA extraction, then PCR, then DNA probes	*Pseudomonas cepacia*: 15–20 copies per cell of naturally occurring DNA sequence	1 (sediment)
Extraction of cells, then PCR	*Rhizobium leguminosarum*: one copy per cell of Tn5 sequence	1–10
Measure total light output	*Escherichia coli*: transformed with luminescence genes	200

Advantages	Disadvantages	Reference
Sensitive, easy to assay many samples	Culturability needed, resistance genes may be lost	115
Sensitive	Culturability needed; time consuming	104
Sensitive, specific; genes are stable and constitutively expressed	Culturability needed	100
Sensitive, specific	Culturability needed; plasmid with genes may be lost	31
Sensitive, specific	Culturability needed for MPN procedure	84
Specific; organism not impaired by added genes	Low sensitivity	92
Very sensitive, specific	Contamination from other amplified DNA possible	143
Sensitive, specific	Needs double PCR amplification	147
Sensitive, specific, can visualize individual cells or total light output	Correlation between light output and number and/or activity not yet established	151

(continued)

Table 1   (Continued)

Method	Organism and marker	Detection limit $(\text{g}^{-1}\ \text{soil})$
Immunofluorescence	*Flavobacterium balustinum*: strain-specific monoclonal antibody	20
Immunofluorescence	*Rhizobium leguminosarum*: strain-specific polyclonal antibody	$1 \times 10^3 - 1 \times 10^4$

of the different soils and bacterial species and numerous (often un-
reported) experimental modifications. It is worth nothing that despite the many drawbacks to plate counts, currently their sensitivity is usually at least equal to that of nucleic acid probe methods
or immunodetection procedures.

Optimizing the extraction of bacteria from soil and minimizing the
variability between replicate samples are essential steps in any viable count method. The main factors to consider are the physical
method used to disperse soil particles and the solution used to dislodge bacteria from soil particles while maintaining their viability.
Commonly used methods for producing soil suspensions include
shaking [83,85–87], blending [88], sonication [89,90], and a combination of blending and centrifugation [91,92]. The recovery efficiency depends on the soil type, as illustrated by Ramsay [90],
who found that there was no difference in the number of indigenous bacteria released from sandy soil by blending, shaking, or
sonication but that sonication extracted more bacteria from a silt
loam soil. Inoculated cells may be more susceptible to injury during extraction than indigenous bacteria, and the period and ampli-
tude of sonication, for instances, must be tested to ascertain which
are appropriate for optimum recovery [89]. Distilled water [86],
phosphate buffer [83,92], Tris buffer [85,93,94], sodium pyrophosphate [95,96], and Ringer's solution [87] are examples of extractants. Ramsay [90] found that the recovery of *Bacillus cereus*
from soil was maximum with Tris buffer as the extractant and 25%
Ringer's solution as the diluent, but it is apparent that for each
soil type and each bacterial species, different extractants need to
be evaluated to discover which is the most appropriate. As a general rule when determining initial recovery rates, the inoculum
should be incubated with the soil for a period of time sufficient

Advantages	Disadvantages	Refer-ence
Sensitive, specific	Time consuming, may count nonviable cells	126
Sensitive, specific	Time consuming	83

for sorption to occur but not long enough for bacterial death or proliferation. Trevors and Berg [97] found that cell numbers of a *Pseudomonas* sp. added to a sterilized loam soil increased during the first 24 h; as a consequence, recovery rates (98.1%) were assessed 1 min after inoculation. The plating medium chosen for enumerating the inoculum is determined by the nutritional requirements of the microorganism in question. In general, low-nutrient agar supports higher counts of soil-inhabiting bacteria than full-strength agar [82,98] because many soil microbes are oligotrophic. However, if the inoculant is a copiotroph, it will require a rich medium containing specific carbon, nitrogen, and energy sources.

The choice of method for adding the soil suspension to the growth medium depends on the sensitivity demanded, the number of samples to be processed, and the degree of selectivity required. Pour plates are good for facultative anaerobes, which can grow both on and within the solid medium and are not killed by the thermal shock of molten agar at 45°C [99]. Spread plates are more appropriate for aerobes. Both these methods allow a relatively large volume of the soil suspension to be applied to each petri dish. Höfte et al. [100] reported the limit of detection of 10 colony-forming units (cfu) g^{-1} soil for a kanamycin-resistant (200 μg ml^{-1}) and β-galactosidase-positive, marked strain of *P. aeruginosa* when 0.5 ml of a concentrated soil suspension was spread on agar containing the antibiotic (the recovery efficiency was 97%). This is probably the lower limit of sensitivity for plate counts. However, when many treatments, replications, and sampling times are involved (as is usually the case in field experiments), spread plate methods demand time-consuming preparation. In this situation, drop plates [101] may be more appropriate. The drops are usually only 20 μl (compared with the 0.1 to 0.5 ml used for spread plates) and, although the limit of detection is not as low, the number of plates needed is reduced compared with the spread plate method. For statistical reliability, the dilution giving the largest number of colonies per drop that do not coalesce is counted; this gives a lower standard error than for spread plates [102]. Further labor, media, and number of petri dishes can be saved using a spiral plater, which yields precision similar to that with drop plates [103] but is particularly efficient if a large number of samples must be processed at frequent intervals [88]. The most probable number (MPN) technique has the highest sensitivity of all the culture methods, with detection levels as low as one cell g^{-1} soil [104]. However, it is time consuming, and this may not be compensated by the increased sensitivity, especially in a large, complex, and prolonged field experiment. Nonetheless, in combination with the newer molecular biological techniques, MPN methods are undergoing a revival (see later).

Counting microcolonies of bacteria using a microscope reduces the incubation period between processing samples and counting colonies visible to the naked eye. Therefore, even though the microcolony method still relies on culturability of the microorganisms, Rodrigues and Kroll [105], in a study of food contamination, were able to enumerate gram-negative bacteria after only a 3-h incubation period and gram-positive bacteria after 6 h. The detection threshold was not changed, compared to conventional colony counting, and for both types of bacteria it was 1×10^3 cfu g^{-1} in a variety of foods. The method should be applicable to the quantification of bacteria in soil.

Detecting and enumerating bacteria by direct observation of soil enables all cells to be counted, including those that would be nonculturable [79]. In addition, the spatial distribution of bacteria in the soil microenvironment is revealed [106], and this may provide important clues to the distribution of the inoculant and its relationship with other microorganisms and with plants. Electron microscopy has been used to good effect with regard to the association of microorganisms within the soil matrix [107]. However, the major drawbacks of direct observation are the time-consuming preparations involved and the tedium of studying and counting a large number of fields (especially if the target species is at low density). With fluorochromes (e.g., acridine orange, fluorescein diacetate, ethidium bromide) and epifluorescence microscopy coupled to image analysis, some degree of automated counting is possible [108] and the process is not so labor intensive. Nonspecific staining and background fluorescence contribute to the inaccuracy of these techniques, but this can be reduced by "gating" if appropriate software is available. For nonselective observation of bacteria in soil, a fluorescent brightener, such as calcofluor white, can be used with incident illumination to observe bacteria on soil particles. However, using calcofluor white with acridine orange as a counterstain, Postma and Altemüller [106] claimed to be able to distinguish introduced bacteria, which were clearly visible on clay particles and organic matter, from indigenous bacteria, which stained at a much lower intensity. Other workers have attempted to overcome the inaccuracies arising from the staining of dead cells by using a so-called vital stain, such as europium chelate [89], which complexes with the nucleic acids of living cells only.

B. Antibiotic Resistance Markers

The dilution plate method, when applied to the identification and enumeration of an inoculant in a soil sample containing many other species, is selective only if the species of interest has some inherently unique property (e.g., pigmentation, unusual nutritional requirement, capacity to grow at temperature extremes). However, this is rarely the case, and induced antibiotic resistance is frequently used as a marker. Many soil bacteria are naturally resistant to low concentrations of single antibiotics, such as streptomycin [109,110]. In these circumstances, for a medium to be selective, the inoculant must be resistant to high concentrations of an individual antibiotic (e.g., 2500 μg ml^{-1} streptomycin [111]) or resistant to an antibiotic that is inhibitory to all other soil bacteria [112]. If the bacteria are resistant to more than one antibiotic, the selectivity of the media is increased [85,98], and, therefore, the best option is to choose a spontaneous or an induced mutant that is resistant to two or three antibiotics [113]. Spontaneous mutants can be used in the field as well as in laboratory experiments and can be detected reliably at 1×10^3 g^{-1} [114] or 1×10^2 g^{-1} soil [115] within an indigenous microbiota.

Generation of antibiotic-resistant bacterial mutants usually depends on the inoculation of actively growing cultures into broth [116] or onto solid media [117] amended with the appropriate antibiotic(s). For broth culture selection, the bacteria are added to flasks with low concentrations of antibiotic, and, if growth is observed, the cells are transferred to media with increasing concentrations of antibiotic. The gradient plate technique [118,119], in which isolates are also selected for survival on progressively higher concentrations of antibiotic, is commonly used to isolate spontaneous mutants. An alternative method of producing bacterial mutants is to introduce an antibiotic resistance gene into the chromosome or via plasmid uptake. However, such recombinant organisms cannot be used in field tests until the purpose and safety of the experiment have been considered by various review committees.

A major drawback to using antibiotic resistance as a marker is that the mutation, whether spontaneous, produced by a mutagen, or genetically engineered, may give rise to a microorganism that is more or less competitive than the wild-type strain. In fact, it cannot be demonstrated easily whether or not the mutant is different because it is difficult to compare it directly with its parent wild type in nonsterile soil. If there is no difference between the inoculant and the wild type, there should be a consistent ratio between them following coinoculation into soil. This relationship could be assessed initially in sterilized soils by adding known concentrations of mutant and wild type and following the populations of the two strains on antibiotic-containing and antibiotic-free media. However, in nonsterile soil (the relevant medium) the survival of the wild type is difficult to monitor unless it possesses a property different from the indigenous bacterial species (see above). Other considerations are that the recovery efficiencies of the mutant and wild type from soil should be comparable [120,121], their growth rates and cell yields should be similar [82,122], and the gross morphology of the mutant and a number of its key biochemical properties should remain unchanged [95,120]. Obviously, the antibiotic resistance gene must be stable, and preferably it should not be carried on a plasmid [123]. Tests for stability of mutants include serial transfers on antibiotic-free medium without loss of resistance (e.g., 25 generations [124]; 50 generations [95]). However, in the field, loss of antibiotic resistance is usually difficult to measure. Halvorson (L. J. Halvorson, Ph.D. thesis, University of Wisconsin, Madison, 1991) showed that antibiotic resistance markers were not maintained in *B. cereus* inoculants in the rhizosphere of field-grown soybeans. Both the wild type and the mutant phenotypes could be monitored in these experiments because the wild-type strains were chosen according to the presence of some other marker such as pigmentation or naturally occurring resistance to tetracycline.

C. Immunological Techniques

Specific immunochemical stains, based on polyclonal antibodies [82, 125] or monoclonal antibodies (MAbs) [126], are being used to visualize and quantify microorganisms in soil. Currently, the detection limit using immunofluorescence techniques for bacteria ranges from 1×10^2 to 1×10^4 cfu g^{-1} soil [127], which is not an improvement over the levels that can be achieved by plate count methods. In fact, immunofluorescence methods may overestimate the numbers of viable cells in a sample because they enumerate cells which are not viable but which still have intact antigens on the outer membrane.

The specificity of the antibodies depends mainly on the uniqueness of the antigen to which the antibodies are raised; e.g., lipopolysaccharide antigens are often strain specific [128]. It is important for the antigen to be located on the bacterial cell surface and to occur at a frequency high enough that a sufficient amount of the antibody is bound to permit direct observation of immuno-fluorescent cells. In addition to the requirement for strain speci-ficity, it is usually desirable that the antigen should be expressed constitutively, so that it can be targeted in all soils regardless of their physical, chemical, or nutritional status. This property may be tested in vitro by growing the inoculant under nutrient-limiting conditions or by starving the cells before exposure to the antibody [126]. It may even be preferable to use starved cells for the immunization procedure, to ensure that an appro-priate epitope is introduced into the rat, mouse, or other ani-mal. Monoclonal and polyclonal antibodies may be equally spe-cific for a particular bacterial strain. However, different MAbs can be developed to recognize variants of the antigens that are expressed during different growth phases or under different en-vironmental conditions. This is an exciting possibility because it may allow monitoring of the physiological condition of the inocu-lant and a number of its properties after introduction into soil. The ability to produce a continual supply of MAbs by tissue culture is a major advantage. On the other hand, production of polyclonal antibodies is usually cheaper. In the future, the spe-cificity and reactivity of antibodies may be ensured by intro-ducing a gene that expresses a unique antigen that is incorpor-ated into the polymers that make up the bacterial cell surface or is part of the chemical that gives the cell its beneficial property (e.g., an extracellular enzyme).

Antibodies have been coupled to fluorochromes for use in di-rect assays (primary antibody coupled directly to stain [129]) or indirect assays (secondary antibody coupled to a stain [79, 130]). The latter approach is usually favored because of its greater sensitivity, which arises from the amplification of the chosen signal (stain or fluorescence) when a secondary antibody is used to bind to the primary antibody. Direct observation of bacteria in soil samples using immunofluorescence can give mis-leading information, as a result of nonspecific sorption of the antibodies to soil colloids and biofilms [99] and the autofluores-cence of other microorganisms. As a consequence, blocking agents, such as bovine serum albumin, and counterstains, such as rhodamine isothiocyanate [131], are usually required in soil preparations. An alternative to enumerating soil bacteria in situ is to extract them before staining and then enumerate them mi-croscopically (possibly with the aid of image analysis [108]) or by using a flow cytometer. Page and Burns [132] compared flow cytometry with other more conventional methods. They in-oculated a loam soil with *Flavobacterium balustinum*, which was labeled with an MAb indirectly conjugated to fluorescein isothio-cyanate. Plate counts and direct counts gave comparable re-sults, but they were one order of magnitude less than predicted

from inoculum levels. Flow cytometry numbers were even less accurate and were approximately two orders of magnitude less than
expected. The major problem was the tenacious association of bacteria with soil particles <30 µm in diameter (a problem common to
other counting methods—see earlier); this reduced the accuracy of
the flow cytometer counts. However, the rapidity with which samples can be processed and the potential for specific detection and
separation of populations suggest that further research into flow
cytometry is justified.

   Microbes may also be recovered (and concentrated) from soil by
immunocapture using magnetic beads coated with the appropriate
antibody. This method has been used to great effect in medical
and diagnostic microbiology [133,134], and recently Morgan et al.
[135] coated magnetic polystyrene beads with a specific MAb and
recovered 15 to 28% of the initial inoculum of *Pseudomonas putida*
from lake water. The use of a similar technique to recover *F.
balustinum* from soil has not been successful (S. Page, personal
communication), again as a result of nonspecific binding to soil
particles, although recovery from aqueous suspensions was 100%.
As a separate approach, extracting and purifying the antigen and
then relating immunofluorescence to a standard curve of antigen
concentration versus bacterial cell numbers may prove useful.

D. Nucleic Acid Probes

Detection of microorganisms with labeled nucleic acid probes depends
on complementarity of single-stranded DNA or RNA species of opposite polarity, where the two nucleic acid sequences are either
identical or very similar. The target sequence is chosen for its
uniqueness to the microorganism and may be part of the genome
(detected by a DNA probe) or a product of gene expression (detected by an RNA probe). The method has the potential to be
species specific, does not depend on expression of the sequence,
and is especially useful for the detection of genetically engineered
organisms that have added DNA sequences [77]. Another advantage of nucleic acid probes is their ability to detect sequences that
may have been transferred from the inoculant to indigenous soil
microorganisms. (Movement of foreign DNA within the soil microbial community is discussed later.) DNA probes can be prepared
by nick translation and RNA probes by transcription; hybridization
efficiency tends to be higher with RNA probes [77]. Nucleic acid
probes can be labeled with radioisotopes (e.g., ^{32}P) or nonradioactive substances (e.g., biotin) for ease of detection. One disadvantage of gene probe methods, in terms of their universal application, is that the procedures used for detection can be time consuming and require specialized equipment. One method of preparing

bacterial samples for labeling with nucleic acid probes is to lyse colonies on nitrocellulose or nylon membranes and then hybridize a labeled probe to the fixed and denatured DNA or RNA [136]. Another widely used method is to directly extract the DNA [137] or RNA from soil and then apply a specific probe.

Using total DNA extracted from soil and a ^{32}P-labeled DNA probe, Holben et al. [92] detected *Bradyrhizobium japonicum* at densities of 4.3 × 10^4 cells g^{-1} soil. Selenska and Klingmüller [138] developed a more efficient extraction method, involving Na$_2$HPO$_4$, centrifugation, and dialysis. They reported the recovery of 100 µg of DNA from 2 g of soil 70 days after inoculation with 1 × 10^7 *Enterobacter agglomerans* g^{-1} soil. At this time, *E. agglomerans* cells were below the limit of detection (10 cells g^{-1} soil). However, this does not mean that the sensitivity of detection is <10 cells g^{-1} soil. This is because the DNA may have consisted of extracellular DNA protected by soil particles [139], that extracted from nonculturable *E. agglomerans* cells, or even indigenous bacterial recipients of the insertion sequence (Tn5). Horizontal genetic exchange in sediments [140] and soils has been demonstrated in a few cases (see Section V,A).

Extraction of RNA and use of RNA probes also have potential for detecting bacteria in environmental samples. Tsai et al. [141] reported recovery rates of 60% of the total RNA of *P. putida* added to sterilized soil. Pitchard and Paul [142] were able to detect a *Vibrio* sp. at 1 × 10^2 cells ml^{-1} seawater with RNA probes to detect the expression of a plasmid-encoded neomycin transferase. Extraction of RNA may be more difficult from soil than from seawater, but the method appears promising, especially to monitor the expression of genes of introduced microorganisms.

The disadvantages of nucleic acid hybridization methods include the possible enumeration of dead cells as well as active cells and the lack of sensitivity in probing for DNA extracted from soil, which may contain very small amounts of the target sequences. In addition, DNA in soil (and in soil extracts) is subject to rapid degradation by nuclease-producing organisms [92], although the use of DNase inhibitors may counteract this effect. The low levels of DNA of a particular bacterial strain in soil samples may be compensated by the polymerase chain reaction (PCR), which amplifies the number of copies of the target DNA sequence [143]. Briefly, the method involves heat denaturing the DNA, annealing oligonucleotide primers to either side of a chosen DNA target sequence, and adding free nucleotides followed by treatment with a thermostable DNA polymerase [144] that extends the annealed primers. Repetition of this process gives an exponential increase of the target DNA sequence with each cycle [145], and, therefore, the amount of the fragment of interest is increased by many orders of magnitude and can be visualized on a stained agarose gel [146]. Using the PCR,

Steffan and Atlas [143] reported the detection of a strain of *P. cepacia* (which had 15 to 20 copies of the target sequence per cell) at one cell g^{-1} sediment that contained high numbers of indigenous microorganisms. Pillai et al. [147] also detected 1 to 10 cells g^{-1} soil based on amplification of a single copy of the target sequence in each cell. However, van Elsas et al. [148] could detect *P. fluorescens* in soil only at populations of 1×10^3 or higher. In contrast, detection in water samples is more sensitive, with 1 to 10 cells per 100 ml being detectable according to Atlas and Bej [149]. Because active, nonactive, and even dead cells containing intact DNA are detected with the PCR method and extracted humic substances may inhibit the polymerase [150], careful interpretation of the data is necessary.

The specificity of nucleic acid probes may be combined successfully with the sensitivity of more traditional culture methods. For example, Fredrickson et al. [84] used a combined MPN-DNA hybridization technique for enumerating *Rhizobium leguminosarum* bv. *phaseoli* and *P. putida* (both carrying the Tn5 transposon) in soil. Dilutions of soil suspensions were incubated in microtiter plates for 4 days, the contents were transferred to membranes, the cells were lysed, DNA-DNA hybridization was carried out on the membranes, and the MPN values were calculated. In nonsterile soil, this method enumerated 100 *R. leguminosarum* cells g^{-1} and as few as five *P. putida* cells g^{-1}.

E. Genetically Engineered Markers

A more recently developed type of marker involves the insertion of a gene into the target bacterium and the use of the unique DNA sequence, either as a direct target for a probe [92] or to express a unique property once the bacterium has been inoculated into soil [151]. Introduction of antibiotic resistance into the genome of a bacterium is an alternative to producing mutants [152]. Plasmid-borne antibiotic resistance genes are frequently used, although plasmids may be lost over time and the growth rate of bacteria which contain additional plasmids may be changed [153]. Larger plasmids may increase survival in soil [154], but survival rates may depend more on the bacterial strain used than on the plasmid size [155]. Tn5 transposons which carry antibiotic resistance genes have often been used to generate antibiotic-resistant mutants for use in soil survival experiments [156]. Pillai and Pepper [121] produced Tn5 mutants of a *Pseudomonas* species that were resistant to kanamycin and neomycin. The parent wild-type strain was also resistant to streptomycin and nalidixic acid, and therefore the survival in soil of both mutant and wild type could be compared. Both strains survived equally well in separate soil inoculations as counted on a medium containing streptomycin and nalidixic acid only, but when the Tn5 mutants were plated on a mecium containing kanamycin and neomycin or on a medium containing streptomycin and nalidixic acid, the counts were significantly lower on the former medium. The resistance coded by the Tn5 transposon appeared to be more sensitive to soil temperature and texture, which would not have been discovered without the comparison with the naturally occurring wild-type markers.

Frequently, a metabolic marker is chosen that will permit growth of the bacterium on highly selective media or give it the capacity to cleave a substrate in a culture medium to yield a product with a distinctive color. Ramos et al. [157] used a strain of *P. putida* which contained the Tol plasmid, enabling it to grow on media containing p-ethylbenzoate. Drahos et al. [31] introduced broad-host-range plasmids containing the *lacZ* (β-galactosidase) and *lacY* (lactose permease) genes from *Escherichia coli* into *P. fluorescens*. The *lacZY* gene gave the cells the unusual property of growth on lactose minimal agar; the selectivity of the medium depended on previous observations that most rhizosphere bacteria had no β-galactosidase activity and, therefore, could not grow on the lactose medium. In addition, the *lacZY* transformants could cleave the chromogenic substrate, X-gal (5-bromo-4-chloro-3-indolyl-β-D-galactopyranoside) with β-galactosidase to give blue-green colonies easily identified in agar plates. Using this marker system, the *lacZY* transformants could be detected at less than 10 cfu g^{-1} soil. Höfte et al. [100] also used *lacZ* genes to identify cells of *P. aeruginosa* introduced into soil and detected approximately 10 cells g^{-1}. Furthermore, they showed that the genes for the *lac* operon were stable and constitutively expressed under the culture conditions used.

Other detection systems based on identification of bacteria by a color change include the introduction of *xylE* genes into *P. putida* [158]. In this case, colonies turned yellow when sprayed with a catechol solution. However, this assumes that all the *P. putida* cells give rise to colony-forming units; if this is not the case direct detection may be preferable. King et al. [159] used the activity of deregulated 2,4-dichlorophenoxyacetate monooxygenase for the detection of *Pseudomonas* species in culture. The monooxygenase converts phenoxyacetate to phenol, which can be measured either chromotographically or spectrophotometrically. The sensitivity of the method was increased from 1×10^4 cells ml^{-1} using gas chromatographic (GC) analysis of phenol to 1×10^3 cells ml^{-1} by reacting the phenol with 4-aminoantipyrine to produce an intensely colored dye that could be measured spectrophotometrically. The β-glucuronidase (GUS) operon is another enzyme marker system that has the potential to be introduced into different bacteria for detection in situ. Wilson et al. [160] cloned the operon into *Rhizobium meliloti* and a *Bradyrhizobium* strain and were able to localize the bacteria within roots using X-gluc (5-chloro-4-bromo-3-indoyl-β-D-glucuronide), a substrate that forms a deep indigo precipitate when cleaved by GUS. The system is stable under a wide variety of conditions and is inducible when the substrate, X-gluc, is added and therefore does not exert a metabolic load on the marked bacteria. Most soil bacteria do not have GUS activity, but bacteria of gut origin (e.g., from dung) have GUS activity and could be present in soil samples. The use of appropriate controls would show if any such bacteria were originally present in the soil.

Light output by bacteria that contain the genes that code for luminescence, cloned from the marine bacterium *Vibrio fischeri*, has also been used to determine the numbers of cells in soil. A major advantage is that measurements can be very rapid (5 min after sampling) as viable cells can be enumerated by luminometry without culturing. In addition, total cell counts can be made by using DNA probes for the *lux* genes, and the spatial arrangement of the luminescent bacteria in soil is revealed by direct observation. However, it may be difficult to determine accurately the numbers of bacteria from total light output measurements, as these measurements depend on how active the bacteria are. Addition of an appropriate substrate to soil to reactivate all cells before assay may overcome this problem. Rattray et al. [151] introduced *lux* genes into *E. coli*. The complete *lux* cassette includes genes *luxA* and *B*, which encode luciferase; *luxC*, *D* and *E*, which encode the production of the aldehyde substrate; and *luxI* and *R*, which encode regulators involved in luciferase production. When the *lux* cassette (complete or with a truncated *luxI* gene) was introduced into *E. coli*, light output was autoinducible, but the light output per cell varied with cell concentration. When only the *luxABE* genes were introduced into *E. coli*, the light output per cell was constitutive but was independent of cell concentration. Thus the constitutive strain was chosen for studies to measure the number of *E. coli* cells in soil. The *E. coli* could be detected at 2×10^2 to 6×10^3 cells g^{-1} soil, a detection level comparable to that achieved by plating or by using nucleic acid probe methods. The sensitive detection of *Erwinia carotovora*, also containing a plasmid coding for constitutive production of luciferase, was achieved following dilution plate counting of soil suspensions [161]. However, efficiency decreased as the proportion of *E. carotovora* in the total indigenous population decreased. If concentrations of the luminescent microorganism were >60% of the total count, accurate enumeration was possible, whereas at <0.03% detection was not possible. de Weger et al. [162] found that constitutive expression of bioluminescence did not occur at high levels in the rhizosphere, most likely because of the energy needed for the cells to synthesize the aldehyde substrate for the luminescence reaction. Addition of nutrients is likely to change the distribution and numbers of cells in the rhizosphere. Therefore, de Weger et al. used either strains in which the *lux* genes were inducible by naphthalene or strains which could produce light only when the luciferase substrate was added. They were then able to monitor the distribution of cells on roots at a detection level of 1×10^3 to 1×10^4 cm^{-1} root. However, in an important field experiment using a genetically engineered, constitutively bioluminescent *Xanthomonas campestris* strain, the Lux assay proved to be as sensitive as serial dilution and plating [163].

There will, undoubtedly, be many more types of markers developed for introduction into bacterial genomes, but it is not likely that one particular method will ever be suitable for detection of all types of bacteria in all habitats [164]. The objectives must be to develop reliable and cost-effective systems that can be used for as broad a range of bacteria as possible and that will detect and quantify active and dormant cells, the gene product(s), and the gene itself. Such data will give information on the numbers of the organisms, their activity, and whether any horizontal transfer of genes has occurred. Ultimately, the development of specific probes for a large number of soil bacteria will be the stimulus necessary to advance our knowledge of microbial community dynamics.

III. SURVIVAL AND DISPERSAL

Microbial inoculants of potential benefit to soil quality and plant growth must survive in order to be effective. Survival depends on many interacting factors, including the physical and chemical properties of the soil; the presence and absence of a rhizosphere; the density, competitiveness, and movement of the inoculants; and the method of inoculation. The influence of these diverse factors makes it difficult to predict how long bacteria will remain active once introduced into soil.

The persistence of introduced bacteria is usually presented in the form of log numbers, but occasionally the uniformity of growth rate and death rate approximates first-order kinetics and allows the expression of survival in terms of half-lives [165]. More often, the decline in numbers with time is sigmoidal, and a value for the reduction of 90, 99, and 99.9% in the original cell concentration may be useful [115], especially if the expressed activity at a particular time is dependent on the number of surviving cells. Changes in bacterial numbers and activity can be represented logarithmically [166], and the exponential decrease of *R. leguminosarum* in soil has been described, using as parameters the initial inoculum density, the final population size, the difference between the initial and final inoculum numbers, the daily reduction factor, and time [114]. Fitting curves to inoculant decay data allows statistical comparisons to be made between the rates of decline observed in laboratory and field experiments [113] or from different starting inoculum densities in the same soil. It may also enable prediction of bacterial survival. However, the decrease (or increase) of an inoculant species will vary according to the soil type, and any mathematical description may be applicable only to individual systems.

In the absence of a specific probe or marker, the survival and movement of inocula have often been studied in sterilized soils. The rationale behind this is that it eliminates biological interactions and allows physical and chemical influences to be identified. Nonetheless, the factors that affect inoculant survival are not merely additive but interactive, and therefore the elimination of the soil microbiota will almost certainly change abiotic properties and can lead to misinterpretation of the data. As a rule, the use of sterilized soil in these studies should be avoided whenever possible.

It is not clear from most published data whether the introduced bacteria merely survive to be counted at the various sampling times or the numbers are composed of introduced and new cells. If the latter is the case, the question arises, do inoculated cells proliferate so that the growth rate is in excess of, equal to, or less than the death rate? This is an important question in relation to predicting bacterial activity in soil, but it is not easy to answer except that when counts become significantly greater than inoculum levels, it is reasonable to assume that proliferation has occurred. However, when numbers remain constant or decline, it is not possible to state what combination of survival, death, and proliferation has contributed to the measured number of cells. Some limited information is provided by measurements of changing respiration rates during prolonged incubation of inoculated sterilized soils [115]. However, if the inoculant has a unique catabolic property, cells may be distinguished from resting and dead cells if the soil is spiked with the specific substrate. These fundamental questions will be answered eventually by using a combination of traditional and novel methods to distinguish total numbers, dead cells, live cells, active cells, and resting cells. Specific stains based on MAbs, new vital strains, genetically engineered markers that indicate activity, and gene probes will all make a contribution toward the understanding of inoculant dynamics in soil.

A. Effect of Soil Properties

Soil texture and water content are major influences on the movement and rate of decline of inoculated bacteria. Most bacteria are associated with particles [167], with more microorganisms attached to the clay size fraction than to silt or sand [168]. Heynen et al. [122] found that survival of *R. leguminosarum* biovar *trifolii* was increased in soils with added bentonite clay. The clay probably

protected the bacteria by clustering around cells to form microniches
that protected the bacteria from bacteriophages and grazing protozoa [169,170]. It is also possible that survival was increased as a
result of concentration of scarce nutrients by the expanding clay
lattices.

Some reports suggest that bacterial inoculants survive better in
drier soils [171–174]. However, wetting cycles generally stimulate
microbial activity [175]. Postma and van Veen [176] found that the
number of *R. leguminosarum* cells significantly decreased when the
moisture content increased in a silty loam from −20 to −5 kPa, although the decline could have been caused by both oxygen limitation and increased protozoan activity. When pore sizes greater
than 3 μm are devoid of water, protozoan activity decreases, either
because the water potential directly affects cell metabolism or because the prey bacteria become inaccessible in the smaller water-
filled pores [177,178]. In the future, the choice of bacteria for
inoculation will probably be made, in part, according to the prevailing soil conditions, and bacteria resistant to desiccation will be
needed for arid soils.

The appropriate temperature range is important for stimulating
the activity of bacteria. Inocula may be sensitive to high temperatures [172], and reports of survival at low temperatures (approximately 4°C) are quite common [85,172]. It is worth remembering
that inoculants may have an optimum temperature range for growth
but, quite possibly, a different temperature for expression of the
desired activity. For instance, Seong et al. [179] showed that a
plant growth-promoting *P. aeruginosa* grew best at 28°C but produced maximum pyoverdin (a siderophore) at 19°C. *P. fluorescens*,
on the other hand, had maximum growth and produced the most
pyoverdin at 12°C. When in soil, *P. fluorescens* survived better
at subzero temperatures, whereas *P. aeruginosa* survived best at
28°C. On roots, *P. aeruginosa* was a better colonizer than *P.
fluorescens* at 30°C, but the reverse was true at 18°C. What
emerges from this type of study is that for each inoculant strain,
the effects of a range of temperatures on survival, growth, and
activity must be determined in laboratory experiments that take
into account the fluctuating temperatures likely to be encountered
in the field.

It is usually desirable that bacterial inoculum be dispersed in
soil, both horizontally and vertically, in order that the beneficial
species colonizes roots and soil not reached directly by the application procedure. For example, a biocontrol agent applied as a
seed dressing will be most effective if it develops in synchrony
with the growing root. If the bacteria are restricted to the seed,
the zone of protection against the target pathogen will be restricted. Reaching all soil or every plant is not feasible for most
field application methods, and it is possible that the value of

inoculant bacteria may be limited because their dispersal is confined
to the soil zone in which the moisture content is close to the water-
holding capacity [180]. Of course, dispersal may not be compatible
with activity, as soils with high water content are likely to be oxygen limited. In laboratory experiments, no passive movement of *R.*
leguminosarum bv. *trifolii* from the source of inoculum occurred
after 14 days at water potentials between -8.9 and -260 kPa [181]
and the presence or absence of roots had no effect. However, in
saturated soil, there was rapid movement within 2 days of inoculation to the bottom of the 15-cm-deep tubes used. Even during
movement, the effective concentration of bacteria will be reduced
as a result of a combination of sorption and filtration coupled with
predation and parasitism. In the absence of an appropriate substrate (which may become available only as plant roots develop),
many bacteria starve and lyse. Vertical movement, whether motile
or nonmotile bacteria are involved, can be expected with water flow
through soil. *P. fluorescens* was not recovered in the rhizosphere
more than 0.5 cm below its inoculation site unless water was added
[182]. Even in flooded soils, metabolites from actively growing surface-applied cyanobacteria did not migrate below the 0.7-cm soil
layer [183]. This underlines the importance of cell movement for
successful inoculation. Transport of bacteria down a root system
does occur in some instances, partly by passive movement with
elongating root tips and partly by active movement in response to
substrate gradients. Chemotactic responses have been described
for rhizobia in soil [184]. Once again, however, the primary determinant is probably the volume of water flow [182]. The downward movement of inoculants into zones that are oxygen limited is
a prerequisite for the in situ bioremediation of subsoils and contaminated aquifers. The transformation of xenobiotics in subsoils
has been reviewed extensively [185].

The rate of movement of bacteria in soil probably depends on
the interaction of various other factors in addition to water availability and flow rate. These include the size and charge of the
cells, the production of extracellular polysaccharides, the presence
of flagella, and the size of soil pores and their distribution [181].
Gannon et al. [186] investigated the movement through soil columns
of 19 different strains belonging to the genera *Enterobacter*, *Pseudo-*
monas, *Arthrobacter*, *Achromobacter*, *Bacillus*, and *Flavobacterium*.
A large number of cell properties were measured to test whether
any were influential on movement. Surprisingly, transport was not
correlated with relative hydrophobicity, net surface charge, flagella, or the presence of capsules, all of which would be expected
to influence retention and restrict bacterial movement in soil. However, with some exceptions, there was a statistical correlation between cell size and movement, with small cells showing the greatest
migration. All experiments were conducted at $3 \pm 1.5°C$ using

sterilized soil, so there was no influence of predators. In laboratory experiments, the downward movement of bacteria may be greater in intact soil cores than in soil columns containing sieved, homogenized, and repacked soil, as macropores and irregularities inherent in undisturbed soil may increase drainage of water. As soil density is decreased, more bacteria are transported to deeper soil layers [156].

Lateral movement of bacterial inoculum over short distances is limited mainly to motile bacteria. On a larger scale, runoff water may transport bacterial inoculants to nearby aquatic environments. *Azospirillum brasilense* does not move horizontally in soil in the absence of plants, but it does move (presumably by chemotaxis) up to 160 cm from the point of inoculation in response to attractants from wheat roots. Plant roots of weed species, between the inoculation point and the wheat, will also act as conduits [187]. Other work suggests that the distance moved by *A. brasilense* through rootless soil to wheat roots may be limited to 15 cm in field soil [188]. The potential for lateral movement of an inoculant needs to be determined in nonsterile soil: Catlow et al. [189] found that both motile and nonmotile mutants of *R. leguminosarum* showed very limited horizontal movement in a nonsterile sand soil, but in a steam-treated sand soil (in which organics may have been solubilized) the motile cells moved at least 16 cm. The comparatively low available organic matter in the nontreated soil may have limited the survival of the introduced species.

B. Competition and Predation

The almost inevitable decline in numbers of bacteria following their introduction into soil has been attributed to a number of biotic and abiotic factors. Prominent among the former are protozoal predation [190,191], but competition [192,193], bacterial parasitism, and production of inhibitory metabolites such as antibiotics and enzymes may all be important in certain soils. The production in situ and significance of antibiotics have long been debated [194], but it is unlikely that secondary metabolites are mere waste products of metabolism. In a highly competitive microenvironment even low concentrations of antibiotics may give the producers some advantage over sensitive microbial neighbors. Abiotic destructive forces include dehydration, salinity, temperature, and pH, factors to which the indigenous population is adapted but which may be alien to the introduced species.

Predation by soil protozoa may cause drastic reductions in populations of introduced bacteria, with the rate of decrease also influenced by the presence of other bacterial competitors. Postma et al. [195] showed that the survival of *Rhizobium* in soil decreased most when not only protozoa but also other bacterial species were added. They suggested that when only *Rhizobium* inoculum and the predators were present, a higher proportion of the rhizobial cells could associate with microniches in the soil and thereby resist predation. Presumably, when other bacteria were included they occupied these protected environments instead. It has been reported that rhizobia associated with microaggregates and in pores <6 μm are protected from protozoal predation [173]. Habte and Alexander [191] also demonstrated the importance of protozoa in determining inoculant survival. When cycloheximide was added selectively to kill the protozoa, the numbers of added rhizobia did not decrease until there was a significant population of cycloheximide-resistant protozoa.

The success of an inoculant strain may be reduced in the presence of a large indigenous population of the same genus or species. A specific example of this is shown by the work of Thies et al. [196], who studied the size of the indigenous rhizobial populations in relation to the effectiveness of introduced rhizobia. They found that the competitive success of the inoculant (1×10^5 to 1×10^7 per pelleted seed) was inversely related to the numbers of indigenous rhizobia. They suggested that the presence of as few as 50 indigenous rhizobial cells g^{-1} soil (estimated by MPN methods) prevented nodulation by the introduced rhizobia. In fact, at these inoculum levels there had to be <10 indigenous cells g^{-1} soil to ensure a significant increase in legume yield. However, it is possible that the inability of inoculants to establish themselves in the presence of an aggressive indigenous microbiota may be overcome by the simultaneous introduction of other species. The additional species would be chosen because they are inhibitory to the indigenous population but do not affect the primary inoculant species. For example, co-inoculation of two antibiotic-producing bacteria (*Streptomyces griseus* and a *Bacillus* sp.) with antibiotic-resistant rhizobia improved nodulation, possibly because of reduced competition by resident bacteria, fungi, and even protozoa [197,198]. The inhibitory coinoculants remained viable on stored seeds for at least 3 months.

C. Rhizosphere Effects

The microbial population of the rhizosphere is a dynamic and interactive community consisting of hundreds (and possibly thousands) of microbial species. Some of the species will be beneficial to the plant and others harmful, and many will have no obvious influence. Inoculation of the rhizosphere is an attempt to adjust the balance between beneficial and harmful microorganisms or to establish a novel bacterium that will offer direct benefits to plants [199]. The bacteria in the root zone have an important influence on plant growth and yield [200,201], and the survival and activity of microbial inoculants are linked with the many aspects of the rhizosphere environment [202]. This reciprocity should be implicit in research aimed at defining the roles of inoculants in agriculture and horticulture. Nonetheless, many researchers see this as impossibly complex and exclude plants from their experiments. The relationships between roots and their associated microbiota are poorly understood, but the contribution made by the rhizosphere to soil fertility and plant growth varies according to plant species, soil type, climate, and the developmental stage and general condition of the plant [203]. Consequently, attempts to manipulate the rhizosphere environment in a rational and predictable way are often unsuccessful or generate data that are difficult to interpret.

The release of organic and inorganic nutrients and energy sources from roots [204] enhances the survival of some bacterial inoculants in the rhizosphere. However, the inoculant must compete with the existing microbiota [193], either on the developing root or within an already established root system. Generally, populations of bacteria are larger in the rhizosphere, with the ratio of numbers in the rhizosphere to numbers in the bulk soil being 10:1 to 50:1 [200,205]. The increased density and the species diversity of the rhizosphere microbiota differ between plant species and with the age of a plant, probably because of changes in root exudates that occur during plant growth [202], and these changes may be controlled by specific host genes [206]. As stated earlier, successful root colonization will also be dependent on some degree of synchronous growth between the developing plant roots and the inoculated bacterium. The ineffectiveness of many putative biological control agents when tested in the presence of a developing plant may be the result of the variable degree of root colonization [207]. Miller et al. [202] reported that the total numbers of bacteria were highest after approximately 8 weeks of growth of spring wheat, with pseudomonads being less than 0.5% of the total. In contrast, pseudomonads have often been reported to constitute a much higher percentage of the total rhizosphere microbiota [201, 208], but in several experiments the relatively rich isolation medium adopted would have precluded isolation of many other bacterial genera. In any case, different inoculant species would be expected to survive differently within the rhizosphere microbiota of a particular host plant, and matching the inoculant with the plant will be an important determinant of survival rates in the rhizosphere. For example, Thompson et al. [115] showed that numbers of introduced *Flavobacterium balustinum* (a rhizosphere, zymogenous strain) declined more slowly in wheat rhizosphere soil than in nonplanted soil. In contrast, the survival of *Arthrobacter globiformis* (a nonrhizosphere, autochthonous strain) was independent of the presence of a plant.

According to some investigators, actinomycetes are dominent in the rhizosphere [202], which may explain the higher incidence of antibiotic-resistant bacteria in rhizosphere soil than in the bulk soil [109]. Therefore, it may be necessary for inoculants to be resistant to certain antibiotics in order to survive and to give them a competitive advantage over a portion of the indigenous rhizosphere inhabitants [209]. Other rhizosphere inhabitants produce measurable concentrations (1–50 nmol) of HCN [210], which not only is detrimental to plants but also may suppress other bacteria [61, 66], including inoculants. The roles of HCN produced by bacteria and of cyanogenic glucosides produced by plant roots are complicated but may be linked to the success of siderophore-producing biocontrol inoculants [211].

The comparatively high microbial numbers (and correspondingly high activity) in the rhizosphere may enhance the microbial degradation of hazardous organics. The potential is illustrated by many reports of accelerated biodegradation of pesticides in the rhizosphere [212–214], and this has been investigated by Walton and Anderson [215] using trichloroethylene (TCE)-contaminated soil. They reported that TCE degradation occurred faster in freshly sampled rhizosphere soil than in nonrhizosphere soil. The importance of the plant-microbe relation in the bioremediation of soils is illustrated by this work, and the possibility exists for augmenting the indigenous rhizosphere population with pollutant-degrading bacteria or planting contaminated soils with tolerant plant species and inoculating with the appropriate rhizobacteria.

IV. DELIVERY AND ACTIVITY OF INOCULANTS

A. Density and Application

The introduction of a sufficiently high number of an inoculant microbial species appears to be essential for achieving the particular beneficial effects associated with that organism. In nonplanted field soil, the population size of the bacterium may be determined by the initial inoculum density, with higher inoculum levels resulting in increased populations after 3 to 4 months [114]. In nonsterile soil microcosms, the initial rate of increase is sometimes higher from low inoculum levels than from high inoculum levels [117], possibly recause the comparatively lower numbers of bacteria are not nutrient limited.

Many interacting factors, in addition to the initial inoculum size, will affect the rate of decline. For example, *P. putida* added to soil (1×10^7 cfu g^{-1} soil) at $-10,000$ kPa to control *Fusarium* wilt could not be detected shortly after inoculation. However, when added to moist soil, which was then dried slowly to $-10,000$ kPa over a period of 5 weeks, the bacteria were able to adapt, and more than 5×10^5 cfu g^{-1} soil could be recovered [117]. This suggests that the physiological state of the species is influenced by the physicochemical condition of the soil and is significant to the subsequent behavior of the inoculant species.

A threshold inoculum number may be necessary, as reported by Iswandi et al. [111], who found that at least 2.3×10^6 pseudomonads per maize seed and 3.6×10^5 per barley seed were needed to establish the bacteria in the rhizosphere. Bashan [216] suggested that the optimum concentration for adding rhizobacteria was 1×10^5 to 1×10^6 cfu ml^{-1} when the method of application was irrigation; presumably, lower rates are necessary when seed inoculations or root dips are used. Other work suggests that the density of root colonization may be independent of the initial inoculum level, and no significant differences were found in root colonization resulting from different inoculum concentrations of rhizosphere—competent *Xanthomonas maltophilia* [98]. When a mixed inoculum, consisting of a *Pseudomonas* sp., a *Mycoplasma* sp., and a *Corynebacterium* sp., were introduced into a barley rhizosphere at 1×10^3, 1×10^5, or 1×10^7 cells mg^{-1} dry root, the maximum colonization density achieved by day 6 was 5×10^7 cells mg^{-1} dry root and was independent of inoculum density [217]. This may be further evidence that in some cases it is the availability of root exudates rather than inoculant numbers themselves that is the primary determinant in the success of a rhizosphere inoculant [218].

Not surprisingly, the effects of inoculum density of rhizobial cells have been much researched. A high inoculum density appears to be important for introduced rhizobia to overcome competition by indigenous rhizobia, which often reduce nodulation by introduced improved strains [119,220]. Many years ago, it was stated that under ideal conditions as few as 10 rhizobia per legume root can initiate nodulation [221]. However, in practice, where there is a requirement for inoculant survival until seed germination has occurred, a minimum of 1×10^4 to 1×10^5 seed^{-1} is more realistic. After germination, the surviving rhizobial cells will respond to root excretions and their numbers will increase [222].

The expression of the activities of rhizosphere inoculants can be manipulated to a degree by ensuring high inoculation densities at the site where most benefit will accrue. In most cases, this means that seed coating or root dipping, where the goal of the treatment is to achieve rapid colonization and, thereby, strongly influence the development of the rhizosphere population. Preemptive establishment is difficult to guarantee, as delivery methods are notoriously variable, and viable counts taken immediately after inoculation may be orders of magnitude lower than expected. Applying an aqueous microbial suspension directly into the seed furrows as the time of planting may be a practical way to promote colonization of the developing rhizosphere [223].

B. Use of Carriers

The use of appropriate methods for preparing, storing, and applying inoculants can greatly increase uniformity and survival in soils
and on seeds and roots. In general, survival in storage is influenced by the original culture medium, the physiological state of the
microorganism at harvest and before storage, the process and rate
of drying, the storage temperature, and whether or not the inoculant is entrapped in a carrier material [224]. A carrier may protect microorganisms from soil physical, chemical, and biological
stresses, as well as releasing viable cells into the soil over a prolonged period of time. Commonly used materials include peat, alginate, clay, oil, and polyacrylamide [225]. Peat may extend inoculant survival to over 6 months in some cases, with the most
important factors being the initial sterility of the peat, moisture,
temperature, and the presence of sticking agents such as methyl
cellulose to make the peat-based inoculant adhere to the seeds.
For example, a *Rhizobium* sp. (*Hedysarum coronarium* L.) survived
in a sterilized peat—gum arabic mixture for 21 days at 25°C [226].
Furthermore, surviving rhizobia were in high enough numbers
(i.e., $>1 \times 10^4$ seed^{-1}) to ensure nodulation. Peat does not have
a defined or constant composition and in some cases may actually
increase the disposition of plants to disease, thereby nullifying
any effect of, for example, a biocontrol agent [227]. Alginate encapsulation has worked well, both to prolong viability during storage [228] and to allow the controlled release of the inoculant into
the soil. Bashan [229] found that for *A. brasilense* and a *Pseudomonas* strain, the rate or release and time of survival of cells in
soil could be controlled by the degree of hardening of alginate
beads. The survival and activity of a chlorophenol-degrading
Flavobacterium sp., entrapped in microcapsules (2.50 μm in diameter) of alginate, agarose, or polyurethane [230], suggest applications in soil bioremediation [231].

A major factor determining the survival of *Rhizobium* species is
pH. In acidic soils the decline of *R. meliloti* inoculum is rapid, but
this may be counteracted by pelleting lucerne seeds with lime [232,
233], which neutralizes the immediate environment of the seedling
as well as supplying essential calcium [234,235].

C. Age and Physiological Condition

There is a temptation to assume that rapidly dividing cells, growing
under optimum conditions and harvested in late exponential phase,
are the most appropriate for inoculating soils. Indeed, the vast
majority of experiments use washed and resuspended cells taken
from a dense culture. However, data arising from this approach

may prove largely irrelevant, because the physiological and nutritional conditions encountered in soil and at the surface of the root and seed are very different from those used for growth in pure culture. Therefore, even if the inoculum survives introduction, a period of time may elapse during which the species becomes acclimated to its new environment. This adjustment may involve such diverse and fundamental processes as the induction of different substrate affinity and transport systems, a switch to alternative metabolic pathways, and adaptation to the physical and chemical stresses encountered in the soil. Consequently, it may be advisable to grow the inoculant in more environmentally realistic conditions (carbon and nitrogen limited, suboptimal temperature and pH, etc.) or to acclimate the bacteria after harvest and before inoculation. The latter goal may be achieved by a period of storage in an appropriate buffer, transfer to a nutrient-limiting medium, or holding the inoculum at a reduced temperature.

Genetic instability of inocula maintained on laboratory media is a recognized problem, so loss of a desired property (e.g., nodulation) during large-scale production or after release must be considered [15]; e.g., laboratory cultures of *R. leguminosarum* bv. *trifolii* have been shown to have significant variability in their symbiotic effectiveness [236].

D. Activity

The primary function of a released bacterium is not that it survives but that it expresses its desired property in soil. As a consequence, the recognition of inoculant activity and the factors that regulate its expression in vivo are important subjects for research. Regrettably, knowledge in this area is increasing slowly, because the heterogeneity of the soil environment and the affinity of bacteria and their metabolites for clay and organic colloidal surfaces [237,238] render many probes useless without further refinements.

Where bioremediation is the objective, a range of responses is possible after inoculation of soil. Concentrations of the xenobiotic may not be sufficient to sustain the growth of a specialized strain raised on high levels of, for example, a pesticide, and even an inoculant chosen specifically for its high affinity for low concentrations of the contaminant may abandon metabolism in favor of a more accessible and less exotic carbon, nitrogen, or energy source. In contrast, cosubstrate metabolism [239] and cometabolism [166] may enable the inoculant to achieve its desired objective, regardless of the target substrate concentration.

For an inoculant to be useful in soil it must either lyse to release its contents, in which case the survival rate of cells is of little consequence, or it must be active (and probably proliferate) and express a beneficial primary or secondary metabolic process. The activity (e.g., enzyme synthesis and secretion) should be constitutive or, if induced, the inducer agents must be present in the environment. A significant problem is that many useful compounds are secondary metabolites (e.g., antibiotics, some enzymes) produced only during late exponential or early stationary phase growth. However, microbes in soils rarely grow exponentially, as nutrients and other factors are rate limiting and the concentration of the beneficial metabolite produced may be very low. Nonetheless, metabolic activities in soil may be influential over millimeter or even micrometer distances, and even nanomolar concentrations of metabolites may prove effective in the microbial microenvironment [237, 240]. In fact, if the direct or indirect desired response is caused by one or a small number of cell products, whose production in soil is unpredictable, then it is worth considering these as amendments in themselves rather than introducing viable cells to produce them. For example, if there was a single rate-limiting step in a biodegradation process to which an environmentally stable enzyme could be applied, then the enzyme might prove a better choice of additive than the cell. Thus, a dehalogenase might initiate rapid metabolism of a pesticide by the indigenous microbiota, enzymic dephosphorylation could increase soluble orthophosphate [241], and the activity of β-glucosidase could accelerate the degradation of agricultural residues [242]. Again, delivery and efficacy of enzymes, antibiotics, polysaccharides, etc., can be manipulated by using various supports (e.g., enzyme adsorption and entrapment) and seed coats (e.g., pellets, films), either as protectants or as a means of ensuring controlled release of the active agent.

The activity of an inoculum is dependent on the extent to which the cells are adapted to living in soil. Introducing genes encoding specific beneficial properties into bacteria isolated from soil will probably be more successful than attempting to establish, in soil, bacteria isolated from other environments (on the basis that they already have some useful function). This is because the genes for "competitive success" are almost certainly greater in number and less easily identified than those for the synthesis and expression of a defined property. A good example of this approach is that of Skot et al. [243], who transferred the gene coding for the polypeptide toxin lethal to various coleopteran larvae from *Bacillus thuringiensis* (Bt) to *E. coli* (using a broad-host-range vector) and then to the soil bacterium *R. leguminosarum* by conjugation. Insect bioassays demonstrated that the toxin was expressed. Most importantly, however, nodulated pea (*Pisum sativum*) and white clover (*Trifolium repens*) suffered less root damage from the larvae of the clover weavil (*Sitona lepidus*) when inoculated with the transformed *Rhizobium* strain. The authors also identified a usual problem in using the construct in the field: the indigenous rhizobia outcompeted the introduced species, thereby producing nodulated plants with no resistance to insect attack. Furthermore, a constitutively expressed toxin may have deleterious side effects on other soil organisms. The latter problem is being addressed by making constructs in which the expression of the toxin gene is controlled by a *nif* gene promoter, thereby restricting toxin production to the nodule. In another experiment, the genes coding for *B. thuringiensis* endotoxins were introduced into a rhizosphere competent strain of *P. fluorescens* [244]. The transformed *P. fluorescens* cells were toxic toward larvae of the malaria mosquito *Anopheles stephensi* and to larvae of leatherjackets, *Tipula oleracea*. However, further research is needed to determine whether the *P. fluorescens* will produce enough toxin in the rhizosphere to control leatherjackets or the malaria mosquito.

An even closer association between Bt and the plant has been achieved by introducing a crystal protein gene into the plant endophytic bacterium *Clavibacter xyli* subsp. *cynodontis* [245]. When seeds coated with this bacterium germinate, the bacterium proliferates in the vascular system of the plant and produces insecticidal toxin that is protected from the environment of the soil. A wide variety of the insecticidal crystal protein (ICP) genes have been cloned and their gene products assessed for activity against various insect pests [246,247].

V. ENVIRONMENTAL EFFECTS

The potential benefits of adding naturally occurring or genetically modified bacteria to soil must be balanced against any deleterious effect on the indigenous microbiota and its activities. The ecological relation between inoculants and the indigenous population is of great significance because inoculant organisms may displace beneficial as well as harmful species.

The primary aim of adding bacteria to soil is to produce an effect on a target organism, accelerate the degradation of a xenobiotic, or stimulate a process beneficial to plant growth. However, researchers also want to determine whether an inoculant adversely alters the balance between species within the nontarget soil microbiota or causes a shift in the overall genetic composition of the community. At present, short- and long-term changes of this type are extremely difficult to measure. There is much debate, but no consensus, about the parameters that should be monitored and how to measure them, and in this regard there are many parallels to the screening of pesticides for side effects. For example, a rhizosphere inoculant might cause a change in the structure of the microbial community that could result in a decline in cellulosic microorganisms. The questions then would be: how should cellulase activity be assayed, for how long, and at what level does any decrease in activity become significant to the breakdown of organic matter and the mineralization of carbon?

An inoculant species must survive in soil in an appropriate location for it to have its intended effect, although this period of survival may be brief or prolonged. However, it is possible that the influence of an inoculant may outlast its survival. If the introduction of a biological control microorganism results in changes in the composition of the indigenous microbiota, these changes may, in turn, include the proliferation of other antagonistic organisms. For example, an introduced strain of *B. cereus* was shown to give prolonged control of damping-off caused by *Phytophthora* root rot of legumes [248]. However, populations of the *B. cereus* strain declined 10 days after planting soybean seeds that had been coated with the inoculant, and it was suggested that early establishment of the inoculant created selection pressures that altered the process of colonization of the seeds and roots by other microorganisms, including *Phytophthora* (G. S. Gilbert, J. Handelsman, and J. L. Parke, Abstr., Phytopath. 80:995, 1990). Such changes in the bacterial community could indirectly or directly result in biological control that was not dependent on the prolonged survival of the original inoculant.

A related phenomenon was revealed in the work of Thompson (R. J. Thompson, Ph.D. thesis, University of Kent, England) as a result of incorporating *Penicillium claviforme* metabolites into sugar beet seed pellets in a project to control damping-off disease caused by *Pythium ultimum* [249]. Although a biocontrol antibiotic was undetectable less than 24 h after seed sowing, protection against both pre- (0 to 3 days) and post- (3 to 10 days) emergence damping-off was maintained. It was suggested that the antibiotic impaired root colonization by antibiotic-sensitive *P. ultimum*, thereby allowing other nonpathogenic microorganisms to establish themselves in the rhizosphere. Once this happened, *P. ultimum* could not compete for a site at the root surface, even though the initial influence (the antibiotic) had disappeared.

A. Horizontal Gene Transfer

The need to monitor the survival and movement of introduced bacteria has been explained. Of equal if not greater importance is the transfer of genes from the inoculant to the indigenous microbiota. The significance of this is twofold. First, it is important to monitor the survival and expression of the novel metabolic potential within the entire soil microbiota, as well as the fate and activity of the inoculant species. Second, there is concern that introduced novel genes may give rise to a deleterious and irreversible change in the microbial ecology of the soil. There is limited evidence for horizontal gene transfer from introduced bacteria to native bacteria [250–252], but it has been suggested that such transfer may be as important a mode of maintenance of recombinant genes in the population as a whole as is the survival of the inoculant organisms [253]. It could be speculated that novel genes may persist if they are transferred to indigenous species that are better adapted to living in soil than the introduced recombinant bacteria. Although it is possible to measure the frequency of gene transfer, it is difficult to design experiments to detect whether such gene transfer will have harmful effects in soil [254]. Gene transfer may occur in natural habitats by transduction, transformation, or conjugation [255] and the rate of transfer is affected, particularly, by the number of donor and recipient organisms that are added [256]. In fact, most of the studies on the rates of gene transfer between different bacteria in soils have involved the introduction of donor and recipient bacteria, and the majority of this work has focused on conjugative transfer [257].

The frequency of plasmid transfer has been shown to depend on the survival of relatively high numbers of donor and recipient cells [258,251]. Thus, environmental factors such as soil type, the presence of plant roots, and nutrient availability, which cause differences in the survival of bacteria, will influence plasmid transfer. van Elsas et al. [259] found higher survival of *P. fluorescens* and greater stability of plasmids in a heavier-textured loam than in a loamy sand and found that plasmids were transferred at a frequency of up to 3.8×10^{-4} between species of *Pseudomonas* in the rhizosphere. Transconjugants were not detected in nonrhizosphere soil [258]. Greater transfer frequency in the rhizosphere could be the result of stimulation of microbial growth in the vicinity of roots and, possibly, the presence of root surfaces at which cell-to-cell contact can occur. Other modes of gene transfer, such as transformation, occur at greater frequencies for bacterial cells attached to sand surfaces than those in suspension [260].

Few studies have convincingly documented the rate of transfer of genes from introduced recombinant bacteria to indigenous bacteria in nonsterile soil. Henschke and Schmidt [252] showed that transfer of plasmids occurred from introduced *Enterobacter aerogenes* to indigenous *P. fluorescens* in nonsterile soil, and evidence that plasmid transfer occurs naturally in soil includes the exchange of *sym* plasmids (which encode genes for nitrogen fixation) between *Rhizobium* populations [261] and the movement of mercury resistance genes between different bacterial species [262]. The frequency at which plasmids are transferred from inoculant cells to indigenous bacteria in soil is not well documented. Using a DNA probe to an entire plasmid, van Elsas et al. [263] detected transfers of a plasmid (conferring resistance to kanamycin and rifampicin) from donor *P. fluorescens* cells to introduced recipient cells in soil. However, the plasmid could not be detected in any of the indigenous bacteria. Furthermore, van Elsas et al. [148] did not detect movement of a plasmid that contained a marker, *pat*, from introduced *P. fluorescens* to other bacteria occurring naturally in soil. In contrast, Henschke and Schmidt [252] demonstrated movement of plasmids from introduced recombinant *E. coli* to indigenous soil bacteria, most of which were fluorescent pseudomonads, and Smit et al. [250] showed that a self-transmissible, broad-host-range plasmid was transferred from *P. fluorescens* to species of *Pseudomonas*, *Enterobacter*, *Comamonas*, and *Alcaligenes*. Such plasmids would be expected to transfer at higher rates than non-self-transmissible plasmids or chromosomally inserted genes. Much further work is needed to document the rates of transfer of genetic material from introduced to indigenous bacteria before the risks, if any, of such transfer can be properly evaluated for useful inoculants.

B. Laboratory Microcosms and Field Experiments

Primary screens to determine whether an inoculant has the potential for beneficial activity are usually conducted under laboratory conditions before evaluation in the field. For most naturally occurring bacteria, field experiments can be conducted with isolates shown to have promise in laboratory screens. However, permission from regulatory authorities is needed before genetically modified inoculants can be tested in field soil, and risk assessments are based on the results of detailed contained microcosm studies. It is, therefore, essential that much thought be given to the design of soil microcosms to ensure that the laboratory model contains enough elements to reflect accurately the terrestrial environment. However, no two microcosms from different laboratories are similar, and whereas some are composed of undisturbed soil profiles, others consist of repacked, air-dried, sterilized, and homogenized soil. Although the latter often show experimental reproducibility, they should not be construed to be realistic models of the soil environment. The widespread use of microcosms in soil biology has been reviewed [264], and it is clear that there is a need for uniformity, if only to allow valid comparisons between experiments.

Usually, laboratory microcosms are maintained at constant temperature and moisture conditions and these conditions might be expected to give better survival of the inoculant in the microcosm than in the field. Bolton et al. [265] found that growth chamber microcosms which simulated field temperatures were better predictors of survival in the field than microcosms incubated continuously at 22°C. Bolton et al. [266] also suggested that the stage of plant growth may be a better reference point than time from inoculation for comparison of samples from laboratory and field experiments.

They showed that the survival of a *Pseudomonas* sp. was prolonged in field lysimeters and field plots when compared with laboratory microcosms. Thompson et al. [113] reported that an *Arthrobacter* sp. (autochthonous) and a *Flavobacterium* sp. (zymogenous) survived longer in the field than in a laboratory experiment. Moreover, the differences in survival observed between these two bacteria were recorded in both laboratory and field experiments. However, in laboratory soil, when bacteria were coinoculated simultaneously or sequentially, the *Flavobacterium* declined more rapidly than when inoculated by itself, whereas the rate of decrease of the *Arthrobacter* was not changed significantly. In contrast, in field soil the rate of decline of the *Flavobacterium* was similar whether it was introduced alone, simultaneously with *Arthrobacter*, or sequentially (i.e., 21 days before, or after, *Arthrobacter*). Such results suggest that caution should be used when predicting survival in the field based on the results of survival in laboratory soil, especially when more than one species of inoculant is involved. Because of the complexity of the soil ecosystem, many intrinsic and extrinsic factors are involved that are not subject to control, and even small-scale field trials may be poor at predicting survival and activity of an inoculant after large-scale application [2]. This suggests that monitoring of inoculant survival should be conducted in experiments of progressively increasing complexity and size.

Most experiments with recombinant organisms have been conducted in laboratory microcosms [30,267]. A recombinant *E. carotovora* was shown to have no significant effect on the numbers of total bacteria in a nonsterile soil microcosm [86], and in another study an endoglucanase gene that was either increased or decreased in copy number in *Pseudomonas solanacearum* had little effect on growth of the bacterium in a rhizosphere microcosm [268]. However, some recombinant bacteria have been introduced into soil in the field, including an *R. meliloti* with increased nitrogen fixation capabilities and *lacZY*-marked pseudomonads [269]. In these cases the inoculum did not migrate to adjacent plants and no detectable genetic transfer took place.

C. Release of Genetically Modified Bacteria

Bacterial genomes have been altered with the aim of introducing a beneficial function, such as biological control of plant pathogens [270], transfer of nitrogen fixation genes to bacteria able to survive in an environment not usually inhabited by nitrogen-fixing bacteria [271], degradation of xenobiotic compounds [45], and solubilization and leaching of minerals [272].

The possible problems associated with release of genetically modified bacteria (GMB) include the potential for unwanted transfer of the new genes to other species [263] and increased competitiveness of the inoculant such that beneficial indigenous species are dis-placed. (It could be argued that field experiments using bacterial mutants produced by traditional techniques, where the exact nature of any modifications made is unknown, are more risky than those resulting from genetic engineering, where the changes are defined. Nonetheless, regulations concerning the environmental release of the former are much less stringent than those that apply to GMB.) Evidence suggests that insertion of new genes should be an extra burden to a microorganism [273], but this could be counteracted if the genes coded for functions leading to increased survival in soil.

At present, there is an emphasis on assessing the environmental impacts of inoculating soil with recombinant bacteria. Where microorganisms are suspected of presenting a risk to human health or causing damage to the environment (for example, a recombinant bacterium that produces the *B. thuringiensis* toxin may proliferate and kill significant numbers of other insects in addition to the target species [2]), experiments are contained and risks are minimized [274]. It is possible that most recombinant organisms pose no more threat to the environment than their wild-type counterparts, but until this is proved, each case for release of GMB should be examined individually [273]. Decisions regarding release should be based on the character of the donor and/or recipient organism and include morphology, physiology, origin, genetic stability, pathogenicity spectrum, genetic properties of the recombinant organism, transfer frequency, expression of the new functions in the environments of interest, survival rates, and growth rates [275]. In other words, the properties of the organism, rather than the way in which it was produced, should be the focus of attention. In addition, microcosm data should be a requirement in an attempt to predict as early as possible whether the organism has improved fitness characteristics that may disrupt the soil ecosystem [276]. In all cases, the genotype and the phenotype need to be tracked, and therefore reliable methods for probing for the gene and detecting the microorganism itself must be developed.

For some uses, it may be desirable for a bacterial inoculant to persist in soil for an extended period. Thus, an antagonistic bacterium that specifically responds to the presence of a pathogen or a particular pollutant would remain dormant until the unique effector was recognized. The species would perform its intended role and then revert to its dormant (e.g., endospore) state. However, from an ecological (and commercial) standpoint, it will generally be preferable that released microorganisms should not survive for long periods in soil. One method for ensuring that a genetically altered bacterium will die out is to include a conditional suicide gene in the recombinant DNA. For example, Molin et al.

[277] transferred a *hok* gene into *E. coli* to restrict its use to a contained system. The *hok* gene product, which induces collapse of the transmembrane potential and causes cells to die, is repressed when tryptophan is present; these conditions can be applied in a bioreactor. If the bacterium should escape into the natural environment, where high levels of tryptophan are absent, the repressor complex is not formed and the cells die. An obvious disadvantage of this approach is that there is strong selection for mutations that cause insensitivity to the lethal protein [271].

VI. CONCLUSIONS

The future uses of microbial inoculants will be driven by commercial and environmental considerations, as well as by scientific advances. Permission to release appropriate microorganisms (including genetically engineered ones) might be forthcoming in certain situations, even in the absence of a complete understanding of the factors that determine survival and activity. Examples are (1) the use of biocontrol agents to replace an environmentally unacceptable and banned pesticide previously effective against an economically significant disease of a food crop and (2) the introduction of a pollutant-degrading bacterial species for the remediation of soil (for subsequent domestic, agricultural or recreational use) and the reduction of pollutant flow to ground water. The comparative urgency and severity of this type of problem, together with risk assessment, will influence the decision on whether or not to release.

The desire to increase agricultural productivity and to conserve the environment has stimulated an enormous volume of research on bacterial introductions into soil. Much of this research has revealed our poor understanding of soil microbial ecology and, therefore, the absence of a body of knowledge on which to base the manipulation of the soil-plant environment. Until we understand how to promote the activity of an inoculant and how to monitor and understand its effects on the soil system, progress toward the wide-scale commercialization of microbial inoculants will be slow.

REFERENCES

1. Fincham, J. R. S., and J. R. Ravetz. 1991. Genetically engineered organisms: benefits and risks. Open University Press, Milton Keynes, U.K.
2. Cavalieri, L.F. 1991. Scaling-up field testing of modified microorganisms. BioScience 41:568–574.
3. Curtiss, R. III. 1988. Engineering organisms for safety: what is necessary? p. 7–20. *In* M. Sussman, C. H. Collins, F. A.

Skinner, and D. E. Stewart-Tull (eds.), The release of ge-
netically-engineered micro-organisms. Academic Press, London.

4. Colwell, R. R., C. Somerville, I. Knight, and W. Straube.
    1988. Detection and monitoring of genetically-engineered mi-
    croorganisms, P. 47–60. *In* M. Sussman, C. H. Collins, F. A.
    Skinner, and D. E. Stewart-Tull (eds.), The release of ge-
    netically-engineered micro-organisms. Academic Press, Lon-
    don.

5. Lambert, B., and H. Joos. 1989. Fundamental aspects of
    rhizobacterial plant growth promotion research. Tibtech 7:
    215–219.

6. Sylvester-Bradley, R. R., R. D. Mosquera, and J. E. Mendez.
    1988. Selection of rhizobia for inoculation of forage legumes in
    savanna and rainforest soils of tropical America, p. 225–234.
    In D. P. Beck and L. A. Materon (eds.), Nitrogen fixation
    by legumes in mediterranean agriculture. Martinus Nijhoff,
    Dordrecht.

7. Giller, K. E., and K. J. Wilson. 1991. Nitrogen fixation in
    tropical cropping systems. CAB International, Wallingford.

8. Alexander, M. 1985. Ecological constraints on nitrogen fix-
    ation in agricultural ecosystems. Adv. Microbial Ecol. 8:163–
    183.

9. Mishustin, E. N. 1970. The importance of non-symbiotic ni-
    trogen fixing microorganisms in agricultural soil. Plant Soil
    32:545-554.

10. Brown, M. E. 1974. Seed and root bacterization. Annu.
    Rev. Phytopathol. 12:181–197.

11. Arshad, M., and W. T. Frankenberger. 1990. Response of
    Zea mays and *Lycopersicon esculentum* to the ethylene precur-
    sors, L-methionine and L-ethionine applied to soil. Plant Soil
    122:219–277.

12. Frankenberger, W. T., A. C. Chang, and M. Arshad. 1990.
    Response of *Raphanus sativus* to the auxin precursor, L-tryp-
    tophan, applied to soil. Plant Soil 129:235–241.

13. Nieto, K. F., and W. T. Frankenburger, Jr. 1990. Microbial
    production of cytokinins, p. 191–248. *In* J.-M. Bollag and G.
    Stotzky (eds.), Soil biochemistry, Vol. 6, Marcel Dekker, New
    York.

14. Dobereiner, J., and F. O. Pedrosa. 1987. Nitrogen-fixing
    bacteria in non-leguminous crop plants. Springer-Verlag,
    Berlin.

15. Giller, K. E., and J. M. Day. 1985. Nitrogen fixation in the
    rhizosphere: significance in natural and agricultural ecosys-
    tems, p. 127–147. *In* A. H. Fitter, D. Atkinson, D. J. Read,
    and M. B. Usher (eds.), Ecological interations in soil: plants,
    microbes and animals. Blackwell, Oxford.

16. Roger, P. A., and I. Watanabe. 1986. Technologies for utilizing biological nitrogen fixation in wetland rice: potentialities, current usage, and limiting factors. Fert. Res. 9:39–77.
17. Witty, J. F. 1979. Algal nitrogen fixation in temperate arable fields. Algal inoculation experiments. Plant Soil 52:165–183.
18. Reynaud, P. A., and B. Metting. 1988. Colonization potential of cyanobacteria on temperature irrigated soils in Washington State, USA. Biol. Agric. Hort. 5:197–208.
19. Rogers, S. L., K. A. Cook, and R. G. Burns. 1991. Microalgal and cyanobacterial soil inoculants and their effect on soil aggregate stability, p. 175–184. *In* W. S. Wilson (ed.), Advances in soil organic matter research: the impact on agriculture and the environment. Royal Soc. Chemistry, Cambridge.
20. Lewin, R. A. 1956. Extracellular polysaccharides of green algae. Can. J. Microbiol. 2:665–672.
21. Barclay, W. R., and R. A. Lewin. 1985. Microalgal polysaccharides as soil-conditioning agents. Plant Soil 88:159–169.
22. Metting, B. 1987. Dynamics of wet and dry aggregate stability from a three-year microalgal soil conditioning experiment in the field. Soil Sci. 143:139–143.
23. Gaur, A. C., and S. Gaind. 1983. Microbial solubilization of insoluble phosphates with particular reference to iron and aluminum phosphate. Sci. Cult. 48:110–112.
24. Sardina, M. G., J. L. Solarde, and R. J. Ertola. 1986. Solubilization of phosphorous from low grade minerals by microbial action. Biotechnology 8:247–252.
25. Gaind, S., and A. C. Gaur. 1991. Thermotolerant phosphate solubilizing microorganisms and their interaction with mung bean. Plant Soil 133:141–149.
26. Azcon, R., R. Rubio, and J. M. Barea. 1991. Selective interactions between different species of mycorrhizal fungi and *Rhizobium meliloti* strains, and their effects on growth, N_2-fixation and nutrition of *Medicago sativa* L. New Phytol. 117:399–404.
27. Abril, M. A., C. Michan, K. N. Timmis, and J. L. Ramos. 1989. Regulator and enzyme specificity of the TOL plasmid-encoded upper pathway for degradation of the substrate range of the pathway. J. Bacteriol. 171:6782–6790.
28. Ramos, J. L., A. Waserfellen, K. Rose, and K. N. Timmis. 1987. Redesigning metabolic routes: manipulation of TOL plasmid pathway for catabolism of alkylbenzoates. Science 235:593–596.
29. Rojo, F., D. Pieper, K. H. Engesser, H. J. Knackmuss, and K. N. Timmis. 1987. Assemblage of ortho cleavage routes for degradation of chloro- and methylaromatics. Science 238:1395–1398.

30. Bentjen, S. A., J. K. Fredrickson, P. van Voris, and S. W. Li. 1989. Intact soil-core microcosms for evaluating the fate and ecological impact of the release of genetically engineered microorganisms. Appl. Environ. Microbiol. 55:198–202.

31. Drahos, D. J., B. C. Hemming, and S. McPherson. 1986. Tracking recombinant organisms in the environment: β-galactosidase as a selectable non-antibiotic marker for fluorescent pseudomonads. Biotechnology 4:439–444.

32. Fulthorpe, R. R., and R. C. Wyndham. 1989. Survival and activity of a 3-chlorobenzoate-catabolic genotype in a natural system. Appl. Environ. Microbiol. 55:1584–1590.

33. Morgan, P., and R. J. Watkinson. 1989. Hydrocarbon degradation in soils and methods for soil biotreatment. CRC Crit. Rev. Biotech. 8:305–333.

34. Shailubhai, K. 1986. Treatment of petroleum industry oil sludge in soil. Tibtech 4:202–206.

35. Clark, C. G., and S. J. L. Wright. 1970. Detoxication of isopropyl N-phenylcarbamate (IPC) and isopropyl N-3-chlorophenylcarbamate (CIPC) in soil, and isolation of IPC-metabolizing bacteria. Soil Biol. Biochem. 2:19–27.

36. Kilbane, J. J., D. K. Chatterjee, and A. M. Chakrabarty. 1983. Detoxification of 2,4,5-trichlorophenoxyacetate acid from contaminated soil by *Pseudomonas cepacia*. Appl. Environ. Microbiol. 45:1697–1699.

37. Kearney, P. C., J. S. Karns, and W. W. Mulbry. 1986. Engineering soil microorganisms for pesticide degradation. Pestic. Sci. Biotechnol. Proc. Int. Congr. Pestic. Chem. 6th, p. 591–596.

38. Barles, R. W., C. G. Daughton, and D. P. H. Hsieh. 1979. Accelerated parathion degradation in soil inoculated with acclimated bacteria under field conditions. Arch. Environ. Contam. Toxicol. 8:647.

39. van der Meer, J. R., W. Roelotson, G. Schraa, and A. J. B. Zehnder. 1987. Degradation of low concentrations of dichlorobenzenes and 1,2,4-bichlorobenzene by *Pseudomonas* sp strain P51 in non-sterile soil columns. FEMS Microbiol. Ecol. 45:333–341.

40. Chaudry, G. R., and L. Cortez. 1988. Degradation of bromacil by a *Pseudomonas* sp. Appl. Environ. Microbiol. 45:1316–1323.

41. Doyle, J. D., K. A. Short, G. Stotzky, R. J. King, and R. H. Olson. 1991. Ecologically significant effects of *Pseudomonas putida* PP0301 (pR0103), genetically engineered to degrad 2,4-dichlorophenoxyacetate, on microbial populations and processes in soil. Can. J. Microbiol. 37:682–691.

42. Short, K. A., J. D. Doyle, R. J. King, R. J. Seidler, G. Stotzky, and R. H. Olsen. 1991. Effects of 2,4-dichlorophenol, a

metabolite of a genetically engineered bacterium, and 2,4-dichlorophenoxyacetate on some microorganism-mediated ecological processes in soil. Appl. Environ. Microbiol. 57:412–418.

43. Salkinoja-Salonen, M., P. Middeldorp, M. Briglia, R. Valo, M. Haggblom, and A. McBain. 1989. Cleanup of old industrial sites, p. 347–367. *In* D. Kamely, A. Chakrabarty, and G. Olemm (eds.), Advances in Applied Biotechnology, Vol.' 4. Gulf Publishing Co., Houston, Texas.

44. Millhomme, H., D. Vega, J. L. Marty, and J. Bastida. 1988. Degradation de l'herbicide chloroprophame dans un sol: role de l'introduction de *Pseudomonas alcaligenes* et de *Pseudomonas cepacia*. Soil Biol. Biochem. 21:307–311.

45. Golovleva, L. A., R. N. Pertsova, A. M. Boronin, V. M. Travkin, and S. A. Kozlovsky. 1988. Kelthane degradation by genetically engineered *Pseudomonas aeruginosa* BS 827 in a soil ecosystem. Appl. Environ. Microbiol. 54:1587–1590.

46. Atlas, R. M. 1977. Simulated petroleum biodegradation. Crit. Rev. Microbiol. 5:371–386.

47. Leahy, J. G., and R. R. Colwell. 1990. Microbial degradation of hydrocarbons in the environment. Microbiol. Rev. 54:305–315.

48. Dibble, J. T., and R. Bartha. 1979. Effect of environmental parameters on the biodegradation of oil sludge. Appl. Environ. Microbiol. 37:729–739.

49. Bewley, R. J. F. 1988. Use of microbial processes for the reclamation of contaminated land. Report for Biototal Ltd. Cardiff. Seesoil 5:1–21.

50. van den Berg, R. J., A. H. M. Verheul, and D. H. Eikelboom. *In situ* biorestoration of an oil contaminated subsoil. Water Sci. Tech. 20:255–256.

51. Davison, J. 1988. Plant beneficial bacteria. Biotechnology 6:282–286.

52. Khachatourians, G. G. 1986. Production and use of biological pest control agents. Tibtech 213:120–123.

53. Peferoen, M. 1991. *Bacillus thuringiensis* in crop protection. Agro-Industry Hi-Tech. 6:59.

54. Moore, L. W., and G. Warren. 1979. *Agrobacterium radiobacter* strain 84 and biological control of crown gall. Annu. Rev. Phytophatol. 17:163–179.

55. Kerr, A., and M. E. Tate. 1980. Agrocins and the biological control of crown gall. Microbiol. Sci. 1:1–14.

56. Simon, A., K. Sivasithamparam, and G. C. Macnish. 1987. Biological suppression of the saprophytic growth of *Gaeumannomyces graminis* var. *tritici* in soil. Can. J. Microbiol. 33: 515–519.

57. Weller, D. M., and R. J. Cook. 1986. Suppression of root diseases of wheat by fluorescent pseudomonads and mechanisms

of action, p. 99—107. *In* T. R. Swinburne (ed.), Iron, siderophores and plant diseases. Plenum Publishing, New York

58. Brisbane, P. G., and A. G. Rovira. 1988. Mechanisms of
inhibition of *Gaeumannomyces graminis* var. *tritici* by fluorescent pseudomonads. Plant Pathol. 37:104—111.

59. Hommer, Y., Z. Sato, F. Hirayama, K. Konno, H. Shirahama,
and T. Suzui. 1989. Production of antibiotics by *Pseudomonas cepacia* as an agent for biological control of soilborne plant
pathogens. Soil Biol. Biochem. 5:723—728.

60. Bakker, A. W., and B. Schippers. 1987. Microbial cyanide
production in the rhizosphere in relation to potato yield reduction and *Pseudomonas* spp.-mediated plant growth stimulations. Soil Biol. Biochem. 19:451—457.

61. Alström, S., and R. G. Burns. 1989. Cyanide production
by rhizobacteria as a possible mechanism of plant growth inhibition. Biol. Fertil. Soils 7:232—238.

62. Becker, J., and R. J. Cook. 1988. Role of siderophores in
suppression of *Pythium* species and production of increased-
growth response of wheat by fluorescent pseudomonads. Phytopathology 78:778—782.

63. Marschner, P., J. S. Ascher, and R. D. Graham. 1991. Effect of manganese-reducing rhizosphere bacteria on the growth
of *Gaeumannomyces graminis* var. *tritici* and on manganese
uptake by wheat (*Triticum aestivum* L). Biol. Fertil. Soils
12:33—38.

64. Crowley, D. E., C. P. P. Reid, and P. J. Szanislo. 1987.
Microbial siderophores as iron sources for plants, p. 375—386.
In G. Winklemann, D. van der Helm, and J. B. Nielands (eds.),
Iron transport in microbes, plants and animals. VCH Publica-
tions, New York.

65. Graham, R. D., and A. D. Rovira. 1986. A role for manganese in the resistance of wheat plants to take-all. Plant
Soil 78:441—444.

66. Défago, G., and D. Haas. 1990. Pseudomonads as antagonists of soilborne plant pathogens: modes of action and
genetic analysis, p. 249—292. *In* J.-M. Bollag and G.
Stotzky (eds.), Soil biochemistry, Vol. 6. Marcel Dekker,
New York.

67. Harman, G. E., and R. D. Lumsden. 1990. Biological disease control, p. 259—280. *In* J. M. Lynch (ed.), The rhizosphere. John Wiley & Sons, Chichester.

68. Burns, R. G., and J. A. Davies. 1986. The microbiology of
soil structure. Biol. Agric. 3:95—113.

69. Foster, S. M. 1990. The role of microorganisms in aggregate
formation and soil stabilization: types of aggregation. Arid
Soil Res. Rehabil. 4:85—98.

70. Piccolo, A., and J. S. C. Mbagwu. 1990. Effects of humic substances and surfactants on the stability of soil aggregates. Soil Sci. 147:47–54.

71. Nadler, A., and J. Letey. 1989. Organic polyanions effect on aggregation of structurally disrupted soil. Soil Sci. 148: 346–360.

72. Painuli, D. K., and I. P. Abrol. 1988. Improving aggregate stability of sodic sandy loam soils by organics. Catena 15: 229–239.

73. Foster, S. M., and T. H. Nicholson. 1981. Aggregation of sand from a maritime embryo sand dune by microorganisms and higher plants. Soil Biol. Biochem. 13:199–203.

74. Metting, B. 1990. Microalgae applications in agriculture. Dev. Ind. Microbiol. 31:265–270.

75. Rao, D. L. N., and R. G. Burns. 1990. The effect of surface growth of blue-green algae and bryophytes on some microbiological, biochemical, and physical soil properties. Biol. Fertil. Soils 9:239–244.

76. Colwell, R. R. 1987. From counts to clones. J. Appl. Bacteriol. Symp. Suppl., p. 1S–6S.

77. Ford, S., and B. H. Olson. 1988. Methods for detecting genetically engineered microorganisms in the environment, p. 45–79. *In* K. C. Marshall (ed.), Advances in microbial ecology, Vol. 10. Plenum, London.

78. Gray, T. R. G. 1990. Methods for studying the microbial ecology of soil, p. 309–343. *In* R. Grigorova and J. R. Norris (eds.), Methods in microbiology. Academic Press, London.

79. Colwell, R. R., P. R. Brayton, D. J. Grimes, D. B. Roszak, S. A. Huq, and L. M. Palmer. 1985. Viable but non-culturable *Vibrio cholerae* and related pathogens in the environment: implications for release of genetically engineered microorganisms. Biotechnology 3:817–820.

80. Roszak, D. B., and R. R. Colwell. 1987. Survival strategies of bacteria in the natural environment. Microbiol. Rev. 51:365–379.

81. Burns, R. G. 1989. Microbial and enzymic activities in soil biofilms, p. 333–349. *In* W. G. Charaklis and P. A. Wilderer (eds.), Structure and function of biofilms. John Wiley & Sons, Chichester.

82. Olsen, R. A., and L. R. Bakken. 1987. Viability of soil bacteria: optimization of plate-counting technique and comparison between total counts and plate counts within different size groups. Microb. Ecol. 13:59–74.

83. Postma, J., J. D. van Elsas, J. M. Govaert, and J. A. van Veen. 1988. The dynamics of *Rhizobium leguminosarum*

biovar *trifolii* introduced into soil as determined by immunofluorescence and selective plating techniques. FEMS Microbiol. Ecol. 53:251–260.

84. Fredrickson, J. K., D. F. Bezdicek, F. J. Brockman, and
S. W. Li. 1988. Enumeration of Tn5 mutant bacteria in soil
by using a most-probable-number-DNA hybridization procedure and antibiotic resistance. Appl. Environ. Microbiol. 54:
446–453.

85. Wessendorf, J., and F. Lingens. 1989. Effect of culture and
soil conditions on survival of *Pseudomonas fluorescens* R1 in
soil. Appl. Microbiol. Biotech. 31:97–102.

86. Orvos, D. R., G. H. Lacy, and J. J. Cairns. 1990. Genetically engineered *Erwinia carotovora*: survival, intraspecific
competition and effects upon selected bacterial genera. Appl.
Environ. Microbiol. 56:1689–1694.

87. Wellington, E. M. H., N. Cresswell, and V. A. Saunders.
1990. Growth and survival of streptomycete inoculants and
extent of plasmid transfer in sterile and non sterile soil.
Appl. Environ. Microbiol. 56:1413–1419.

88. Hartel, P. G., J. W. Williamson, and M. A. Schell. 1990.
Growth of genetically altered *Pseudomonas solanacearum* in
soil and rhizosphere. J. Soil Sci. Soc. Am. 54:1021–1025.

89. Anderson, J. R., and J. M. Slinger. 1975. Europium chelate
and fluorescent brightener staining of soil propagules and
their photomicrographic counting. 1. Methods. Soil Biol.
Biochem. 7:205–209.

90. Ramsay, A. J. 1984. Extraction of bacteria from soil: efficiency of shaking or ultrasonication as indicated by direct
counts and autoradiography. Soil Biol. Biochem. 16:475–481.

91. Bakken, L. R. 1985. Separation and purification of bacteria
from soil. Appl. Environ. Microbiol. 49:1482–1487.

92. Holben, W. E., J. K. Jansson, B. K. Chelm, and J. M.
Tiedje. 1988. DNA probe method for the detection of spe-
cific microorganisms in the soil bacterial community. Appl.
Environ. Microbiol. 54:703–711.

93. Griffiths, B. S., and K. Ritz. 1988. A technique to extract,
enumerate and measure protozoa from mineral soils. Soil Biol.
Biochem. 20:163–173.

94. Hartel, P. G., J. W. Billingsley, and J. W. Williamson. 1989.
Styrofoam cup-membrane assembly for studying microorganism-
root interactions. Appl. Environ. Microbiol. 55:1291–1294.

95. van Elsas, J. D., A. F. Dijkstra, J. M. Govaert, and J. A.
van Veen. 1986. Survival of *Pseudomonas fluorescens* and
Bacillus subtilis introduced into two soils of different texture
in field microplots. FEMS Microbiol. Ecol. 38:151–160.

96. Henschke, R. B., and F. R. J. Schmidt. 1989. Survival,
distribution, and gene transfer of bacteria in a compact soil
microcosm system. Biol. Fertil. Soils 8:19–24.

97. Trevors, J. T., and G. Berg. 1989. Conjugal RP4 transfer between pseudomonads in soil and recovery of RP4 plasmid DNA from soil. Syst. Appl. Microbiol. 11:223–227.
98. Juhnke, M. E., D. E. Mathre, and D. C. Sands. 1989. Relationship between bacterial seed inoculum density and rhizosphere colonization of spring wheat. Soil Biol. Biochem. 21: 591–595.
99. Herbert, R. A. 1990. Methods for enumerating microorganisms and determining biomass in natural environments, p. 1–39. *In* R. Grigorova and J. R. Norris (eds.), Methods in microbiology. Academic Press, London.
100. Höfte, M., M. Mergeay, and W. Verstraete. 1990. Marking the rhizopseudomonas strain 7NSK$_2$ with a mu d(*lac*) element for ecological studies. Appl. Environ. Microbiol. 56:1046–1052.
101. Miles, A. A., and S. S. Misra. 1938. The estimation of the bactericidal power of blood. J. Hyg. 38:732–748.
102. Hartman, P. A. 1968. Miniaturized microbiological methods. Academic Press, New York.
103. Kramer, J. M., M. Kendall, and R. J. Gilbert. 1979. Evaluation of the spiral plate and laser colony counting techniques for the enumeration of bacteria in foods. Eur. J. Appl. Microbiol. Biotech. 6:289–299.
104. Scanferlato, V. S., D. R. Orvos, G. H. Lacy, and J. Cairns. 1990. Enumerating low densities of genetically engineered *Erwinia carotovora* in soil. Lett. Appl. Microbiol. 10:55–59.
105. Rodrigues, U. M., and R. G. Kroll. 1988. Rapid selective enumeration of bacteria in foods using a microcolony epifluorescence microscopy technique. J. Appl. Bacteriol. 64:65–78.
106. Postma, J., and H. J. Altemüller. 1990. Bacteria in thin soil sections stained with the fluorescent brightener calcofluor white M2R. Soil Biol. Biochem. 22:89–96.
107. Foster, R. C. 1988. Microenvironments of soil microorganisms. Biol. Fertil. Soils 6:189–203.
108. Wynn-Williams, D. D. 1988. Television image analysis of microbial communities in Antarctic fellfields. Polarforschung 58:239–249.
109. Dobereiner, J., and V. L. D. Baldini. 1979. Selective infection of maize roots by streptomycin-resistant *Azospirillum lipoferum* and other bacteria. Can. J. Microbiol. 25:1264–1269.
110. van Elsas, J. D., and M. T. P. R. R. Pereira, 1986. Occurrence of antibiotic resistance among bacilli in Brazilian soils and the possible involvement of resistance plasmids. Plant Soil 94:213–226.
111. Iswandi, A., P. Bossier, J. Vandenabeele, and W. Verstraete. 1987. Influence of the inoculation density of the rhizopseudomonad strain 7NSK2 on the growth and the composition of the

root microbial community of maize (*Zea mays*) and barley (*Hordeum vulgare*). Biol. Fertil. Soils 4:119–123.

112. Norelli, J. L., T. J. Burr, A. M. Lo Cicero, M. T. Gilbert, and B. H. Katz. 1991. Homologous streptomycin resistance gene present among diverse gram-negative bacteria in New York State apple orchards. Appl. Environ. Microbiol. 57: 486–491.

113. Thompson, I. P., C. S. Young, K. A. Cook, G. Lethbridge, and R. G. Burns. 1992. Survival of two ecologically distinct bacteria (*Flavobacterium* and *Arthrobacter*) in unplanted and planted rhizosphere soil: field studies. Soil Biol. Biochem. 24:1–14.

114. Postma, J., C. H. Hok-a-Hin, and J. H. Oude Voshaar. 1990. Influence of the inoculum density on the growth and survival of *Rhizobium leguminosarum* biovar *trifolii* introduced into sterile and non-sterile loamy sand and silt loam. FEMS Microbiol. Ecol. 73:49–58.

115. Thompson, I. P., K. A. Cook, G. Lethbridge, and R. G. Burns. 1990. Survival of two ecologically distinct bacteria (*Flavobacterium* and *Arthrobacter*) in unplanted and rhizosphere soil: laboratory studies. Soil Biol. Biochem. 22: 1029–1037.

116. Danso, S. K. A., M. Habte, and M. Alexander. 1973. Estimating the density of individual bacterial populations introduced into natural ecosystems. Can. J. Microbiol. 19:1450–1451.

117. Dupler, M., and R. Baker. 1984. Survival of *Pseudomonas putida*, a biological control agent, in soil. Phytopathology 74:195–200.

118. Bryson, V., and W. Szybalsky. 1952. Microbial selection. Science 116:45–51.

119. Carlton, B. C., and B. J. Brown. 1981. Gene mutation, p. 222–242. *In* G. Gerhardt (ed.), Manual of methods for general bacteriology. American Society for Microbiology, Washington, D.C.

120. Reddy, M. S., and J. E. Rahe. 1989. Growth effects associated with seed bacterization not correlated with populations of *Bacillus subtilis* inoculant in onion seedling rhizospheres. Soil Biol. Biochem. 21:373–378.

121. Pillai, S. D., and I. L. Pepper. 1991. Transposon Tn5 as an identifiable marker in rhizobia: survival and genetic stability of Tn5 mutant bean rhizobia under temperature stressed conditions in desert soils. Microb. Ecol. 21:21–33.

122. Heynen, C. E., J. D. van Elsas, P. J. Kuikman, and J. A. van Veen. 1988. Dynamics of *Rhizobium leguminosarum* biovar *trifolii* in soil: the effect of bentonite clay on predation by protozoa. Soil Biol. Biochem. 20:483–488.

123. Griffiths, R. P., C. L. Moyer, B. A. Caldwell, C. Ye, and R. Y. Morita. 1990. Long term starvation-induced loss of antibiotic resistance in bacteria. Microb. Ecol. 19:251–258.
124. Dijkstra, A. F., J. M. Govaert, G. H. N. Scholten, and J. D. van Elsas. 1987. A soil chamber for studying the bacterial distribution in the vicinity of roots. Soil Biol. Biochem. 19:351–352.
125. Dewey, F. M. 1988. Development of immunological diagnostic assays for fungal plant pathogens. Brighton crop protection conference—pests and diseases 2:777–786.
126. Mason, J., and R. G. Burns. 1990. Production of a monoclonal antibody specific for a *Flavobacterium* species isolated from soil. FEMS Microbiol. Ecol. 73:299–308.
127. Trevors, J. T., and J. D. van Elsas. 1989. A review of selected methods in microbial genetics. Can. J. Microbiol. 35:895–902.
128. Wright, S. F. 1990. Production and epitope analysis of monoclonal antibodies against a *Rhizobium leguminosarum* biovar *trifolii* strain. Appl. Environ. Microbiol. 56:2262–2264.
129. Leeman, M., J. M. Raaijmakers, P. A. H. M. Bakker, and B. Schippers. 1991. Immunofluorescence colony staining for monitoring pseudomonads introduced into soil, p. 374–379. In A. B. R. Beemster, G. J. Bollen, M. Gerlagh, M. A. Ruissen, B. Schippers, and A. Tempel (eds.), Biotic interactions and soil-borne diseases. Proceedings of the first conference of the European Foundation for Plant Pathology. Elsevier, Amsterdam.
130. Desmonts, C., J. Minet, R. R. Colwell, and M. Cormier. 1990. Fluorescent-antibody method useful for detecting viable but non-culturable *Salmonella* spp. in chlorinated wastewater. Appl. Environ. Microbiol. 56:1448–1452.
131. Bohlool, B. B., and E. L. Schmidt. 1980. The immunofluorescence approach in microbial ecology. Microb. Ecol. 4:203–241.
132. Page, S., and R. G. Burns. 1991. Flow cytometry as a means of enumerating bacteria introduced into soil. Soil Biol. Biochem. 23:1025–1028.
133. Johne, B., and J. Jarp. 1988. A rapid assay for protein-A in *Staphylococcus aureus* strains, using immunomagnetic mono-sized polymer particles. Acta Pathol. Microbiol. Scand. 96:1–7.
134. Lund, A. A., L. Helleman, and F. Vartdal. 1988. Rapid isolation of K88[+] *Escherichia coli* by using immunomagnetic particles. J. Clin. Microbiol. 26:2572–2575.
135. Morgan, J. A. W., C. Winstanley, R. W. Pickup, and J. R. Saunders. 1991. Rapid immunocapture of *Pseudomonas putida* cells from lake water by using bacterial flagella. Appl. Environ. Microbiol. 57:503–509.

136. Trevors, J. T. 1985. DNA probes for the detection of specific genes in bacteria isolated from the environment. Trends Biotech. 3:291–293.

137. Steffan, R. J., J. Goksoyr, A. K. Bej, and R. M. Atlas. 1988. Recovery of DNA from soils and sediments. Appl. Environ. Microbiol. 54:2908–2915.

138. Selenska, S., and W. Klingmüller. 1991. DNA recovery and direct detection of Tn5 sequences from soil. Lett. Appl. Microbiol. 13:21–24.

139. Lorenz, M. G., and W. Wackernagel. 1989. Adsorption of DNA to sand and variable degradation rates of adsorbed DNA. Appl. Environ. Microbiol. 53:2945–2952.

140. Stewart, G. J., and C. D. Singalliano. 1990. Detection of horizontal gene transfer by natural transformation in native and introduced species of bacteria in marine and synthetic sediments. Appl. Environ. Microbiol. 56:1818–1824.

141. Tsai, Y. L., M. J. Park, and B. H. Olson. 1991. Rapid method for direct extraction of mRNA from seeded soils. Appl. Environ. Microbiol. 57:765–768.

142. Pitchard, S. L., and J. H. Paul. 1991. Detection of gene expression in genetically engineered microorganisms and natural phytoplankton populations in the marine environment by mRNA analysis. Appl. Environ. Microbiol. 57:1721–1727.

143. Steffan, R. J., and R. M. Atlas. 1988. DNA amplification to enhance detection of genetically enginnered bacteria in environmental samples. Appl. Environ. Microbiol. 54:2185–2191.

144. Saiki, R. F., D. H. Gelfand, S. Stoffel, S. S. Scharf, and R. Higuchi. 1988. Primer-directed enzymatic amplification of DNA with thermostable DNA polymerase. Science 239:487–491.

145. Arnheim, N. T. White, and W. E. Rainey. 1990. Application of PCR: organismal and population biology. BioScience 40:174–182.

146. Schesser, K., A. Luder, and J. M. Henson. 1991. Use of polymerase chain reaction to detect the take-all fungus, *Gaeumannomyces graminis*, in infected wheat plants. Appl. Environ. Microbiol. 57:553–556.

147. Pillai, S. D., K. L. Josephson, R. L. Bailey, C. P. Gerba, and I. L. Pepper. 1991. Rapid method for processing soil samples for polymerase chain reaction amplification of specific gene sequences. Appl. Environ. Microbiol. 57:2283–2286.

148. van Elsas, J. D., L. S. van Overbeek, and R. Fouchier. 1991. A specific marker, *pat*, for studying the fate of introduced bacteria and their DNA in soil using a combination of detection techniques. Plant Soil 138:49–60.

149. Atlas, R. M., and A. K. Bej. 1990. Detecting bacterial pathogens in environmental water samples by using PCR and gene probes, p. 399–406. *In* M. A. Innis, D. H. Gelfand, J. J. Sninsky, and T. J. White (eds.), PCR protocols, a guide to methods and applications. Academic Press, San Diego.

150. Tsai, Y. L., and B. H. Olson. 1992. Detection of low numbers of bacterial cells in soils and sediments by polymerase chain reaction. Appl. Environ. Microbiol. 58:754–757.

151. Rattray, E. A. S., J. I. Prosser, K. Killham, and L. A. Glover. 1990. Luminescence-based nonextractive technique for *in situ* detection of *Escherichia coli* in soil. Appl. Environ. Microbiol. 56:3368–3374.

152. Atlas, R. M., and G. S. Sayler. 1988. Tracking microorganisms and genes in the environment, p. 31–45. *In* G. S. Omenn (ed.), Environmental biotechnology: reducing risks from environmental chemicals through biotechnology, Plenum Publishing, New York.

153. Hong, Y., J. J. Pasternak, and B. R. Glick, 1991. Biological consequences of plasmid transformation of the plant growth promoting rhizobacterium *Pseudomonas putida* GR12-2. Can. J. Microbiol. 37:796–799.

154. Chao, W. L., and R. L. Feng. 1990. Survival of genetically engineered *Escherichia coli* in natural soil and river water. J. Appl. Bacteriol. 68:319–325.

155 Devanas, M. A., D. Rafaeli-Eshkol, and G. Stotzky. 1986. Survival of plasmid-containing strains of *Escherichia coli* in soil: effect of plasmid size and nutrients on survival of hosts and maintenance of plasmids. Curr. Microbiol. 13: 269–277.

156. van Elsas, J., T. Trevors, and L. S. van Overbeek. 1991. Influence of soil properties on the vertical movement of genetically-marked *Pseudomonas fluorescens* through large soil microcosms. Biol. Fertil. Soils 10:249–255.

157. Ramos, J. L., E. Duque, and M. I. Ramos-Gonzalez. 1991. Survival in soils of an herbicide resistant *Pseudomonas putida* strain bearing a recombinant TOL plasmid. Appl. Environ. Microbiol. 57:260–266.

158. Winstanley, C., J. A. W. Morgan, R. W. Pickup, and J. R. Saunders. 1991. Use of *xylE* marker gene to monitor survival of recombinant *Pseudomonas putida* populations in lake water by culture on nonselective media. Appl. Environ. Microbiol. 57:1905–1913.

159. King, R. J., K. A. Short, and R. J. Seidler. 1991. Assay for detection and enumeration of genetically engineered microorganisms which is based on the activity of a deregulated

2,4-dichlorophenoxyacetate monooxygenase. Appl. Environ. Microbiol. 57:1790–1792.

160. Wilson, K. J., K. E. Giller, and R. A. Jefferson. 1991. β-Glucuronidase (GUS) operon fusions as a tool for studying plant-microbe interactions. Adv. Mol. Gen. Plant-Microbe Interact. 1:226–229.

161. Grant, F. A., L. A. Glover, K. Kilham, and J. J. Prosser. 1991. Luminescence-based viable cell enumeration of *Erwinia carotovora* in soil. Soil Biol. Biochem. 23:1021–1024.

162. de Weger, L. A., P. Dunbar, W. F. Mahafee, B. J. Lugtenberg, and G. S. Sayler. 1991. Use of bioluminescence markers to detect *Pseudomonas* spp. in the rhizosphere. Appl. Environ. Microbiol. 57:3641–3644.

163. Shaw, J. J., F. Dane, D. Geiger, and J. W. Kloepper. 1992. Use of bioluminescence for detection of genetically engineered microorganisms released into the environment. Appl. Environ. Microbiol. 58:267–273.

164. Pickup, R. W. 1991. Development of molecular methods for the detection of specific bacteria in the environment. J. Gen. Microbiol. 137:1009–1019.

165. Jensen, E. S., and L. H. Sorensen. 1987. Survival of *Rhizobium leguminosarum* in soil after addition as inoculant. FEMS Microbiol. Ecol. 45:221–226.

166. Cork, D. J., and J. D. Kreuger. 1991. Microbial transformations of herbicides and pesticides. Adv. Appl. Microbial. 36:1–66.

167. Postma, J., C. H. Hok-a-Hin, J. M. T. Schotman, C. A. Wijffelman, and J. A. van Veen. 1991. Population dynamics of *Rhizobium leguminosarum* Tn5 mutants with altered cell surface properties introduced into sterile and nonsterile soils. Appl. Environ. Microbiol. 57:649–654.

168. Paul, E. A., and F. E. Clark. 1989. Soil microbiology and biochemistry. Academic Press, London.

169. Roper, M. M., and K. C. Marshall. 1978. Effect of clay particle size on clay–*Escherichia coli*–bacteriophage interactions. J. Gen. Microbiol. 106:187–189.

170. Roper, M. M., and K. C. Marshall. 1978. Effects of a clay mineral on microbial predation and parasitism of *Escherichia coli*. Microb. Ecol. 4:279–298.

171. Moffet, M. L., J. E. Giles, and B. A. Wood. 1983. Survival of *Pseudomonas solanacearum* biovars 2 and 3 in soil: effect of moisture and soil type. Soil Biol. Biochem. 15:587–591.

172. Park, E. W., and S. M. Lim. 1985. Overwintering of *Pseudomonas syringae* pv. *glycinia* in the field. Phytopathology 75:520–524.

173. Postma, J., J. A. van Veen, and S. Walter. 1989. Influence of different initial soil moisture contents on the distribution and population dynamics of introduced *Rhizobium leguminosarum* biovar *trifolii*. Soil Biol. Biochem. 21:437–442.
174. Heijnen, C. E., and J. A. van Veen. 1991. A determination of protective microhabitats for bacteria introduced into soil. FEMS Microbiol. Ecol. 85:73–90.
175. Wardle, D. A., and D. Parkinson. 1990. Interactions between microclimatic variables and the soil microbial biomass. Biol. Fertil. Soils 9:273–280.
176. Postma, J., and J. A. van Veen. 1990. Habitable pore space and survival of *Rhizobium leguminosarum* biovar *trifolii* introduced into soil. Microb. Ecol. 19:149–161.
177. Rutherford, P. M., and N. G. Juma. 1992. Influence of texture on habitable pore space and bacterial-protozoan populations in soil. Biol. Fertil. Soils 12:221–227.
178. Kuikman, P. J., A. G. Jansen, and J. A. van Veen. 1991. ^{15}N-nitrogen mineralization from bacteria by protozoan grazing at different soil moisture regimes. Soil Biol. Biochem. 23:193–200.
179. Seong, K. Y., M. Hofte, J. Boelens, and W. Verstraete. 1991. Growth, survival, and root colonization of plant growth beneficial *Pseudomonas fluorescens* ANP15 and *Pseudomonas aeruginosa* 7NSK2 at different temperatures. Soil Biol. Biochem. 23:423–428.
180. Parke, J. L., R. Moen, A. D. Rovira, and G. D. Bowen. 1986. Soil water flow affects the rhizosphere distribution of a seed-borne biological control agent, *Pseudomonas fluorescens*. Soil Biol. Biochem. 15:583–588.
181. Worral, V., and R. J. Roughley. 1991. Vertical movement of *Rhizobium leguminosarum* bv. *trifolii* in soil as influenced by soil water potential and water flow. Soil Biol. Biochem. 23:485–486.
182. Liddell, C. M., and J. L. Parke. 1989. Enhanced colonization of pea taproots by a fluorescent pseudomonad biocontrol agent by water infiltration into soil. Phytopathology 79:1327–1332.
183. Rao, D. L. N., and R. G. Burns. 1990. The effect of surface growth of blue-green algae and bryophytes on some microbiological biochemical and physical soil processes. Biol. Fertil. Soils 9:239–244.
184. Dowling, D. N., and W. T. Broughton. 1986. Competition for nodulation of legumes. Annu. Rev. Microbiol. 40:131–157.
185. Ghiorse, W. C., and J. T. Wilson. 1988. Microbial ecology of the terrestrial subsurface. Adv. Appl. Microbiol. 33:107–172.

186. Gannon, T. J., V. B. Manilal, and M. Alexander. 1991. Relationship between cell surface properties and transport of bacteria through soil. Appl. Environ. Microbiol. 57:190–193.
187. Bashan, Y., and H. Levanony. 1987. Horizontal and vertical movement of *Azospirillum brasilense* Cd in the soil and along the rhizosphere of wheat and weeds in controlled and field environments. J. Gen. Microbiol. 133:3473–3480.
188. Chamblee, D. S., and R. D. Warren. 1990. Movement of rhizobia between alfalfa plants. Agron. J. 82:283–286.
189. Catlow, H. Y., A. R. Glenn, and M. J. Dilworth. 1990. The use of transposon-induced non-motile mutants in assessing the significance of motility of *Rhizobium leguminosarum* biovar *trifolii* for movement in soils. Soil Biol. Biochem. 22:331–336.
190. Acea, M. J., and M. Alexander. 1988. Growth and survival of bacteria introduced into carbon-amended soil. Soil Biol. Biochem. 5:703–709.
191. Habte, M., and M. Alexander. 1977. Further evidence for the regulation of bacterial populations in soil by protozoa. Arch. Microbiol. 113:181–183.
192. Ramirez, C., and M. Alexander. 1980. Evidence suggesting protozoan predation on *Rhizobium* associated with germinating seeds and in the rhizosphere of beans (*Phaseolus vulgaris* L.). Appl. Environ. Microbiol. 40:492–499.
193. Mazzola, M., and R. J. Cook. 1991. Effects of fungal root pathogens on the population dynamics of biocontrol strains of fluorescent pseudomonads in the wheat rhizosphere. Appl. Environ. Microbiol. 57:2171–2178.
194. Williams, S. T., and J. C. Vickers. 1986. The ecology of antibiotic production. Microb. Ecol. 12:43–52.
195. Postma, J., C. H. Hok-a-Hin, and J. A. van Veen. 1990. Role of microniches in protecting introduced *Rhizobium leguminosarum* biovar *trifolii* against competition and predation in soil. Appl. Environ. Microbiol. 56:495–502.
196. Thies, J. E., P. W. Singleton, and B. B. Bohlool. 1991. Influence on the size of indigenous rhizobial populations on establishment and symbiotic performance of introduced rhizobia on field-grown legumes. Appl. Environ. Microbiol. 57:19–28.
197. Li, D. M., and M. Alexander. 1988. Co-inoculation with antibiotic-producing bacteria to increase colonization and nodulation by rhizobia. Plant Soil 108:211–219.
198. Li, D. M., and M. Alexander. 1990. Factors affecting co-inoculation with antibiotic-producing bacteria to enhance rhizobial colonization and nodulation. Plant Soil 129:195–202.
199. Burns, R. G. 1985. The rhizosphere: microbial and enzymic gradients and prospects for manipulation. Pedologie 35:283–295.

200. Bazin, M. J., P. Markham, E. M. Scott, and J. M. Lynch. 1990. Population dynamics and rhizosphere interactions, p. 99–128. *In* J. M. Lynch (ed.), The rhizosphere, John Wiley & Sons, Chichester.

201. de Freitas, J. R., and J. J. Germida. 1990. Plant growth promoting rhizobacteria for winter wheat. Can. J. Microbiol. 36:265–272.

202. Miller, H. J., E. Liljeroth, G. Henken, and J. A. van Veen. 1990. Fluctuations in the fluorescent pseudomonad and actinomycete populations of rhizosphere and rhizoplane during the growth of spring wheat. Can. J. Microbiol. 36:254–258.

203. Rovira, A. D., G. D. Bowen, and R. C. Foster. 1983. The significance of rhizosphere microflora and mycorrhizas in plant nutrition, p. 61–93. *In* A. Lauchli and R. L. Bieski (eds.), Inorganic plant nutrition. Springer-Verlag, Berlin.

204. Whipps, J. M. 1990. Carbon economy, p. 59–98. *In* J. M. Lynch (ed.), The Rhizosphere. John Wiley & Sons, Chichester.

205. Kloepper, J. W., and K. L. Bowen. 1991. Quantification of the geocarposphere and rhizosphere effect of peanut (*Arachis hypogea* L.). Plant Soil 136;103–109.

206. Kremer, R. J., M. F. T. Begonia, L. Stanley, and E. T. Lanham. 1990. Characterization of rhizobacteria associated with weed seedlings. Appl. Environ. Microbiol. 56:1649–1655.

207. Weller, D. M. 1988. Biological control of soilborne plant pathogens in the rhizosphere with bacteria. Annu. Rev. Phytopathol. 26:379–407.

208. Dijkstra, A. F., G. H. N. Scholten, and J. A. van Veen. 1987. Colonization of wheat seedling (*Triticum aestivum*) roots by *Pseudomonas fluorescens* and *Bacillus subtilis*. Biol. Fertil. Soils 4:41–46.

209. Schippers, B., A. W. Bakker, P. A. H. M. Bakker, and R. van Peer. 1990. Beneficial and deleterious effects of HCN-producing pseudomonads on rhizosphere interactions. Plant Soil 129:75–83.

210. Dartnall, A. M., and R. G. Burns. 1987. A sensitive method for measuring cyanide and cyanogenic glucosides in sand culture and soil. Biol. Fertil. Soils 5:141–147.

211. Bakker, A. W., and B. Schippers. 1987. Microbial cyanide production in the rhizosphere in relation to potato yield reduction and *Pseudomonas* spp.-mediated plant growth-stimulation. Soil Biol. Biochem. 19:451–457.

212. Hsu, T. S., and R. Bartha. 1979. Accelerated mineralization of two organophosphate insecticides in the rhizosphere. Appl. Environ. Microbiol. 37:36–41.

213. Reddy, B. R., and N. Sethunathan. 1983. Mineralization of parathion in the rice rhizosphere. Appl. Environ. Microbiol. 45:826–829.

214. Sandmann, E. R. I. C., and M. A. Loos. 1984. Enumeration of 2,4-D-degrading microorganisms in soils and crop plant rhizospheres using indicator media; high populations associated with sugarcane (*Saccharum officinarum*). Chemosphere 13:1073–1084.

215. Walton, B. T., and T. A. Anderson. 1990. Microbial degradation of trichloroethylene in the rhizosphere: potential application to biological remediation of waste sites. Appl. Environ. Microbiol. 56:1012–1016.

216. Bashan, Y. 1986. Significance of timing and level of inoculation with rhizosphere bacteria on wheat plants. Soil Biol. Biochem. 18:297–301.

217. Bennett, R. A., and J. M. Lynch. 1981. Colonization potential of bacteria in the rhizosphere. Curr. Microbiol. 6: 137–138.

218. van Vuurde, J. W. L., and B. Schippers. 1980. Bacterial colonisation of seminal wheat roots. Soil Biol. Biochem. 12: 559–565.

219. Singleton, P. W., and J. W. Tavares. 1986. Inoculation response of legumes in relation to the number and effectiveness of indigenous rhizobium populations. Appl. Environ. Microbiol. 51:1013–1018.

220. Bohlool, B. B., and E. L. Schmidt. 1973. Persistence and competition aspects of *Rhizobium japonicum* observed in soil by immunofluorescence microscopy. Soil Sci. Soc. Am. Proc. 37:561–564.

221. Purchase, H. F., and P. S. Nutman. 1957. Studies on the physiology of nodule formation. VI. The influence of bacterial numbers in the rhizosphere on nodule initiation. Ann. Bot. 11:439–454.

222. van Egeraat, A. W. S. M. 1975. The growth of *Rhizobium leguminosarum* on the root surface and in the rhizosphere of pea seedlings in relation to root exudates. Plant Soil 42: 367–379.

223. Zablotowicz, R. M., E. M. Tipping, F. M. Scher, M. Ijzerman, and J. W. Kloepper. 1991. In furrow spray as a delivery system for plant growth–promoting rhizobacteria and other rhizosphere-competent bacteria. Can. J. Microbiol. 37: 632–636.

224. Fages, J. 1990. An optimized process for manufacturing an *Azospirillum* inoculant for crops. Appl. Microbiol. Biotechnol. 32:473–478.

225. van Elsas, J. D., and C. E. Heijnen. 1990. Methods for the introduction of bacteria into soil: a review. Biol. Fertil. Soils 10:127–133.

226. Rodrigues-Navarro, D. N., F. Temprano, and R. Orive. 1991. Survival of *Rhizobium* sp (*hedysarum coronarium*) on

peat-based inoculants and inoculated seeds. Soil Biol. Biochem. 23:375–379.

227. Huber, D. M., H. El-Nasshar, L. W. Moore, D. E. Mathre, and J. E. Wagner. 1989. Interactions between a peat carrier and bacterial seed treatments evaluated for biological control of the take-all diseases of wheat (*Triticum aestivum* L.). Biol. Fertil. Soils 8:166–171.

228. Sougoufara, B., H. G. Diem, and Y. R. Dommergues. 1989. Response of field grown *Casuarina equisetifolia* to inoculation with *Frankia* strain ORS 02001 entrapped in alginate beads. Plant Soil 118:133–137.

229. Bashan, Y. 1986. Alginate beads as synthetic inoculant carriers for slow release of bacteria that affect plant growth. Appl. Environ. Microbiol. 51:1–89–1098.

230. O'Reilly, K. T., and R. L. Crawford. 1989. Degradation of pentachlorophenol by polyurethane-immobilized *Flavobacterium* cells. Appl. Environ. Microbiol. 55:2113–2118.

231. Stormo, K. E., and R. L. Crawford. 1992. Preparation of encapsulated microbial cells for environmental applications. Appl. Environ. Microbiol. 58:727–730.

232. Deinum, B., and J. Eleveld. 1986. Effect of liming and seed pelleting on the growth of lucerne (*Medicago sativa* L.) on sandy soils, p. 270–273. *In* F. M. Borba and J. M. Abreau (eds.), Grassland facing the energy crisis. Proc. 11th Gen. Meeting European Grassl. Fed., Lisbon, Portugal.

233. Pijnenborg, J. W. M., and T. A. Lie. 1990. Effect of lime-pelleting on the nodulation of lucerne (*Medicago sativa* L.) in an acid soil: a comparative study in the field, in pots and in rhizotrons. Plant Soil 121:225–234.

234. Pijnenborg, J. W. M., T. A. Lie, and A. J. B. Zehnder. 1990. Inhibition of nodulation of lucerne (*Medicago sativa* L.) by calcium depletion in acid soil. Plant Soil 127:31–39.

235. Pijnenborg, J. W. M., T. A. Lie, and A. J. B. Zehnder. 1990. Nodulation of lucerne (*Medicago sativa* L.) in an acid soil: pH-dynamics in the rhizosphere of seedlings growing in rhizotrons. Plant Soil 126:161–168.

236. Gibson, A. H., D. H. Demezas, R. R. Gault, T. V. Bhuvaneswari, and J. Brockwell. 1990. Genetic stability in rhizobia in the field. Plant Soil 127:37–44.

237. Stotzky, G., and R. G. Burns. 1982. The soil environment: clay-humus-microbe interactions, p. 105–133. *In* R. G. Burns and J. H. Slater (eds.), Experimental microbial ecology. Academic Press, London.

238. Burns, R. G. 1986. Interaction of enzymes with soil minerals and organic colloids, p. 429–451. *In* P. M. Huang and M. Schnitzer (eds.), Interaction of soil minerals with natural

organics and microbes. Soil Science Society of America, Madison, Wisconsin.

239. Schmidt, S. K., and M. Alexander. 1985. Effects of dissolved organic carbon and second substrates on the biodegradation of organic compound at low concentrations. Appl. Environ. Microbiol. 49:822–827.

240. Burns, R. G. 1983. Extracellular enzyme-substrate interactions in soil, p. 249–298. *In* J. H. Slater, R. Whittenbury and J. W. T. Wimpenny (eds.), Microbes in their natural environment. Cambridge University Press, Cambridge.

241. Burns, R. G., and J. Ladd. 1985. Stability of immobilized phosphates in soil. Soc. Gen. Microbiol. Quart. 12:17.

242. Sarkar, J. M., and R. G. Burns. 1984. Synthesis and properties of β-D-glucosidase—phenolic copolymers as analogues of soil humic-enzyme complexes. Soil Biol. Biochem. 16:619–625.

243. Skot, L. S., D. Harrison and A. Nath. 1990. Expression of insecticidal activity in *Rhizobium* containing the δ-endotoxin gene from *Bacillus thuringiensis* subsp. *tenebrionis*. Plant Soil 127:285–295.

244. Waalwijk, C., A. Dullemans, and C. Maat. 1991. Construction of a bioinsecticidal rhizosphere isolate of *Pseudomonas fluorescens*. FEMS Microbiol. Lett. 77:257–264.

245. Dimock, M. B., R. M. Beach, and P. S. Carlson. 1989. Biological pesticides and novel plant-pest resistance for insect pest management, p. 88–101. *In* D. W. Roberts and R. R. Granadas (eds.), Biotechnology. Boyce Thompson Institute for Plant Research, Ithaca, New York.

246. Höfte, H., and H. R. Whitely. 1989. Insecticidal crystal proteins of *Bacillus thuringiensis*. Microbiol. Rev. 53:242–255.

247. Masson, L., W. J. Moar, K. van Frankenhuyzen, M. Bossé, and R. Broussen. 1992. Insecticidal properties of a crystal protein gene product isolated from *Bacillus thuringiensis* subsp. *kenyae*. Appl. Environ. Microbiol. 58:642–646.

248. Handelsman, J., S. Raffel, E. H. Mester, L. Wunderlich, and C. R. Grau. 1990. Biological control of damping-off of alfalfa seedlings with *Bacillus cereus* UW85. Appl. Environ. Microbiol. 56:713–718.

249. Thompson, R. J., and R. G. Burns. 1989. Control of *Pythium ultimum* with antagonistic fungal metabolites incorporated into sugar beet seed pellets. Soil Biol. Biochem. 32:745–748.

250. Smit, E., J. D. van Elsas, J. A. van Veen, and W. M. de Vos. 1991. Detection of plasmid transfer from *Pseudomonas fluorescens* to indigenous bacteria in soil by using bacteriophage φR2f for donor counterselection. Appl. Environ. Microbiol. 57:3482–3488.

251. van Elsas, J. D., J. M. Govaert, and J. A. van Veen. 1987. Transfer of plasmid pFT30 between bacilli in soil as influenced by bacterial population dynamics and soil conditions. Soil Biol. Biochem. 19:639–647.

252. Henschke, R. B., and F. R. J. Schmidt. 1990. Plasmid mobilization from genetically engineered bacteria to members of the indigenous soil microflora in situ. Curr. Microbiol. 20:105–110.

253. Coughter, J. P., and G. J. Stewart. 1989. Genetic exchange in the environment. Antonie van Leeuwenhoek 55:15–22.

254. Stotzky, G. 1990. Ecological considerations related to the release of genetically engineered microorganisms in the environment, p. 145–157. Advances in Biotechnology, Proc. Int. Conf., Stockholm.

255. Trevors, J. T., T. Barkay, and W. Bourquin. 1987. Gene transfer among bacteria in soil and aquatic environments: a review. Can J. Microbiol. 33:191–198.

256. Stotzky, G. L., R. Zeph, and M. A. Devanas. 1991. Factors affecting the transfer of genetic information among microorganisms in soil, p. 95–121. *In* L. R. Ginzberg (ed.), Assessing ecological risks of biotechnology. Butterworth-Heinemann, Boston.

257. Stotzky, G., and H. Babich. 1986. Survival of, and genetic transfer by, genetically engineered bacteria in natural environments. Adv. Appl. Microbiol. 31:93–138.

258. van Elsas, J. D., J. T. Trevors, and M. E. Starodub. 1988. Bacterial conjugation between pseudomonads in the rhizosphere of wheat. FEMS Microb. Ecol. 53:299–306.

259. van Elsas, J. D., J. T. Trevors, L. S. van Overbeek, and M. E. Starodub. 1989. Survival of *Pseudomonas fluorescens* containing plasmids RP4 or PRK2501 and plasmid stability after introduction into two soils of different texture. Can. J. Microbiol. 35:951–959.

260. Lorenz, M. G., B. W. Aardemia, and W. Wackernagel. 1988. Highly efficient genetic transformation of *Bacillus subtilis* attached to sand grains. J. Gen. Microbiol. 134:107–112.

261. Schofield, P. R., A. H. Gibson, W. F. Dodman, and J. M. Watson. 1987. Evidence for genetic exchange and recombination of *Rhizobium* symbiotic plasmids in a soil population. Appl. Environ. Microbiol. 53:2942–2947.

262. Kelly, W. J., and D. C. Reanney. 1984. Mercury resistance among soil bacteria: ecology and transferability of genes encoding resistance. Soil Biol. Biochem 16:1–8.

263. van Elsas, J. D., M. Nikkel, and L. S. van Overbeek. 1989. Detection of plasmid RP4 transfer in soil and rhizosphere and

the occurrence of homology to RP4 in soil microorganisms. Curr. Microbiol. 19:375–381.

264. Burns, R. G. 1988. Laboratory models in the study of soil microbiology, p. 51–98. *In* J. W. T. Wimpenny (ed.), Handbook of laboratory model systems for microbial ecosystem research. CRC Press, Boca Raton, Florida.

265. Bolton, H., J. K. Fredrickson, J. M. Thomas, S. W. Li, D. J. Workman, S. A. Bentjen, and J. L. Smith. 1991. Field calibration of soil-core microcosms: ecosystem structural and functional comparisons. Microb. Ecol. 21:175–189.

266. Bolton, H., J. K. Fredrickson, S. A. Bentjen, D. J. Workman, S. W. Li, and J. M. Thomas. 1991. Field calibration of soil-core microcosms: fate of a genetically altered rhizobacterium. Microb. Ecol. 21:163–173.

267. Trevors, J. T., J. D. van Elsas. L. S. Overbeek, and M. E. Starodub. 1990. Transport of a genetically engineered *Pseudomonas fluorescens* strain through a soil microcosm. Appl. Environ. Microbiol. 56:401–408.

268. Williamson, J. W., and P. G. Hartel. 1991. Rhizosphere growth of *Pseudomonas solanacearum* genetically altered in extracellular enzyme production. Soil Biol. Biochem. 5:453–458.

269. Drahos, D. J. 1991. Field testing of genetically engineered microorganisms. Biotech. Adv. 9:157–171.

270. Pimentel, D. 1985. Using genetic engineering for biological control: reducing ecological risks, p. 129–140. *In* H. O. Halvorson, D. Palmer, and M. Rogul (eds.), Engineered organisms in the environment: scientific issues. American Society for Microbiology, Washington, D.C.

271. Lindow, S. E., N. J. Panopoulos, and B. L. McFarland. 1989. Genetic engineering of bacteria from managed and natural habitats. Science 244:1300–1307.

272. Nicolaidis, A. A. 1987. Microbial mineral processing: the opportunities for genetic manipulation. J. Chem. Tech. Biotechnol. 38:167–185.

273. Brill, W. J. 1985. Safety concerns and genetic engineering in agriculture. Science 227:381–384.

274. Kuenzi, M., F. Assi, A. Chmiel, C. H. Collins, M. Donikian, J. B. Dominguez, L. Financsek, L. M. Fogarty, W. Frommer, F. Hasko, J. Hovland, E. H. Houwink, J. L. Mahler, A. Sandkvist, K. Sargeant, C. Sloover, and T. Tiujnenborg Muijs. 1985. Safe biotechnology: general considerations. Appl. Microbiol Biotechnol. 21:1–6.

275. Domsch, K. H., A. J. Briesel, W. Goebel, W. Lindenmaier, W. Lotz, H. Reber, and F. Schmidt. 1988. Considerations on release of gene-technologically engineered microorganisms into the environment. FEMS Microbiol. Ecol. 53:261–272.

276. Cairns, J., and J. R. Pratt. 1986. Ecological consequence assessment; effects of bioengineered microorganisms. Water. Res. Bull. 22:171–182.
277. Molin, S., P. Klemm, L. K. Poulsen, H. Biehl, K. Gerdes, and P. Andersson. 1987. Conditional suicide system for containment of bacteria and plasmids. Biotechnology 5: 1315–1318.

2

White Rot Fungi and Their Potential Use in Soil Bioremediation Processes

JOHN A. BUMPUS *University of Notre Dame, Notre Dame, Indiana*

I. INTRODUCTION

The white rot fungus *Phanerochaete chrysosporium* has the ability to degrade a wide variety of environmentally persistent compounds, many of which are toxic organopollutants [1–35]. Compounds degraded by this fungus include 1,1,1-trichloro-2,2-bis(4-chlorophenyl)ethane (DDT), 2,4,5-trichlorophenoxyacetic acid (2,4,5-T), 2,3,7,8-tetra-chlorodibenzo(p)dioxin, 2,4,6-trinitrotoluene (TNT), pentachlorophe-nol (PCP), benzo[a]pyrene, and other polycyclic aromatic hydrocarbons (PAHs). It is of interest that carbon dioxide has been shown to be the final oxidation product in many of these degradations. Chemicals degraded by this fungus are listed in Table 1.

Because of its unique biodegradative abilities, *P. chrysosporium* is a potentially useful microorganism for the bioremediation of contaminated soils. This chapter reviews the use of this fungus in soil bioremediation, as well as selected aspects of soil biochemistry that have general or specific relevance to the use of this fungus or other white rot fungi in soil bioremediation. Selected aspects of lignin degradation are also summarized, as this system appears to be responsible, at least in part, for the ability of *P. chryso-sporium* to degrade organopollutants.

A. White Rot Fungi

The hyphae of certain wood-rotting fungi are able to penetrate wood and cause white-rot, "a light colored spongy mass (white rot)

65

Table 1 Partial List of Organic Compounds Degraded by *Phanerochaete chrysosporium*[a]

Polycyclic aromatic compounds	Chlorinated alkylhalides
Benzo[*a*]pyrene	Lindane
Biphenyl	Chlordane
2–Methylnaphthalene	Mirex
Phenanthrene	
Benz[*a*]anthracene	Biopolymers
Pyrene	Lignin
Anthracene	Cellulose
Perylene	Kraft lignin
Dibenzo[*p*]dioxin	4–Chloroaniline–lignin conjugate
Fluorene	3,4–Dichloroaniline–lignin conjugate
9–Fluorenone	Humic acids
1,4–Naphthoquinone	
	Triphenylmethane dyes
Chlorinated aromatic compounds	Crystal violet
4–Chlorobenzoic acid	Pararosaniline
Dichlorobenzoic acid (unknown congener)	Cresol red
2,4,6–Trichlorobenzoic acid	Bromphenol blue
4,5–Dichloroguaiacol	

		Other compounds
		Metolachlor
		Cyanide
		2,4,6-Trinitrotoluene
	Ethyl violet	
	Malachite green	
	Brilliant green	
	Polymeric dyes	
	Poly B	
	Poly R	
	Poly Y	Azo dyes
		Congo red
		Orange II
		Tropaeolin O
	Heterocyclic dyes	
	Azure B	
6-Chlorovanillin		
4,5,6-Trichloroguaiacol		
Tetrachloroguaiacol		
Pentachlorophenol		
2,4-Dichlorophenol		
4-Chloroaniline		
3,4-Dichloroaniline		
2,4,5-Trichlorophenoxyacetic acid		
Polycyclic chlorinated aromatic compounds		
DDT [1,1,1-trichloro-2,2-bis(4-chloro-phenyl)ethane]		
2,3,7,8-Tetrachlorodibenzo-p-dioxin		
3,4,3',4'-Tetrachlorobiphenyl		
2,4,5,2',4',5'-Hexachlorobiphenyl		
Aroclor 1254		
Aroclor 1242		
2-Chlorodibenzo[p]dioxin		
Dicofol [2,2,2-trichloro-1,1-bis(4-chlorophenyl)ethanol]		
DDE[1,1-dichloro-2,2-bis(4-chlorophenyl)ethene]		

[a]Compiled from references 1–35.

containing white pockets or streaks separated by thin areas of firm wood" [36]. Fungi with this ability are collectively known as white-rot fungi. Although most of the fungi that cause white rot are members of the subdivision Basidiomycotina, the term *white-rot fungi* is a functional rather than a systematic classification as many fungi that cause white-rot have phylogenetic affiliations only at the level of the order. For example, *Armillaria mellea*, *Pleurotus ostreatus*, and *Panus tigris* are members of the Agaricales, whereas *Trametes versicolor*, *Fomes fomentarius*, and *P. chrysosporium* are members of the Aphyllophorales.

B. Lignin Degradation: An Overview

Fungi, along with bacteria, are the predominant microorganisms responsible for decomposition of organic carbon in the biosphere. Of particular interest and importance in carbon turnover is the biodegradation of lignin [37]. Lignin may be described as a complex, water-insoluble, nonrepeating heteropolymer whose function is to provide structural support in plants. It is composed of a variety of aromatic monomers that are connected by at least 12 different types of carbon-carbon and carbon-oxygen linkages [37]. Adding to its structural complexity is the fact that lignin is a large three-dimensional polymer whose chiral carbons exist in both the D and L configurations. Because of its complexity, relatively few types of microorganisms have the ability to degrade lignin to carbon dioxide. This, coupled with the fact that lignin is the most abundant renewable aromatic compound on Earth, suggests that lignin turnover may be the rate-limiting step in the carbon cycle [37]. White-rot fungi cause extensive biodegradation of lignin, and of this group *P. chrysosporium* is, unquestionably, the most thoroughly studied with respect to its lignin-degrading abilities [38–40]. Although the manner in which lignin degradation (especially the initial stages) occurs has been studied for many years, it was only in the early 1980s that researchers discovered that the initial oxidations were mediated by two families of peroxidases (lignin peroxidases and Mn peroxidases) that are secreted by *P. chrysosporium* under nutrient-limited conditions [41–44]. The action of lignin and Mn peroxidases appears to result in the oxidative depolymerization of lignin, forming smaller and more soluble metabolites that can then be internalized and further metabolized to carbon dioxide by the fungus to complete the mineralization process. More detailed accounts of lignin degradation by *P. chrysosporium* are available in several reviews [38–40]. The lignin-degrading system is of importance in organopollutant degradation as several lines [1–35] of indirect and direct evidence demonstrate that this system is also involved in the degradation of many of the compounds listed in Table 1. Examples of compounds in which initial oxidations are catalyzed by lignin peroxidases are presented in Table 2.

II. SOIL AS A HABITAT FOR FUNGI: NUTRITIONAL ASPECTS

Soils are extremely heterogeneous, and it is difficult, if not impossible, to make assertions that apply in all cases to all soils. Indeed, an attempt merely to provide a description of soil has resulted in the following unwieldy definition:

> Unconsolidated mineral matter on the surface of the Earth that
> has been subjected to and influenced by genetic and environ-
> mental factors of: parent material, climate (including moisture
> and temperature effects), macro- and microorganisms, and top-
> ology, all acting over a period of time and producing a product-
> soil that differs from the material from which it is derived in
> many physical, chemical and biological properties and charac-
> teristics. [45]

Despite its complexities, some reasonable assessments can be made concerning the suitability of soil as a habitat for fungi. The following is a brief overview of selected nutritional aspects that affect the growth of fungi in soil. Topics which may affect the growth and biodegradative capabilities of white-rot fungi in soil are emphasized.

In general, soils (or the many microhabitats within soils) may be regarded as lacking available carbon and incapable of maintaining sustained microbial growth [46–48]. This is to be expected in some mineral soils, where the organic content (i.e., humus) may be less than 1%. However, even in organic soils (where the total organic content may be between 60 and 95%), the types of carbon compounds present are typically poor growth substrates for most fungi. The effect of this on fungi in soil is that they generally exist as quiescent spores or in a vegetative state in which minimal metabolism and growth occur. The effect on fungi to be used in bioremediation processes is that if the fungus cannot use an organopollutant as a source of carbon and energy (i.e., if degradation occurs by cometabolism), an exogenous growth substrate must be added for biodegradation of the organopollutant to occur. An exception to this characterization of soil as lacking in nutritionally available carbon is the area immediately surrounding plant roots (i.e., the rhizosphere). Similarly, dead plant and animal matter incorporated into soil naturally or by agricultural practices provides transient increased amounts of available carbon and

Table 2 Representative Organic Compounds Oxidized by Lignin Peroxidases from *Phanerochaete chrysosporium*[a]

Substrate	Product	Reaction type	Reference
Benzo[a]pyrene	Benzo[a]pyrene-1,6-, -3,6-, and 6,12-quinones	Oxygenation	30
Pyrene	Pyrene-1,6- and -1,8-diones	Oxygenation	31
N,N,N',N'',N''-Hexamethylpararosaniline (crystal violet)	N,N,N',N'',N''-Pentamethyl-pararosaniline	N-Demethylation	15
Thianthrene	Thianthrene monosulfoxide	S-Oxygenation	32
2,4-Dichlorophenol	2-Chloro-1,4-benzoquinone	Oxidative dechlorination	34
Pentachlorophenol	2,3,5,6-Tetrachloro-2,5-cyclohexadiene-1,4-dione	Oxidative dechlorination	3, 33
1,4-Dimethoxybenzene	1,4-Benzoquinone	O-Demethylation	35
4-Chloroaniline	Complex mixture of dimers, trimers, and tetramers	Oxidative polymerization	Chang and Bumpus (unpublished)
Oxalic acid	Carbon dioxide	Decarboxylation	54

[a]In addition to the above, the following compounds have also been reported to be oxidized by lignin peroxidases from *P. chrysosporium*: benz[a]anthracene, anthracene, perylene, dibenzo[p]dioxin, 2-chlorodibenzo[p]dioxin, sodium azide, 3-amino-1,2,4-triazole, azure B, orange II. tropaeolin O, pararosaniline, cresol red, bromphenol blue, ethyl violet, malachite green, and brilliant green (references 10, 15, 31, and unpublished observations).

other nutrients, which results in microbial activity that is orders of magnitude greater than that observed in soils with no exogenous organic matter.

Inorganic constituents in soil are also important for fungal nutrition. In general, most soils contain adequate amounts of iron and smaller amounts of calcium, magnesium, potassium, manganese, sodium, nitrogen, phosphorus, and sulfur [46,49,50]. Substantial amounts of aluminum and silicon may also be found [49,50]; however, these elements are of little nutritional value to most fungi. Although they may be present in substantial quantities, the bioavailability of inorganic nutrients may be often limited. Among the factors which influence the bioavailability of inorganic nutrients are soil pH, cation exchange capacity of the soil, the soil's clay mineral content, and the ability of fungi to produce and secrete substances which sequester inorganic nutrients and facilitate their uptake by the cell [46,49,50]. As a case in point, iron is often found in adequate amounts in soils, especially in acid soils. However, iron deficiencies can arise in alkaline soils due to the formation of insoluble iron hydroxides and oxides [50]. In many cases, fungi, like bacteria, have evolved mechanisms to cope with such deficiencies. For example, siderophores, compounds that sequester iron, are secreted by some bacteria and fungi in response to a deficiency in available iron. Fekete et al. [51] have shown that several members of the Basidiomycotina (including *P. chrysosporium* and *Phanerochaete sordida*) produce siderophores: the 10 species screened included representatives of the Agaricales and the Aphyllophorales. The fact that all 10 species produced siderophores suggests that the production of siderophores is widespread among wood-rotting fungi. In addition to siderophores, fungi also secrete a variety of chelating compounds that are able to solubilize calcium, magnesium, silicon, manganese, aluminum, sodium, and other elements that exist in an insoluble form in soil [49]. The sequestration and uptake of calcium are especially important for *P. chrysosporium* because this fungus requires a relatively high concentration of calcium for optimal lignin-degrading activity [52]. Many fungi accumulate calcium by secreting oxalic acid, which forms calcium oxalate that is taken up by these fungi [53]. The report that *P. chrysosporium* secretes oxalic acid [54] suggests that this fungus may accumulate calcium in this manner.

Fungi use inorganic nitrogen (nitrates, nitrites, and ammonia) as well as organic nitrogen for growth. However, unlike bacteria, fungi do not appear to use atmospheric nitrogen (via nitrogen fixation) as a major source of nutrient nitrogen [55].

Although many fungi do utilize nitrates, some do not. The inability to use nitrates is common in the Blastocladiales, the Saprolegiaceae, yeasts, and the higher Basidiomycotina [55]. In the case of *P. chrysosporium*, nitrate is utilized, even when both nitrate and ammonia are present [56]. *P. chrysosporium* is also able to utilize amino acids and urea as sole sources of nitrogen [57].

Soils often become deficient in available nitrogen, which can limit fungal growth. Nitrogen deficiency is easily remedied in agriculture by application of nitrogen fertilizers. However, in bioremediation processes using *P. chrysosporium* too much nitrogen, whether organic or inorganic, would be expected to suppress the expression of the biodegradative abilities of this fungus [2,57]. The role of nitrogen in the degradation of organopollutants by *P. chrysosporium* is further discussed later in this chapter.

III. OCCURRENCE OF FUNGI IN SOIL

A. Estimation of Fungal Mass in and Isolation of Fungi from Soil

On a mass basis, fungi are as abundant as or more abundant than bacteria in many soils [49,50]. They are typically abundant in the upper 4 inches of soil. Relatively few fungi are able to survive at depths below 12 inches [55]. In addition to soil depth, other environmental factors influence the number and type of fungal species found in a given soil, e.g., pH, temperature, plant cover, salinity, season, moisture content, and clay mineral content [46, 49,50,55,58].

The pH of soil is of special concern. Individual species can be found that grow at pH extremes as low as pH 2 and as high as pH 11, and fungi, as a group, tend to have an advantage, relative to bacteria, in acidic soils and account for much of the microbial activity in such soils [49,58,59].

Estimation of the population size and diversity of fungi in soil is complex, and no single technique has been developed that accurately assesses these parameters. Plate counts of soil dilutions on agar, such as those used to enumerate bacteria, originally suffered from the fact that bacteria, particularly actinomycetes, are usually abundant in soil and often outcompete fungi on nutrient agars [49,59]. This limitation has been addressed by using agars of low pH (pH ~4) on which few bacteria can grow or by using agars containing broad-spectrum bacteria-specific antibiotics, such as streptomycin and penicillin [49,59]. The combination of low pH and broad-spectrum bacteria-specific antibiotics is also used. Although these procedures allow fungi to be grown on nutrient agars, it has been noted that accurate relative numbers of individuals of a given species cannot be determined, as even gentle agitation can cause shearing of mycelium, individual fragments of which may be able to give rise to individual colonies [49]. Similarly, spore germination from profusely sporulating fungi, such as *Penicillium* sp.

and *Aspergillus* sp., tends to give results that overestimate their abundance in soil [49]. At best, plate count dilution methods give only an estimate of the number of fungal spores and viable mycelium fragments in a given soil sample. An alternative approach to the quantification of fungi in soil is to estimate the total mass of fungal hyphae per unit of soil mass using the soil-agar film method, the membrane filter method, or the nylon mesh method, each of which has unique strengths and drawbacks, as described in a recent review [59].

In addition to quantification of total fungal mass in soil, a knowledge of the relative abundance of individual species is important in many ecological and agricultural studies. Like the quantification of total fungal biomass, estimation of species diversity in soils is not a trivial task. No single isolation medium has yet been described that is suitable for isolating all fungi and, indeed, it is unlikely that such a medium will ever be developed, given the varied nutritional requirements that exist among different species of fungi. The slow growth of some important fungi also makes their isolation difficult. Thus, it has become common practice to use nutrient media that allow as many fungal species as possible to be isolated from a given soil, with the tacit understanding that not all species present may be isolated. Although useful, this approach has a number of shortcomings, the most obvious of which is that large and important groups of fungi may be underrepresented by such procedures. For example, the Basidiomycotina have often been underestimated in studies of forest and prairie soils, despite the fact that macroscopic fruiting bodies and microscopic clamp-bearing hyphae (indicative of Basidiomycotina) are frequently observed in such soils [59]. Thus, when slow-growing or nutritionally fastidious fungi are to be studied, selective media for the fungi in question must be used. In studies which require a thorough census of the fungal species present, a variety of selective media would be required to ensure representative sampling.

Even with their acknowledged limitations, the foregoing techniques have been used to document many important quantitative and qualitative changes in fungal populations and the effect of these changes on naturally occurring biodegradation processes. It should be possible to adapt these techniques to study fungal occurrence, survival of introduced species, and population changes in bioremediation processes in which fungi may have an active role. For example, composting of contaminated soils has often been suggested as a bioremediation technique. A large literature exists on the composting of agricultural residues, yard wastes, sewage sludge, and by-products from the lumbering industry [60–62]. However, relatively few thorough studies have been published concerning composting of hazardous wastes or contaminated soils, and those that have been published tend to focus on biodegradation

by bacteria in such systems. Other bioremediation processes in
which fungal isolation and identification will be necessary include
those in which contaminated soils are amended with appropriate
growth substrates and fungi that have known biodegradative capabilities. It will be critical in such cases to show that the introduced fungus is able to survive and thrive in such soils. Of particular interest in this chapter are the selective media that have
been developed for wood-rotting fungi and other members of the
Basidiomycotina. Isolation of wood-rotting fungi from soils and
other materials is often complicated by the fact that many species
are slow growing and are overwhelmed by other fungi and bacteria.
The observation that some fungal species are more resistant than
others to certain fungicides has led to the development of selective
media useful in isolating a variety of wood-rotting fungi. In these
selective media, fungicides, such as benomyl, pentachloronitroben-
zene, and *ortho*-phenylphenol, are included at concentrations that
are sublethal to the fungus of interest but inhibitory or lethal to
fungi that would normally be faster growing [63–70]. Such inhibitors may be used alone or in conjunction with an antibacterial compound. Dietrich and Lamar [63] have shown that 15 mg/L of
benomyl and 500 mg/L of streptomycin sulfate in 2% malt agar
proved to be an effective selective medium for the isolation of *P.
chrysosporium* from nonsterile forest soil.

B. Relative Occurence of Fungi in Soil

Among the fungi most commonly found in soils are species of *Aspergillus, Penicillium,* and *Trichoderma* [49,55]. For this reason,
members of the Deuteromycotina have been considered to be among
the most abundant fungi generally present in soils. This is somewhat misleading, however, as the Deuteromycotina is an artificial
subdivision of the Eumycota, with membership being based on the
fact that the ~15,000 known species in this group lack a sexual
state and, thus, cannot be placed in other fungal subdivisions [55,
59]. Most species of the Deuteromycotina are thought to be anamorphs of species belonging to the Ascomycotina, with a smaller
number of species belonging to the Basidiomycotina [59].
    Kendrick and Parkinson [59] have analyzed the list of the taxa
of fungi isolated from soil that appears in the *Compendium of soil
fungi* [71] with the intent of assessing the relative likelihood of an
isolated species belonging to one of the major fungal classifications.
Of the 461 species listed, the overwhelming majority (97.2%) were
members of the division Amastigomycota. Of these, the Ascomycotina and their anamorphs accounted for 85.2% of the total, and the
Zygomycotina and filamentous yeasts accounted for 10 and 0.7%,
respectively. The Basidiomycotina accounted for only 1.1% of the
species listed, and species of the division Mastigomycota represented

2.8% of the total. Consequently, the likelihood of an isolate belonging to the Basidiomycotina appears strikingly small and is clearly not indicative of their actual occurrence in soils. The lack of appropriate culture media, coupled with the difficulty of identifying these fungi in the absence of macroscopic fruiting bodies, has contributed to this situation.

IV. BIODEGRADATION OF ORGANOPOLLUTANTS IN SOIL BY *P. CHRYSOSPORIUM*

Because of its ability to degrade a wide variety of environmentally persistent organopollutants, several studies have focused on the development of biotreatment systems using *P. chrysosporium* immobilized on a variety of supports to treat aqueous wastes [72–75]. In principle, these systems can be further developed to treat contaminated soil. Further development, however, is not a trivial matter, as several problems will need to be addressed before such systems can be put in place. One of the most important issues is that of competition by other microorganisms, which would be expected to be present in abundance in soil. In our experience with *P. chrysosporium* in bioreactors, with either agitated or stationary cultures, contamination and subsequent competition by other microorganisms are rarely a problem. This is not to say that competition never occurs, as, on occasion, some cultures have been contaminated with other fungi (yeasts and *Aspergillus* sp.) and bacteria. In a soil slurry system or in a system to treat wastewater from a soil-washing operation, the material to be decontaminated may require a pretreatment to reduce competition by other microorganisms. Ultraviolet (UV) ozonolysis is one approach that might be useful in this respect. This procedure would also be expected to cause some oxidation of the compounds to be degraded. Furthermore, the high oxygen content of the effluent resulting from ozonolysis might promote the growth of *P. chrysosporium*, as it has been established that this fungus is tolerant of highly oxygenated growth media [57, 76].

The effect of shear forces on fungal mycelia and on lignin peroxidases may prove to be a problem in developing slurry reactors using *P. chrysosporium*. Studies have shown that lignin peroxidases are susceptible to inactivation by agitation in aqueous cultures [77,78] and therefore would be expected to be adversely affected by shear forces created by stirring in soil slurry cultures. Furthermore, shear forces combined with the scouring effect of suspended soil particles would be expected to dislodge, at least partially, fungal mycelia in immobilized systems. However, such problems are not insurmountable. For example, certain detergents protect lignin peroxidases from inactivation in agitated or stirred bioreactors [77,78]. The development of bioreactors, which use this fungus in slurries for soil bioremediation, in which shear forces are minimized to acceptable levels may, however, prove exceedingly difficult.

Treatment of soils by microorganisms in situ is, conceptually, a seemingly more straightforward approach to decontaminating problem soils. In practice, in situ treatment presents its own set of variables and complications. For example, it is convenient, but incorrect, to regard a given soil type as homogeneous. In reality, soils exist as "discrete microhabitats, the chemical, physical, and biological characteristics of which undoubtedly differ in time and space" [46]. As an obvious example, it is clear that the microorganisms in the first few centimeters of a soil are exposed to quite different environmental conditions (moisture, temperature, nutrient availability, oxygen availability, among others) than microorganisms which exist only a few centimeters farther below the surface. Thus, bioremediation strategies which require introduction of an exogenous microorganism must deal with the fact that the introduced species must be able to compete successfully with a variety of indigenous microorganisms that are adapted to different microhabitats. Indeed, the supposition that introduced microorganisms will not be able to compete with indigenous microorganisms is one of the arguments against using such procedures. However, it should be noted that the successful introduction of exogenous microorganisms into soil is a common agricultural practice. *Rhizobium* spp. are often added to agricultural soils to promote nitrogen fixation by legumes [49]. Typically, suspensions of *Rhizobium* sp., grown in liquid culture, are mixed with humus, peat or peat-charcoal combinations. Such mixtures are then mixed with seeds and planted. Similar procedures have also been adapted for introduction of *P. chrysosporium* into soil in laboratory studies (see references in Table 3) and field trials [79]. Typically, corncobs, wheat straw, peat, or wood chips are inoculated with the fungus, which is allowed to grow on the growth substrate for some time before introduction into the contaminated soil.

Another problem encountered during soil bioremediation in situ is that the concentration of contaminants is typically not uniform throughout the soil. This problem is often addressed by thoroughly mixing (e.g., rototilling) the soil. This, however, is only a partial remedy, as in most cases mixing to homgeneity is not technically or economically feasible. Even in what appears to be well-mixed soil, there may be "hot spots" in which high concentrations of an organopollutant may be present. Similarly, some areas may be present in which the concentration of organopollutant is well below the mean concentration for the soil. Thus, it is critical to design and use procedures which ensure that replicate samples are obtained from multiple sites throughout the soil during the bioremediation process. Analysis of such samples will allow a statistically representative assessment of changes in pollutant concentration during the bioremediation process. Thorough mixing of soil with fungal inoculum should also help to establish the introduced fungus in soil. As noted above, soils consist of many microhabitats, each occupied by a population of microorganisms that are adapted to each microhabitat [46]. It is reasonable to suggest that thorough mixing coupled with the introduction of an exogenous fungus (growing on a good growth substrate for the fungus) may alter some soil microhabitats such that the introduced fungus may compete successfully with indigenous microorganisms.

Soil bioremediation using *P. chrysosporium* has been the subject of considerable research effort [11,12,21–23,79,80,82–84,86–90]. Table 3 is a summary of the compounds that have been shown to be degraded by *P. chrysosporium* in soil, and the following is a synopsis of individual pollutants that have been degraded by this fungus in soils.

A. DDT

Although the use of DDT has been prohibited in the United States since 1972, it is still used in a number of developing countries, and residual DDT in soils still presents a remediation problem at several sites in the United States. *P. chrysosporium* is able to degrade DDT to carbon dioxide in axenic culture [1,7,16]. Unlike the major pathway proposed for DDT degradation by bacterial consortia, the initial oxidation of DDT involves hydroxylation to form dicofol [2,2,2-trichloro-1,1-bis(4-chlorophenyl)ethanol], followed by conversion to FW-152 [2,2-dichloro-1,1-bis(4-chlorophenyl)-ethanol] and, subsequently, to DBP (4,4'-dichlorobenzophenone) [1, 16]. DBP is then thought to undergo aromatic ring cleavage (or dechlorination followed by aromatic ring cleavage) to form unidentified metabolites that are eventually degraded to carbon dioxide. Fernando et al. [19] have shown that DDT adsorbed onto a silt loam soil is also degraded to carbon dioxide. Fungal growth and DDT degradation were dependent on the presence of an appropriate growth substrate (ground corncobs, in this case) and an adequate moisture content. A moisture content of approximately 40% (w/w) of the amount required to saturate the soil-corncob mixture was found to result in the greatest levels of DDT mineralization. Although DDT was converted to carbon dioxide, the extent of this conversion in soil cultures amended with corncobs (10% mineralization in 60 days) was only about one-third of that observed in liquid cultures (30% mineralization in 60 days) containing cellulose and glucose as carbon sources. Mass balance analysis of the ^{14}C-labeled metabolites derived from [^{14}C]DDT in soil cultures amended with ground corncobs showed that 5% of the DDT metabolites were

Table 3 Environmental Pollutants Degraded by *Phanerochaete chrysosporium* in Soil

Compound	Initial concentration (mg/kg)	Soil type (pH)
2,4,5-T	1.5	Silt loam, pH 6.4
2,4,6-TNT	10,000	Silt loam, pH 6.4
PCP	50	Sandy loam, pH 6.4
	50	Sandy loam, pH 7.0
	50	Sandy clay loam, pH 4.8
	0.33	Silt loam, pH 6.4
	250–400	A strongly alkaline, calcarious soil, pH 9.6
Chlordane	2.1	Silt loam, pH 6.4
Lindane	1.4	Silt loam, pH 6.4
Mirex	2.8	Silt loam, pH 6.4
Aldrin	10.1	Silt loam, pH 6.4
Dieldrin	0.2	Silt loam, pH 6.4
Heptachlor	10.1	Silt loam, pH 6.4
Aroclor 1242[b]	1000	Horticultural sand, pH 6.1
Aroclor 1254	0.4	Silt loam, pH 6.4
Fluorene	75	Silt loam, pH 4.4
9-Fluorenone	70	Silt loam, pH 4.4
1,4-Naphthoquinone	65	Silt loam, pH 4.4
Phenanthrene	1	Silt loam, pH 6.4
Benzo[a]pyrene	100	Silt loam, pH 7.2
DDT	0.4	Silt loam, pH 6.4
Creosote	3,396	Clay loam
	200–860	Sandy loam, pH 5.0

[a]NR = not reported.
[b]This system also contained PCB-degrading bacteria.

Mineral-ization (%)	Disappear-ance (%)	Incubation period (days)	Amendment	Refer-ence
28—32	NR[a]	30	Corncobs	21
18.4	85	30	Corncobs	11
<3	~98	56	Wood chips	89
<3	~98	56	Wood chips	89
<3	~98	56	Wood chips	89
32	NR	45	Corncobs	22
NR	~90	45.5	Wood chips and peat moss	79
14.9	28	30	Corncobs	12
2.2	35	30	Corncobs	12
4.0	15	30	Corncobs	12
0.8	85.7	30	Corncobs	12
0.2	28	30	Corncobs	12
2.2	19.5	30	Corncobs	12
NR	40	140	Straw	90
11.5	NR	30	Corncobs	22
NR	80—85	14	Nutrient medium	84
NR	52	14	Nutrient medium	84
NR	94	14s	Nutrient medium	84
11.9	NR	45	Corncobs	22
NR	80	100	Corncobs	82,83
10	60	60	Corncobs	19
NR	90	30	Corncobs	86
NR	85—95	112	Corncobs	86

water soluble, 40% were upgraded DDT, 10% comigrated with DDT metabolites (dicofol, DDD, DDE, and DBP) during thin-layer chromatography (TLC), and 18% were unextractable and, possibly, covalently bound to the soil-corncob matrix. Approximately 15% of the radiolabeled metabolites of DDT recovered following TLC were unidentified and were more polar than dicofol.

In contrast to the results of Fernando et al. [19], Nelson and Kinsella [80] concluded that in bench-scale studies, *P. chrysosporium* was not effective in the bioremediation of soil from the Leetown Pesticide Site that was contaminated with DDT, DDD, and DDE. It is well documented that *P. chrysosporium* is able to degrade DDT [1,2,4,7,16,19,23,81]. Apparently, characteristic differences in soils may have profound effects on DDT degradation in a particular soil. Of interest in the study of Nelson and Kinsella [80] is the observation that even when growth of *P. chrysosporium* was apparent, biodegradation, as assessed by [^{14}C]DDT mineralization, DDT disappearance, and metabolite formation, was minimal or nonexistent. Inasmuch as DDT is degraded by *P. chrysosporium* under ligninolytic or nonligninolytic conditions [7, 16,81] (under nonligninolytic conditions DDD is the major metabolite), substantial disappearance of DDT would have been expected. That it was not suggests that DDT may be tightly bound to the Leetown soil in such a manner that it is protected from biodegradation by the fungus.

B. Benzo[*a*]pyrene

Benzo[*a*]pyrene is degraded to carbon dioxide by *P. chrysosporium* [1,17] via a pathway in which the initial oxidation products have been shown to be the benzo[*a*]pyrene-1,6-, -3,6-, and -6,12-quinones [17,30. Interestingly, these benzo[*a*]pyrene quinones have been shown to be formed in vitro by the oxidation of benzo[*a*]pyrene catalyzed by lignin peroxidase, which represents the first direct evidence for involvement of the lignin-degrading system in the biogradation of an organopollutant [30]. In nitrogen-limited liquid cultures of *P. chrysosporium*, benzo[*a*]pyrene (initial concentration 252 µg/L) underwent at least a 99.5% disappearance during 54 h of incubation [17].

McFarland and his associates studied the effects of *P. chrysosporium* and amendment with ground corncobs on the bioremediation of a soil contaminated with benzo[*a*]pyrene [82,83]. This treatment significantly enhanced the removal rates of benzo[*a*]-pyrene in a nonsterile silt loam Kidman soil. During a 120-day incubation, approximately 80% of the benzo[*a*]pyrene initially present (initial concentration 100 mg/kg) was removed relative to uninoculated control soil in which 40% disappearance was observed.

A major mechanism for benzo[a]pyrene removal was that of immobilization, as approximately 37% of the [^{14}C]benzo[a]pyrene derivatives were bound as solvent-unextractable residues. Benzo[a]-pyrene removal, however, was significantly different from that observed in control cultures only when cultures were flushed with oxygen. Even though the Kidman soil had a slightly alkaline pH, fungal growth and benzo[a]pyrene disappearance still occurred. This is of interest because it has been established that lignin peroxidases are barely active above pH 5.0 [38]. The authors [82, 83] suggested that the fungus may be able to regulate the pH of its microenvironment such that a favorable pH would be maintained in the vicinity of developing hyphae, which could allow secreted lignin peroxidases to remain active.

C. Fluorene, 9-Fluorenone, and 1,4-Naphthoquinone

George and Neufeld [84] studied the ability of *P. chrysosporium* to degrade fluorene, 9-fluorenone, and 1,4-naphthoquinone in a sterile silt loam soil. Their results demonstrated that 80 to 85% of the fluorene, 52% of the 9-fluorenone, and 94% of the 1,4-naphthoquinone (having initial concentrations of 75, 70, and 65 mg/kg, respectively) were degraded during a 14-day incubation. These results were compared with those obtained with nonsterile soils not amended with *P. chrysosporium* in which 35 to 50% of the fluorene and 38% of the 9-fluorenone were degraded by indigenous soil microorganisms. The degradation of 1,4-naphthoquinone by soil microorganisms was not presented. In an effort to increase the rates of biodegradation, as assessed by fluorene disappearance, experiments were conducted in which sterile soil was inoculated with *P. chrysosporium* and supplemented with veratryl alcohol and hydrogen peroxide [84]. Veratryl alcohol partially protects lignin peroxidases from inactivation [85], and, in some cases, veratryl alcohol radical is thought to act as a co-oxidant [54]. Hydrogen peroxide is the oxidizing cosubstrate used by lignin peroxidase in vivo [38]. The addition of veratryl alcohol (0.2 to 50 mmol/50 g soil) did not enhance biodegradation of fluorene, and concentrations higher than 50 mmol/50 g soil often inhibited growth of the fungus. The addition of hydrogen peroxide (20 and 40 mmol/50 g soil) resulted in marginally better rates of fluorene degradation. In other experiments, George and Neufeld [84] assessed the ability of *P. chrysosporium* to degrade fluorene in a sterile silt loam soil in which fines (soil particles smaller than 0.074 mm) were removed. This procedure resulted in a substantial increase in the rate of fluorene degradation. The authors [84] suggested that fine removal opened up the pore space, resulting in increased oxygen availability, which, in turn, may have had a role in increasing fluorene degradation by this fungus.

D. Mixtures of Polycyclic Aromatic Hydrocarbons

Brooks and Mills (S. C. Brooks and A. L. Mills, Abstr. Annu. Meet. Am. Soc. Microbiol. 1991, Q77, p. 289) studied the biodegradation of PAHs in soil contaminated with creosote and were unable to demonstrate any difference in the disappearance of individual PAHs between nonsterile soils amended or not amended with *P. chrysosporium*. In constrast, Stroo et al. [86] reported that *P. chrysosporium* was effective in increasing the rate and extent of the biodegradation of PAHs in creosote-contaminated soils. In one highly contaminated soil, the total content of PAHs was reduced 90%, from 3396 mg/kg to 329 mg/kg, during a 30-day incubation (the soil was amended with corncobs as a growth substrate for the fungus). Even PAHs containing five and six rings were degraded.

In bench-scale pilot studies, Leuschner and Snyder [87] showed that an 85 to 95% disappearance of PAHs (including five- and six-ring compounds) occurred in two of three coal tar—contaminated sandy loam soils during a 16-week incubation in which the soils were amended with *P. chrysosporium* and ground corncobs. In one soil, only moderate (35%) disappearance of PAHs was observed and it was suggested that the levels of cyanide and metals in this soil may have inhibited the biodegradative system of the fungus. In these studies, no rigorous attempt was made to differentiate disappearance resulting from fungal action from degradation resulting from the action of indigenous microorganisms. However, the authors pointed out that in previous work with one of the soils, virtually no degradation of PAHs resulting from bacterial action was observed during a 16-week incubation and only 50% disappearance was observed during incubation for 25 weeks [87]. In any case, the levels of disappearance of PAH in soils amended with the fungus and corncobs were deemed sufficiently encouraging to suggest that this system be used in field trials.

It appears that, as with DDT, the results of studies on the degradation of PAHs by *P. chrysosporium* are mixed. One reason for this may be that some soils that are contaminated with coal tar may also be contaminated with box wastes. Box wastes are by-products of a scrubbing step that was used during coal gasification (of which coal tar is also a by-product) to remove sulfur and nitrogen compounds. Typically, the scrubbers (or "boxes") contained iron filings or wood chips. In the case of the iron filings, highly toxic ferricyanides were produced. High concentrations of such materials in soils would be expected to be inhibitory to the growth or biodegradative enzymes of many microorganisms, including *P. chrysosporium*.

E. Pentachlorophenol

The wood preservative PCP undergoes extensive biodegradation by
P. chrysosporium [3,33,75]. In some in vitro experiments, more
than 50% of this compound was degraded to carbon dioxide, and
lignin peroxidases have been shown to catalyze the initial oxidation
of PCP to form 2,3,5,6-tetrachloro-2,5-cyclohexadiene-1,4-dione [3,
33]. Biodegradation of PCP by *P. chrysosporium* in soils has also
been reported [22,23,79,88,89]. For example, 32% of the [^{14}C]-
PCP (333 µg/kg soil) adsorbed onto a silt loam soil amended with
ground corncobs was mineralized during 45 days of incubation [23].
Similarly, [^{14}C]PCP adsorbed on sand was reported to be mineralized; however, addition of peat, which was added as a potential
carbon source, inhibited the degradation of [^{14}C]PCP to [^{14}C]car-
bon dioxide [22].

The most thorough reported investigation of PCP degradation by
white rot fungi in soil has been performed by Lamar and Dietrich
[79]. In field studies, it was shown that treatment of PCP-contami-
nated soil with *P. chrysosporium* or *P. sordida* resulted in an 88 to
91% decrease in PCP concentration (initial PCP concentrations 250
to 400 mg/kg soil). In these studies, field plots were sterilized by
fumigation with BROM-O-GAS (98% methyl bromide, 2% chloropicrin)
and amended with wood chips and peat 3.35 and 1.93% (w/w), respectively, relative to the dry weight of the soil. After extraction with *n*-hexane—acetone (1:1) and gas chromatographic examination of the treated soil, pentachloroanisole was the only transformation product found. Based on previous findings that degradation of PCP to carbon dioxide and volatilization of PCP account
for only 2 to 3% of the PCP disappearance observed in soil [89],
it was suggested that most of the disappearance of PCP may result from the conversion of PCP to nonextractable soil-bound products. These conclusions are in contrast to those of Bumpus et al.
[22] and Aust and Bumpus [23], who showed that for several
chlorinated aromatic compounds, including [^{14}C]PCP, mineralization accounts for a substantial amount of the biodegradation observed. These differences in results between investigators may
have been caused by the fact that different soils were used and
that the soil used by Lamar and Dietrich [79] was from a "real-
world" site rather than a soil that was artificially contaminated in
the laboratory. It is also possible that the relatively high concentrations of growth substrates and the relatively low levels of [^{14}C]-
PCP used by Bumpus et al. [22] combined to promote the higher
levels of mineralization observed in their study.

F. Alkyl Halides

Kennedy et al. [12] tested the ability of *P. chrysosporium* to degrade six common alkyl halide insecticides (lindane, aldrin, dieldrin,

heptachlor, chlordane, and mirex). Of these, only [^{14}C]lindane
and [^{14}C]chlordane were extensively degraded, as shown by the
fact that 9 and 23%, respectively, of these compounds were mineralized in nutrient-limited liquid cultures of the fungus. A small
amount (~2%) of [^{14}C]mirex was mineralized, and little or no [^{14}C]-
heptachlor, [^{14}C]aldrin, or [^{14}C]dieldrin was mineralized. In general, the same pattern was observed when these compounds were
adsorbed onto a silt loam soil. Approximately 15 and 23%, respectively, of the [^{14}C]chlordane and [^{14}C]lindane were mineralized.
Four percent of the [^{14}C]mirex was mineralized, and less than 0.8%
of the [^{14}C]heptachlor, [^{14}C]aldrin, or [^{14}C]dieldrin adsorbed on
soil was mineralized.

G. 2,4,5-Trichlorphenoxyacetic Acid

Ryan and Bumpus [21] have shown that ^{14}C-labeled 2,4,5-trichloro-
phenoxyacetic acid ([^{14}C]2,4,5-T) undergoes extensive mineralization (62% in 30 days) in nitrogen-limited liquid cultures of *P. chrysosporium*. They also showed that substantial (28 to 32%) amounts
of [^{14}C]2,4,5-T were mineralized in a [^{14}C]2,4,5-T—contaminated
silt loam soil amended with ground corncobs and that there was no
statistical difference in the mineralization of [^{14}C]2,4,5-T in soils
that were sterilized or not sterilized before inoculation with *P. chrysosporium*. However, a statistically significant lower level (18%) of
mineralization (caused by indigenous microorganisms) was observed
in soils that were not amended with corncobs or inoculated with *P.
chrysosporium*. *P. chrysosporium* was found to be a good competitor in this soil, as evidenced by the fact that it became the
most prevalent fungus in nonsterile soils that also contained *Aspergillus niger* and *Fusarium* sp.

H. 2,4,6-Trinitrotoluene (TNT)

Contamination of water and soil with TNT is a major problem at
sites where this explosive was manufactured. Although resistant
to degradation by many microorganisms, [^{14}C]TNT was readily degraded by *P. chrysosporium* [11]. In liquid cultures (initial [^{14}C]-
TNT concentration 100 mg/L), 85% of the [^{14}C]TNT initially present
disappeared during a 90-day incubation. Approximately 18% of the
[^{14}C]TNT was degraded to [^{14}C]carbon dioxide. In a nonsterile
silt loam (initial [^{14}C]TNT concentration 10,000 mg/kg) amended
with *P. chrysosporium* growing on corncobs, 85% of the [^{14}C]TNT
present disappeared during a 90-day incubation. Approximately
19% of the [^{14}C]TNT initially present was degraded to [^{14}C]carbon
dioxide. Although the relative percentages of [^{14}C]TNT disappearance and mineralization in liquid culture and soil culture are very

similar, it should be noted that the soil cultures contained 100-fold more [14C]TNT than was present in liquid cultures [11].

I. Polychlorinated Biphenyls

The ability of *P. chrysosporium* to degrade PCBs has been reported [1,4,14]. Eaton [14] showed that [14C]Aroclor 1254 was mineralized (8% in 22 days) by ligninolytic cultures of this fungus, and packed-column gas chromatography suggested that extensive disappearance of all PCB congeners in this PCB mixture had occurred. Bumpus and Aust [4] have also shown that ligninolytic cultures of this fungus mineralized [14C]Aroclor 1254 and [14C]Aroclor 1242 (18 and 20.5%, respectively in 60 days). However, when pure PCB congeners ([14C]3,4,3',4'-tetrachlorobiphenyl and [14C]2,4,5,2',4',5'-hexachlorobiphenyl) were assayed, only 1 to 2% mineralization was observed, suggesting that some PCB congeners are degraded rather slowly [1,4]. In preliminary studies, Viney and Bewley [90] showed that *P. chrysosporium* was more effective than selected bacteria in degrading several of the more recalcitrant PCB congeners in liquid culture. In studies in which Aroclor 1242 (1000 mg/kg) was adsorbed on sand, a reduction of approximately 40% was achieved during a 20-week incubation by a mixed culture containing *P. chrysosporium* and PCB-degrading bacterium. The bulk of the disappearance, however, was attributed to metabolism of the less highly chlorinated PCB congeners [90]. Taken together, these studies demonstrate that PCBs can, indeed, be degraded by *P. chrysosporium*. They also suggest that, similar to the situation with bacteria, some PCB congeners are more difficult to degrade than others. Unfortunately, a systematic study using capillary gas chromatography to assess the ability of this fungus to degrade individual PCB congeners is not available. Such studies will be required before serious attempts can be made to use this fungus for bioremediation of soils contaminated with PCBs.

V. FATE OF ORGANOPOLLUTANTS FOLLOWING INCORPORATION INTO THE INSOLUBLE FRACTION OF SOIL AND PLANTS

In addition to biodegradation, incorporation into the insoluble matter of plants or the humic fraction of soils is a major fate for many organopollutants. The role of oxidases in these processes has been extensively studied by Bartha and by Bollag and their associates [91–97]. Some of these investigations have focused on the use of laccases from *T. versicolor*, which is also a white rot fungus and, like *P. chrysosporium*, produces lignin peroxidases

[98] and Mn peroxidases [90]. Once covalently incorporated into soil, organopollutants, such as chlorophenols, can be considered "detoxified." Although their possible rerelease into the environment presents a potential hazard, studies indicate that covalently bound organopollutants may be further degraded to carbon dioxide or rereleased, largely unchanged, only at low concentrations and so slowly that they are considered to be unlikely to cause adverse effects [91].

The ability of *P. chrysosporium* to degrade organopollutants covalently incorporated in insoluble material has also been investigated. Arjmand and Sandermann [8] have shown that ~60% of $[^{14}C]$4-chloroaniline and $[^{14}C]$3,4-dichloroaniline that had been covalently bound to a synthetic lignin was mineralized during a 30-day incubation with ligninolytic cultures of *P. chrysosporium*. Similar findings were obtained when wheat lignin, with covalently bound $[^{14}C]$4-chloroaniline or $[^{14}C]$3,4-dichloroaniline, was incubated with *P. chrysosporium* [100]. Sandermann et al. [101] have also shown that ~66% of such covalently bound chloroanilines ingested by rats are bioavailable during digestion; i.e., they are released as the free chloroanilines from the chloroaniline-lignin conjugates. This is of significance in agriculture, where large quantities of plant material containing bound chloroanilines may be ingested by livestock. In contrast, the ability of soil microorganisms to effect the release of large quantities of free chloroanilines from chloroaniline-lignin conjugates in soil does not appear to be a problem. Only 0.5 and 0.2% of 4-chloroaniline and 3,4-dichloroaniline, respectively, were released from chloroaniline-lignin conjugates during a 90-day incubation [101].

The ability of *P. chrysosporium* to degrade organopollutants covalently incorporated in soil, or, more specifically, in the humic acid fraction has been studied by Haider and Martin [102], who showed that several ^{14}C-labeled organopollutants covalently bound to humic acids underwent between 13 and 56% degradation to $[^{14}C]$carbon dioxide during 18 days of incubation in ligninolytic cultures of *P. chrysosporium*. Furthermore, ^{14}C-labeled humic acids were also degraded to $[^{14}C]$carbon dioxide at comparable rates. These results are of interest, as the turnover of humic substances in soil is generally very slow [103]. Thus, covalent incorporation of organopollutants into humic substances would normally be expected to make these materials less bioavailable as the result of immobilization. However, by developing methods by which *P. chrysosporium* or other microorganisms enhance the biodegradation of humic substances, the degradation of organopollutants covalently incorporated into these substances may also be enhanced [102].

VI. PROBLEMS, STRATEGIES, AND POTENTIAL ADVANTAGES USING *P. CHRYSOSPORIUM* FOR THE BIOREMEDIATION OF SOILS

The use of microorganisms for the bioremediation of contaminated soils is, in many cases, an attractive, economical, and environmentally compatible alternative to incineration or land filling. Indeed, numerous sites worldwide have been successfully bioremediated (in situ or in slurry reactors) using such processes. References [104–106] are selected examples that have been documented in the literature. Typically, such processes have been successful when used to treat soils that have been contaminated with pollutants that are relatively easy to degrade, such as the components of gasoline, kerosene, and one- to four-ring nonhalogenated aromatic compounds. Unfortunately, long-chain aliphatics (carbon chain length ~30 or greater), PAHs of five rings or more, and polyhalogenated aromatic compounds (DDT, PCBs, chlorodioxins, among others) still present formidable problems and have been resistant to most bioremediation efforts. Most soil bioremediation processes rely on the action of indigenous microorganisms, enrichment bacterial cultures derived from contaminated soils, or induced mutations. Although various interesting bacteria with unique biodegrative abilities have been produced by recombinant DNA techniques, these microorganisms have not been used in real-world bioremediation efforts.

There is a sizable literature ([107] and references therein for examples) on the biodegradative abilities of fungi. However, most bioremediation efforts have focused on the use of bacteria to degrade organopollutants in contaminated soils. In some respects, this is to be expected, because bacteria, in general, possess a wide spectrum of biodegradative abilities, are often (but not always) easy to culture and manipulate in the laboratory, and generally adapt with relative ease to soil bioremediation systems, such as slurry reactors.

Compared to bacteria, fungi are generally slower-growing microorganisms, have a more complex genetic makeup, and have more complex life cycles, and many, because of their size, might not be amenable for use in slurry reactors. This would be an especially difficult problem for filamentous fungi, whose mycelial network would be disrupted by shear forces in such systems. In spite of these problems, a greater effort to promote the use of fungi in bioremediation efforts is warranted. As noted earlier, fungi are as abundant as or more abundant than bacteria on a mass basis in some soils. This abundance and the biodegradative abilities of fungi suggest that they are undoubtedly responsible for a substantial amount of the natural biodegradation observed in soils.

In the specific case of P. chrysosporium, we have suggested
that this microorganism may be useful in soil bioremediation systems [1], especially composting and land farming. P. chrysosporium has the ability to degrade a wide variety of environmentally
persistent compounds [1—35]. In many cases, carbon dioxide has
been identified as the final metabolite, demonstrating that this microorganism possesses a complete pathway for the biodegradation
of such compounds. The lignin-degrading system has been implicated indirectly and directly in the biodegradation of many organopollutants by P. chrysosporium [1—35]. In nature, white rot
fungi are responsible, in large measure, for lignin turnover, and
it is thought that lignin must eventually undergo complete turnover (i.e., degradation). Thus, by analogy, it is reasonable to
suggest that the organopollutants that are attacked by P. chrysosporium may eventually undergo complete degradation.

Lignin is an extremely complex heteropolymer [37]; and any microbial system capable of degrading lignin must, by definition, be
nonspecific. Similarly, the lignin-degrading system is relatively
nonstereoselective in action [38,39]. Such a nonspecific degradation system might well be expected to have application in the treatment of hazardous organochemical waste as such wastes often occur
as mixtures.

Typically, the initial oxidation of an organopollutant by P. chrysosporium results in a metabolite that is more water soluble than the
parent compound. In the case of DDT [1,16], the initial oxidation
to form dicofol represents a 667-fold increase in water solubility
(the water solubilities of DDT and dicofol are 1.2 and 800 µg/L,
respectively [108,109]). The fact that the lignin-degrading system
normally attacks a substrate that is water insoluble and difficult to
degrade (i.e., lignin) is significant because many environmental
pollutants (e.g., DDT, PCB, chlorodioxins) are resistant to biodegradation, in part because they are insoluble in water and, therefore, are generally not readily available for attack.

A number of microorganisms have been discovered by taking
advantage of the fact that, on exposure to a given organopollutant, microbial populations are often enriched with organisms
that have the ability to metabolize the compound (or compounds)
of interest. Often, contaminated sites are excellent sources of
such microorganisms. Other techniques include the selection of
microbial populations following exposure to a mutagen or the use
of recombinant DNA techniques to construct microorganisms with
desired capabilities. In contrast, P. chrysosporium is a naturally
occurring microorganism that has been found in habitats from the
Sonoran dessert in Arizona [110] to central Siberia [72]. The
fact that wild-type strains of P. chrysosporium are available for
use in waste treatment systems should not be discounted as a
possible operational advantage given the difficulties associated

with the introduction of laboratory-constructed strains of microorganisms into the environment. Furthermore, biodegradative enzymes of many adapted, mutant, or recombinant strains must be induced by the presence of the organopollutant to be degraded. In *P. chrysosporium*, biodegradative abilities are expressed in response to nutrient depletion. Although carbon and sulfur depletion [111] may be used, most degradation strategies using this fungus rely on procedures that render the fungal cultures nutrient nitrogen depleted. In general, a requirement for nutrient nitrogen—limited conditions should not present a problem, as in most soils the relative amounts of bioavailable carbon are substantially greater than the amounts of bioavailable nitrogen. Although C/N ratios vary from soil to soil, ratios of 10 to 20:1 or even greater are not uncommon. Moreover, not all of the nitrogen present in soil is necessarily available for use by the fungus. For instance, some of the ammonium ions may be tightly bound to anionic components of certain soils [46,50]. Similarly, nitrogen may be in an organic form that is not immediately available for use by the fungus. For example, most of the nitrogen in humus consists of bound amino acids and amino sugars (e.g., glucosamine and galactosamine) [49]. Although free amino acids have been found in soil, their concentrations are typically low [49]. Another consideration is that *P. chrysosporium* does not appear to use most organopollutants (or lignin for that matter) as growth substrates. This, coupled with the fact that soils do not possess sufficient nutrients for sustained fungal growth, suggests that a logical strategy for promoting growth and biodegradation would be to supplement contaminated soils with suitable growth substrates. As noted previously in this chapter, abundant and relatively inexpensive agricultural residues, such as ground corncobs, peat, wood chips and wheat straw, have been used for this purpose. Such supplemental material would have the added effect of increasing the C/N ratio, as the amount of carbon in such material is substantially greater than the amount of nitrogen. For example, the C/N ratio of wheat straw is ~80:1, whereas the C/N ratio in wood chips can be 350:1 or greater [58]. In some cases where soils possess excessive nitrogen, this procedure might lead to establishment of C/N ratios that favor the expression of degradative enzymes in this fungus.

The fact that peroxidases are involved in the initial oxidations of lignin and of many organopollutants is important. In general, peroxidases are relatively nonspecific, and lignin peroxidases have been shown to be even less specific than some other peroxidases. For example, Hammel et al. [31] showed that horseradish peroxidase was able to oxidize PAHs having ionization potentials up to 7.35 eV, whereas a lignin peroxidase could catalyze PAHs having ionization potentials up to 7.55 eV. The nonspecific nature of peroxidase oxidations is due in part to the fact that some compounds that are not substrates (or are poor substrates) for the peroxidase are oxidized via cooxidation. In this process, a substrate of the peroxidase undergoes a one-electron oxidation to generate a free radical intermediate that is able to catalyze the oxidation of compounds that are not substrates for the peroxidase. Evidence for such cooxidations has been presented for some lignin peroxidase—mediated reactions. Popp et al. [54] presented evidence implicating the veratryl alcohol radical in the cooxidation of malonic acid and oxalic acid, and Tuisel et al. (H. Tuisel, T. Grover, J. A. Bumpus, and S. D. Aust, Abstr. 1990. NIEHS Superfund Basic Research Conference on Biodegradation of Hazardous Wastes, April 9—10, Utah State University) have presented data implicating the veratryl alcohol radical as a cooxidant for 3-amino-1,2,4-triazole. Similarly, Mn^{3+} complexes, formed by Mn peroxidases, catalyze a variety of cooxidations [112]. The significance and extent of such lignin peroxidase— and Mn peroxidase—mediated cooxidations in the degradation of organopollutants by *P. chrysosporium* are topics of current research interest.

Because peroxidases are integral components of the degradative systems of *P. chrysosporium*, there is the possibility that bacterial catalases might interfere with these degradative processes in soil, as they would be expected to eliminate quickly hydrogen peroxide, the oxidizing cosubstrate required for these enzymes. This, however, is probably not of great concern, as wood-rotting fungi, in vivo, typically produce their own hydrogen peroxide [113,114] via a variety of enzyme-mediated reactions. In *P. chrysosporium*, hydrogen peroxide is generated by at least two enzymes, glyoxal oxidase and glucose oxidase [115—117]. Glucose oxidase is located in the periplasmic space, in close proximity to the extracellular peroxidases [115].

VI. GENERAL OBSERVATIONS
AND CONCLUSIONS

The ability of *P. chrysosporium* to degrade a wide variety of environmentally persistent organopollutants was discovered in the mid-1980s. Despite the obvious appeal of using a naturally occurring microorganism with such capabilities, the development of practical bioremediation systems using this fungus has been slow, as the result of at least two factors. First, although biodegradation of many environmentally persistent compounds (DDT, chlordane, benzo[a]pyrene) occurs in fungus-amended soil, the rates of degradation are low and degradation is incomplete. Furthermore, although degradation of the relatively more easily degraded compounds (e.g., PCP, 2,4,5-T, PAHs having four rings or less) is often more rapid in fungus-amended soils, degradation is often also accomplished by more conventional microbial systems. Second, *P. chrysosporium* is not thought of as a typical soil microorganism. Thus, there has been a reluctance to try to use this microorganism in soil systems, based on the premise that established microorganisms typically possess a competitive advantage in their "ecological niches" and may prevent or restrict the growth of introduced microorganisms. This is a valid concern, and when *P. chrysosporium* is used in practical bioremediation systems it is necessary to supplement these systems with substantial quantities of a suitable growth substrate (e.g., wood chips, wheat straw, corncobs). This is required as the fungus cannot usually use most pollutants as growth substrates, and most soils do not possess sufficient available nutrients to support sustained growth of the fungus. Supplementing soils with such materials and mixing should have the effect of altering the ecological niche, such that *P. chrysosporium* will, in some cases, be competitive with indigenous microorganisms. If the growth substrate is inoculated with *P. chrysosporium* before addition to soil, the chances for successful establishment of this fungus may be further increased. Another approach to the competition problem is soil sterilization. Gas sterilization of soils before inoculation is feasible. However, technical problems and expense may limit the use of this procedure. Perhaps the greatest obstacle to the use of microorganisms in general, and of *P. chrysosporium* in particular, in soil bioremediation is a lack of understanding and/or appreciation of the complexities of the microbial ecology of soil systems. In the case of *P. chrysosporium*, a picture is emerging which suggests that the problem of microbial competition can be dealt with. However, the most formidable problem to be addressed is that of discovering how to use this fungus to increase rates and extents of biodegradation of the compounds that are most environmentally persistent.

ACKNOWLEDGMENTS

Research cited that was conducted by the author was supported by Cooperative Agreements CR-814448, CR-814162 and CR-813369 with the United States Environmental Protection Agency (P. R. Sferra, Project Officer), Grant ES04922 from the National Institute of Environmental Health Sciences, and Grant 14-08-0001-61733 from the United States Geological Survey. The author thanks Professor Steven D. Aust (Biotechnology Center, Utah State University), who directed some of this research, for his continuing interest, advice, and collaboration.

REFERENCES

1. Bumpus, J. A., M. Tien, D. Wright, and S. D. Aust. 1985. Oxidation of persistent environmental pollutants by a white rot fungus. Science 228:1434–1436.

2. Bumpus, J. A., and S. D. Aust. 1987. Biodegradation of environmental pollutants by the white rot fungus *Phanerochaete chrysosporium*: involvement of the lignin degrading system. BioEssays 6:166–167.

3. Mileski, G. J., J. A. Bumpus, M. A. Jurek, and S. D. Aust. 1988. Biodegradation of pentachlorophenol by the white rot fungus *Phanerochaete chrysosporium*. Appl. Environ. Microbiol. 54:2885–2889.

4. Bumpus, J. A., and S. D. Aust. 1986. Biodegradation of chlorinated organic compounds by *Phanerochaete chrysosporium*, p. 340–349. *In* J. H. Exner (ed.), Solving Hazardous Waste Problems, ACS Symposium Series 338. American Chemical Society, Washington, DC.

5. Bumpus, J. A., and S. D. Aust. 1987. Mineralization of recalcitrant environmental pollutants by a white rot fungus, p. 146–151. *In* Proceedings of the national conference on hazardous wastes and hazardous materials. Hazardous Materials Control Research Institute, Silver Spring, MD.

6. Bumpus, J. A., T. Fernando, G. J. Mileski, and S. D. Aust. 1987. Biodegradation of organopollutants by *Phanerochaete chrysosporium*: practical considerations, p. 411–418. *In* Land disposal, remedial action, incineration and treatment of hazardous waste. Proceedings of the thirteenth annual research symposium, Cincinnati. EPA/600/9-87/015.

7. Bumpus, J. A., and S. D. Aust. 1985. Studies on the biodegradation of organopollutants by a white rot fungus, p. 404–410. *In* Proceedings of the international conference on new frontiers for hazardous waste management. EPA/600/9-85/025.

8. Arjmand, M., and H. Sandermann, Jr. 1985. Mineralization of chloroaniline/lignin conjugates and of free chloroanilines by the white rot fungus *Phanerochaete chrysosporium*. J. Agic. Food Chem. 33:1055–1060.

9. Bumpus, J. A. 1989. Biodegradation of polycyclic aromatic hydrocarbons by *Phanerochaete chrysosporium*. Appl. Environ. Microbiol. 55:154–158.

10. Cripps, C., J. A. Bumpus, and S. D. Aust. 1990. Biodegradation of azo and heterocyclic dyes by *Phanerochaete chrysosporium*. Appl. Environ. Microbiol. 56:1114–1118.

11. Fernando, T., J. A. Bumpus, and S. D. Aust. 1990. Biodegradation of TNT (2,4,6-trinitrotoluene) by *Phanerochaete chrysosporium*. Appl. Environ. Microbiol. 56:1666–1671.

12. Kennedy, D. W., S. D. Aust, and J. A. Bumpus. 1990. Comparative biodegradation of alkyl halide insecticides by the white rot fungus, *Phanerochaete chrysosporium* (BKM-F-1767). Appl. Environ. Microbiol. 56:2347—2353.

13. Aust, S. D., H. Tuisel, C.-W. Chang, and J. A. Bumpus. 1990. Oxidation of environmental pollutants by lignin peroxidases from white rot fungi, p. 454—464. *In* C. C. Reddy, G. A. Hamilton, and K. M. Madyastha (eds.), Biological oxidation systems, Vol. I. Academic Press, San Diego.

14. Eaton, D. C. 1985. Mineralization of polychlorinated biphenyls by *Phanerochaete chrysosporium*: a ligninolytic fungus. Enzyme Microb. Technol. 7:194—196.

15. Bumpus, J. A., and B. J. Brock. 1988. Biodegradation of crystal violet by the white rot fungus *Phanerochaete chrysosporium*. Appl. Environ. Microbiol. 54:1143—1150.

16. Bumpus, J. A., and S. D. Aust. 1987. Biodegradation of DDT (1,1,1-trichloro-2,2-bis(4-chlorophenyl)ethane) by the white rot fungus *Phanerochaete chrysosporium*. Appl. Environ. Microbiol. 59:2001—2008.

17. Sanglard, D. M. S. A. Leisola, and A. Fiechter. 1986. Role of extracellular ligninases in biodegradation of benzo[a]pyrene by *Phanerochaete chrysosporium*. Enzyme Microb. Technol. 8:209—212.

18. Huynh, V.-B., H.-m. Chang, T. W. Joyce, and T. K. Kirk. 1985. Dechlorination of chloro-organics by a white rot fungus. TAPPI J. 68(7):98—102.

19. Fernando, T., S. D. Aust, and J. A. Bumpus. 1989. Effects of culture parameters on DDT (1,1,1-trichloro-2,2-bis(4-chlorophenyl)ethane) biodegradation by *Phanerochaete chrysosporium*. Chemosphere 19:1387—1398.

20. Liu, S.-Y., A. J. Freyer, and J. M. Bollag. 1991. Microbial dechlorination of the herbicide metolachlor. J. Agric. Food Chem. 39:631—636.

21. Ryan, T. P., and J. A. Bumpus. 1989. Biodegradation of 2,4,5-trichlorophenoxyacetic acid in liquid culture and in soil by the white rot fungus *Phanerochaete chrysosporium*. Appl. Microbiol. Biotechnol. 31:302—307.

22. Bumpus, J. A., T. Fernando, M. A. Jurek, G. J. Mileski, and S. D. Aust. 1989. Biological treatment of hazardous wastes by *Phanerochaete chrysosporium*, p. 176—183. *In* G. Lewandowski, P. Armenante, and B. Baltzis (eds.), Biotechnology applications in hazardous waste treatment. Engineering Foundation, New York.

23. Aust, S. D., and J. A. Bumpus. 1987. Biodegradation of halogenated hydrocarbons. 5 p. Environmental Research Brief. U.S. Environmental Protection Agency. EPA/600/M-87/012.

24. Morgan, P., S. T. Lewis, and R. J. Watkinson, 1991. Comparison of abilities of white-rot fungi to mineralize selected xenobiotic compounds. Appl. Microbiol. Biotechnol. 34:693–696.
25. Aitken, M. D., R. Venkatadri, and R. L. Irvine. 1989. Oxidation of phenolic pollutants by a lignin degrading enzyme from the white rot fungus *Phanerochaete chrysosporium.* Water Res. 23:443–450.
26. Eriksson, K.-E., and M.-C. Kolar. 1985. Microbial degradation of chlorolignins. Environ. Sci. Technol. 19:1086–1089.
27. Glenn, J. K., and M. H. Gold. 1983. Decolorization of several polymeric dyes by the lignin-degrading basidiomycete *Phanerochaete chrysosporium.* Appl. Environ. Microbiol. 45:1741–1747.
28. Aust, S. D. 1990. Degradation of environmental pollutants. Microb. Ecol. 20:197–209.
29. Blondeau, R. 1989. Biodegradation of natural and synthetic humic acids by the white rot fungus, *Phanerochaete chrysosporium.* Appl. Environ. Microbiol. 55:1282–1285.
30. Haemmerli, S. D., M. S. A. Leisola, D. Sanglard, and A. Fiechter. 1986. Oxidation of benzo(a)pyrene by extracellular ligninases of *Phanerochaete chrysosporium:* veratryl alcohol. and stability of ligninase. J. Biol. Chem. 261:6900–6903.
31. Hammel, K. E., B. Kalyanaraman, and T. K. Kirk. 1986. Oxidation of polycyclic aromatic hydrocarbons and dibenzo-(p)dioxins by *Phanerochaete chrysosporium.* J. Biol. Chem. 261:16948–16952.
32. Schreiner, R. P., S. E. Stevens, Jr., and M. Tien. 1988. Oxidation of thianthrene by the ligninase of *Phanerochaete chrysosporium.* Appl. Environ. Microbiol. 54:1858–1860.
33. Hammel, K. E., and P. J. Tardone. 1988. The oxidative 4-dechlorination of polychlorinated phenols is catalyzed by extracellular fungal lignin peroxidases. Biochemistry 27:6563–6568.
34. Valli, K., and M. H. Gold. 1991. Degradation of 2,4-dichlorophenol by the lignin-degrading fungus *Phanerochaete chrysosporium.* J. Bacteriol. 173:345–352.
35. Kersten, P. J., M. Tien, B. Kalyanaraman, and T. K. Kirk. 1985. The ligninase of *Phanerochaete chrysosporium* generates cation radicals from methoxybenzenes. J. Biol. Chem. 260:2609–2612.
36. Agrios, G. N. 1978. Plant pathology. Academic Press, New York.
37. Crawford, R. 1981. Lignin biodegradation and transformation. John Wiley & Sons, New York.
38. Tien, M. 1987. Properties of ligninases from *Phanerochaete chrysosporium* and their possible applications. CRC Crit. Rev. Microbiol. 15:141–168.

39. Kirk, T. K., and R. L. Farrell. 1987. Enzymatic "combustion": the microbial degradation of lignin. Annu. Rev. Microbiol. 41:465–505.

40. Gold, M. H., H. Wariishi, and K. Valli. 1989. Extracellular peroxidases involved in lignin degradation by the white rot basidiomycete *Phanerochaete chrysosporium*, p. 127–140. *In* J. R. Whitaker and P. E. Sonnet (eds.), Biocatalysis in agricultural biotechnology. ACS Symposium Series No. 389. American Chemical Society, Washington, D.C.

41. Tien, M., and T. K. Kirk. 1983. Lignin-degrading enzyme from the hymenomycete *Phanerochaete chrysosporium* Burds. Science 221:661–663.

42. Tien, M., and T. K. Kirk. 1984. Lignin-degrading enzyme from *Phanerochaete chrysosporium*: purification, characterization, and catalytic properties of a unique H_2O_2-requiring oxygenase. Proc. Natl. Acad. Sci. USA 81:2280–2284.

43. Glenn, J. K., M. A. Morgan, M. B. Mayfield, M. Kuwahara, and M. H. Gold. 1983. An extracellular H_2O_2-requiring enzyme preparation involved in lignin degradation by the white rot basidiomycete *Phanerochaete chrysosporium*. Biochem. Biophys. Res. Commun. 114:1077–1083.

44. Kuwahara, M., J. K. Glenn, M. A. Morgan, and M. H. Gold. 1984. Separation and characterization of two extracellular H_2O_2-dependent oxidases from ligninolytic cultures of *Phanerochaete chrysosporium*. FEBS Lett. 169:247–250.

45. Glossary of soil science terms. 1979. Soil Science Society of America, Madison, Wisconsin.

46. Stotzky, G. 1972. Activity, ecology and population dynamics of microorganisms in soil. Crit. Rev. Microbiol. 2:59–137.

47. Lynch, J. M. 1982. Limits to microbial growth in soil. J. Gen. Microbiol. 128:405–410.

48. Wainwright, M. 1988. Metabolic diversity of fungi in relation to growth and mineral cycling in soil. A review. Trans. Br. Mycol. Soc. 90:159–170.

49. Alexander, M. 1977. Introduction to soil microbiology, 2nd ed. John Wiley & Sons, New York.

50. Foth, H. D. 1984. Fundamentals of soil science, 7th ed. John Wiley & Sons, New York.

51. Fekete, F. A., V. Chandhoke, and J. Jellison. 1989. Iron-binding compounds produced by wood-decaying basidiomycetes. Appl. Environ. Microbiol. 55:2720–2722.

52. Letham, G. F. 1986. The ligninolytic activities of *Letinus edodes* and *Phanerochaete chrysosporium*. Appl. Microbiol. Biotechnol. 24:51–58.

53. Cromack, K., P. Sollins, R. L. Todd, R. Fogel, A. W. Todd, W. M. Fender, M. E. Crossley, and D. A. Crossley. 1977. The role of oxalic acid and bicarbonate in calcium cycling by

fungi and bacteria. Some possible implications for soil animals. Ecol. Bull. (Stockh.) 25:246–252.

54. Popp, J. L., B. Kalyanaraman, and T. K. Kirk. 1990. Lignin peroxidase oxidation of Mn^{2+} in the presence of veratryl alcohol, malonic or oxalic acid and oxygen. Biochemistry 29: 10475–10478.

55. Moore-Landecker, E. 1990. Fundamentals of the fungi, 3rd ed. Prentice-Hall, Englewood Cliffs, New Jersey.

56. Lekkerkerk, L., H. Lundkvist, G. I. Agren, G. Ekbohm, and E. Bosatta. 1990. Decomposition of heterogeneous substrates: an experimental investigation of a hypothesis on substrate and microbial properties. Soil Biol. Biochem. 22:161–167.

57. Kirk, T. K., E. Schultz, W. J. Connors, L. F. Lorenz, and J. G. Zeikus. 1978. Influence of culture parameters on lignin metabolism by *Phanerochaete chrysosporium*. Arch. Microbiol. 117:277–285.

58. Deacon, J. W. 1984. Introduction to modern mycology, 2nd ed. Blackwell Scientific Publications, Oxford.

59. Kendrick, W. B., and Parkinson, D. 1990. Soil fungi, p. 49–68. *In* D. L. Dindal (ed.), Soil biology guide, John Wiley & Sons, New York.

60. Frankland, J. C., J. N. Hedger, and M. J. Swift. 1982. Decomposer basidiomycetes: their biology and ecology. Cambridge University Press, Cambridge.

61. Biddlestone, A. J., and K. R. Gray. 1985. Composting, p. 1059–1070. *In* M. Moo-Young (ed.), Comprehensive biotechnology, Vol. II. Pergamon Press, New York.

62. Finstein, M. S., F. C. Miller, P. F. Strom, S. T. MacGregor, and K. M. Psarianos. 1983. Composting ecosystem management for waste treatment. Biotechnology 1:347–353.

63. Dietrich, D. M., and R. T. Lamar. 1990. Selective medium for isolating *Phanerochaete chrysosporium* from soil. Appl. Environ. Micorbiol. 56:3088–3092.

64. Russel, P. 1956. A selective medium for the isolation of basidiomycetes. Nature 177:1038–1039.

65. Uscuplic, M., and R. G. Pawsey. 1970. A selective medium for the isolation of *Polyporous schweinitz*. Trans. Br. Mycol. Soc. 55:161–163.

66. Vaartaja, O. 1960. Selectivity of fungicidal materials in agar cultures. Phytopathology 50:870–873.

67. Cary, J. K., and A. V. Hull. 1987. A selective medium for the isolation of wood-rotting basidiomycetes. *In* Biodeterior. Bull. 25:373–376.

68. Martin, J. P. 1950. Use of acid, rose bengal and streptomycin in the plate method for estimating soil fungi. Soil Sci. 69: 215–232.

69. Maloy, O. C. 1974. Benomyl-malt agar for the purification of cultures of wood decay fungi. Plant Dis. Rep. 58:902—904.

70. Hale, M. D. C., and J. G. Savory. 1976. Selective agar media for the isolation of basidiomycetes from wood-a review. *In* Biodeterior. Bull. 12:112—115.

71. Domsch, K. H., W. Gams, and T. Anderson. 1980. Compendium of soil fungi, Vols. 1 and 2. Academic Press, London.

72. Chang, H-m., T. W. Joyce, A. G. Campbell, E. D. Gerrad, V-B. Huynh, and T. K. Kirk. 1983. Fungal discolorization of bleach plant effluents, p. 257—268. *In* T. Hicuchi, H-m. Chang, and T. K. Kirk (eds.), Recent advances in lignin biodegradation. Uni Publishers, Tokyo.

73. Lewandowski, G. A., P. M. Armenante, and D. Pak. 1990. Reactor design for hazardous waste treatment using a white rot fungus. Water Res. 24:75—82.

74. Combustion Engineering, Inc. 1987. Biotreatment of pink-water with a fungal-based treatment method—summary report.

75. Lin, J-E., H. Wang, and R. F. Hickey. 1990. Degradation kinetics of pentachlorophenol by *Phanerochaete chrysosporium*. Biotechnol. Bioengin. 35:1125—1134.

76. Reid, I. D., and K. A. Seifert. 1982. Effect of an atmosphere of oxygen on growth, respiration and lignin degradation by white rot fungi. Can. J. Bot. 60:252—260.

77. Jager, A., S. Croan, and T. K. Kirk. 1985. Production of ligninases and degradation of lignin in agitated submerged cultures of *Phanerochaete chrysosporium*. Appl. Environ. Microbiol. 50:1274—1278.

78. Venkatadri, R., and R. L. Irvine. 1990. Effect of agitation on ligninase and ligninase production by *Phanerochaete chrysosporium*. Appl. Environ. Microbiol. 56:2684—2691.

79. Lamar, R. T., and D. M. Dietrich. 1990. *In situ* depletion of pentachlorophenol from contaminated soil by *Phanerochaete chrysosporium*. Appl. Environ. Microbiol. 56:3093—3100.

80. Nelson, M. J., and J. V. Kinsella. 1990. Bench-scale treatability study—Leetown pesticide site. REM III program-EPA contract 68-01-7250. Ecova Corp., Redmond, Washington.

81. Kohler, A., A. Jager, H. Willerhausen, and H. Graf. 1988. Extracellular ligninase of *Phanerochaete chrysosporium* Burdsall has no role in the degradation of DDT. Appl. Microbiol. Biotechnol. 29:618—620.

82. McFarland, M. J., X. J. Qiu, W. A. April, and R. C. Sims. 1990. Biological composting of petroleum waste organics using the white rot fungus *Phanerochaete chrysosporium*. *In* Pap. Inst. IGT Symp. Gas, Oil, Coal Environ. Biotechnol., 2nd, pp. 37—57, Inst. Gas Technol., Chicago, Illinois.

83. McFarland, M. J., and X. J. Qiu. 1991. Bound residue for-
     mation in PAH contaminated soil composting using *Phanero-
     chaete chrysosporium*. J. Hazard. Wastes Hazard. Mat. 8:
     115–126.

84. George, E. J., and R. D. Neufeld. 1989. Degradation of
     fluorene in soil by fungus *Phanerochaete chrysosporium*. Bio-
     technol. Bioeng. 33:1306–1310.

85. Wariishi, H., and M. H. Gold. 1990. Lignin peroxidase com-
     pound. III: Mechanism of formation and decomposition. J.
     Biol. Chem. 265:2070–2077.

86. Stroo, H. F., M. A. Jurek, J. A. Bumpus, M. F. Torpy, and
     S. D. Aust. 1989. Bioremediation of wood preserving wastes
     using the white rot fungus *Phanerochaete chrysosporium*, p.
     1–7. Proc. Am. Wood. Preservers Assoc., San Francisco.

87. Leuschner, A., and S. Snyder. 1991. White rot fungus
     treatment of soils contaminated with manufactured gas plant
     chemicals of interest. Report prepared by Remediation Tech-
     nologies for New York State Electric and Gas Co.

88. Lamar, R. T., M. J. Larsen, and T. K. Kirk. 1990. Sensi-
     tivity to and degradation of pentachlorophenol by *Phanero-
     chaete* spp. Appl. Environ. Microbiol. 56:3519–3526.

89. Lamar, R. T., J. A. Glaser, and T. K. Kirk. 1990. Fate of
     pentachlorophenol (PCP) in sterile soils inoculated with the
     white rot basidiomycete *Phanerochaete chrysosporium*: min-
     eralization, volatilization and depletion of PCP. Soil Biol.
     Biochem. 22:433–440.

90. Viney, I., and R. J. F. Bewley. 1990. Preliminary studies
     on the development of a microbiological treatment for poly-
     chlorinated biphenyls. Arch. Environ. Contam. Toxicol.
     19:789–796.

91. Dec, J., and J.-M. Bollag. 1988. Microbial release and deg-
     radation of catechol and chlorophenols bound to synthetic hu-
     mic acids. Soil Sci. Soc. Am. J. 52:1366–1371.

92. Bollag, J.-M., and W. B. Bollag. 1990. A model for enzy-
     matic binding of pollutants in the soil. Int. J. Environ. Anal.
     Chem. 39:147–157.

93. Bollag, J.-M., and C. Myers. 1992. Detoxification of aquatic
     and terrestrial sites through binding of pollutants to humic
     substances. Sci. Total Environ. 117/118:357–366.

94. Nannipieri, P., and J.-M. Bollag. 1991. Use of enzymes to
     detoxify pesticide-contaminated soils and waters. J. Environ.
     Qual. 20:510–517.

95. Bartha, R., and L. Bordeleau. 1969. Cell-free peroxidases
     in soil. Soil Biol. Biochem. 1:139–143.

96. Bartha, R. 1971. Fate of herbicide-derived chloroanilines in
     soil. J. Agric. Food Chem. 19:385–387.

97. Bartha, R., I.-S. You, and A. Saxena. 1983. Humus-bound residues of phenylamide herbicides: their nature, persistence, and monitoring, p. 345–350. *In* J. Miyamoto et al. (eds.), IUPAC pesticide chemistry, human welfare and the environment. Pergamon Press, New York.

98. Jonsson, L., T. Johansson, K. Sjostrom, and P. O. Nyman. 1987. Purification of ligninase isozymes from the white-rot fungus *Trametes versicolor*. Acta Chem. Scand. B41:766–769.

99. Johansson, T., and P. O. Nyman. 1987. A manganese (II)-dependent extracellular peroxidase from the white-rot fungus *Trametes versicolor*. Acta Chem. Scand. B41:762–765.

100. Argmand, M., and H. Sandermann, Jr. 1986. Plant biochemistry of xenobiotics. Mineralization of chldoroaniline/lignin-metabolites from wheat by the white rot fungus *Phanerochaete chrysosporium*. Z. Naturforsch. C. 41C:206–214.

101. Sandermann, H. S., Jr., M. Arjmand, I. Gennity, R. Winkler, C. B. Struble, and P. W. Aschbacher. 1990. Animal bioavailability of defined xenobiotic lignin metabolites. J. Agric. Food Chem. 38:1877–1880.

102. Haider, K., and J. P. Martin. 1988. Mineralization of ^{14}C-labelled humic acids and of humic-acid bound ^{14}C-xenobiotics by *Phanerochaete chrysosporium*. Soil Biol. Biochem. 20:425–429.

103. Jenkinson, D. S., and J. H. Rayner. 1977. The turnover of soil organic matter in some of the Rothamsted classical experiments. Soil Sci. 123:298–305.

104. Valo, R., and M. Salkinoja-Salonen. 1986. Bioreclamation of chlorophenol-contaminated soil by composting. Appl. Microbiol. Biotechnol. 25:68–75.

105. Sims, J. L., R. C. Sims, and J. E. Matthews. 1990. Approach to bioremediation of contaminated soil. J. Hazard. Wates Hazard. Mater. 7:117–149.

106. Duffy, J. J., W. C. Ying, R. M. Bewley, and T. Balba. 1989. Biological remediation of hazardous waste sites, p. 63–79. *In* G. Lewandowski, P. Armenante, and B. Baltzis (eds.), Biotechnology applications in hazardous waste treatment. Engineering Foundation, New York.

107. Rochkind, M. L., J. W. Blackburn, and G. S. Sayler. 1986. Microbial decomposition of chlorinated aromatic compounds. Publ. EPA/600/2-86/090. U.S. Environmental Protection Agency Hazardous Waste Engineering Research Laboratory, Cincinnati.

108. Bowman, M. C., F. Acrea, Jr., and M. K. Corbett. 1960. Solubility of carbon-14 DDT in water. J. Agric. Food Chem. 8:406–410.

109. Walsh, P. R., and R. A. Hites. 1979. Dicofol solubility and
      hydrolysis in water. Bull. Environm. Contam. Toxicol. 22:
      305–311.
110. Burdsall, H. H. Jr., and W. E. Eslyn. 1974. A new *Phan-
      erochaete* with a *chrysosporium* imperfect state. Mycotaxon
      1:123–133.
111. Jeffries, T. W., S. Choi, and T. K. Kirk. 1981. Nutritional
      regulation of lignin degradation by *Phanerochaete chrysospor-
      ium*. Appl. Environ. Microbiol. 42:290–296.
112. Paszczynski, A., V.-B. Huynh, and R. Crawford. 1986.
      Comparison of ligninase-I and peroxidase M2 from the white
      rot fungus *Phanerochaete chrysosporium*. Arch. Biochem.
      Biophys. 244:750–765.
113. Koenigs, J. W. 1972. Production of extracellular hydrogen
      peroxide by wood-rotting fungi. Phytopathology 62:100–110.
114. Koenigs, J. W. 1974. Production of hydrogen peroxide by
      wood rotting fungi in wood and its correlation with weight
      loss, depolymerization, and pH change. Arch. Microbiol.
      99:129–145.
115. Forney, L. J., C. A. Reddy, and S. Pankratz. 1982. Ul-
      trastructural localization of hydrogen peroxide production in
      ligninolytic *Phanerochaete chrysosporium* cells. Appl. En-
      viron. Microbiol. 44:732–736.
116. Kelley, R. L., and C. A. Reddy. 1986. Identification of
      glucose oxidase activity as the primary source of hydrogen
      peroxide production in ligninolytic cultures of *Phanerochaete
      chrysosporium*. Arch. Microbiol. 144:248–253.
117. Kersten, P. J. 1990. Glyoxal oxidase of *Phanerochaete
      chrysosporium*: Its characterization and activation by lignin
      peroxidase. Proc. Natl. Acad. Sci. USA 87:2936–2940.

3

Biodegradation and Humification Processes in Forest Soils

INGRID KÖGEL-KNABNER *University of Bochum, Bochum, Germany*

I. INTRODUCTION

Soil organic matter represents a major component of the surface carbon reserves of the world. The total carbon in dead organic matter on the forest floor and in the underlying mineral soil has been estimated to be 1450×10^9 t C, exceeding the amount stored in living vegetation by a factor of two or three [1–3]. Mineralization and humification represent important processes in the terrestrial carbon cycle. Schlesinger [4] estimated that about 0.7% of the annual terrestrial net primary production is sequestered in the carbon of refractory humic substances. Organic matter of forest soils is composed of a mixture of above- and belowground plant residues (primary resources), microbial residues (secondary resources), and humic compounds [5]. Humic compounds are formed concomitantly during the microbial decomposition of primary and secondary resources. The conceptual view of soil organic matter in this chapter is that of a continuum ranging from fresh plant litter to humic substances, the final products of humification. A complete separation of plant remains and humic compounds is not possible, because organic matter at all stages of degradation and humification is present simultaneously in natural soils, albeit in different amounts. Undisturbed soils have a characteristic distribution of undecomposed and decomposed litter and humus on and within their surface [3]. As the horizon sequence in forest humus is directly related to the degree of humification, the pathway of formation of humic substances can be delineated by following the chemical evolution in plant-derived biomacromolecules and in humic substances with depth. It is now generally accepted that humification in soils involves a number of processes and that more than one humification process is active in a particular soil. The predominance of specific pathways of formation of humic substances in a particular environment presumably depends on the type of precursor material and on the environmental conditions [6,7]. The objective of this chapter is to delineate the major humification processes that operate in forest soils of temperate climates.

II. GENESIS OF FOREST HUMUS LAYERS

A. The Humus Profile

Before discussing the biodegradation and humification of different classes of chemical compounds in forest humus layers, the various factors that influence the distribution of specific compounds with depth should be considered. The parent material for the humification processes is derived from plant residues (primary resources) and microbial and animal products (secondary resources). The composition of forest humus layers depends on the input and composition of primary and secondary resources and on the decomposition rate of the different classes of compounds [5].

The total process of litter decomposition involves comminution, catabolism, and leaching [5]. Figure 1 gives an overview of the different processes that are operating. Fragmentation of plant litter by soil animals strongly affects the release of dissolved organic carbon [8]. The individual processes mentioned above are reiterated at different stages of decomposition [9], leading to the accumulation of organic matter in distinctive forest floor horizons. They represent various stages of decomposition and humification, from fresh litter material (L horizon), to transitional horizons (Of), and to completely humidified forest floor (Oh) and mineral soil horizons (Ah). The sequence of organic and mineral horizons constitutes the humus profile. Depending on the decomposition rate, the forest humus type mull, moder, or mor develops. Carbon turnover rates are controlled by three main groups of factors: the site-specific environment (climatic factors, such as water regime and temperature, and interactions with the soil matrix) results in a definate resource quality (chemical composition of litter), and both factors, in turn, control the nature of the decomposer community [5,10].

With depth in the forest floor, the organic matter is gradually mineralized and humified, resulting in a mixture of macromorphologically identifiable plant residues and morphologically unstructured

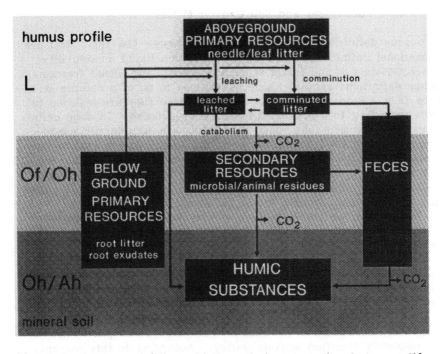

Figure 1 Processes of litter biodegradation operating between different compartments of carbon in forest soils. The horizon designations of the German classification system L, Of, and Oh, correspond to the Oi (L), Oe (Of), and Oa (Oh) horizon designations of the American classification.

humic compounds. These morphological changes are directly related to the degree of humification [11]. However, reworking and transport of the organic material can occur through the action of the soil fauna, especially earthworms, which are of major importance in mull humus profiles [10,12]. Additionally, the input of root litter and the transport of organic material into the A horizon must be considered [13]. In deeper humus horizons and in the mineral soil, the contribution of root litter is significant.

Organic matter is transported in the mineral soil horizon by the action of animals and also by leaching and precipitation of water-soluble organic matter. Babel [14] and Ohta et al. [15] point out that the transport of particulate organic matter into the A horizon is also of importance in some forest soils.

B. Humic Substances and Soil Organic Matter

Numerous definitions exist for humic substances [16–19]. Most are operational definitions, such as the definition based on solubility characteristics of the fulvic acid, humic acid, and humin fractions. Other definitions require humic substances to be of a definite structure (e.g., aromatic, phenolic rich), because they are assumed to form according to a presumed humification process. In some definitions it is assumed that humic substances are structurally nondefined compounds, although the basis for these requirements is possibly the result of limitations in analytical techniques suitable for the characterization of humic substances. The number of definitions reflects the problems associated with the separation of plant, animal, or microbial residues from humic substances.

In this chapter humic substances are viewed as an integrated component of soil organic matter. The soil organic matter compartment itself is composed of plant, animal, and microbial residues at different stages of decomposition and of humic substances, but excluding live roots, animals, and microorganisms. No suitable technique exists that allows a complete separation of plant remains and humic substances. This can probably not be achieved in principle, because soils contain a continuum of organic matter at different stages of decomposition and humification, from fresh plant remains to completely humified organic matter. According to this concept, only two major compartments of carbon exist in soil. One is the "living" compartment, consisting of plant roots, animals, and microorganisms, and the other is the "dead" compartment or soil organic matter compartment. With increasing depth and decomposition in forest soils, the relative amount of morphologically and chemically identifiable detritus decreases and the amount of amorphous material formed during humification increases. This definition of humic substances includes a large range of compounds and also allows the use of different separation and fractionation techniques. In contrast, the operational definition used by the International Humic Substances Society [17] is based on the solubility behavior in aqueous acids and bases, and humic substances are studied exclusively in the form of fulvic acids, humic acids, and humin. This limited definition prevents the use of additional separation techniques, which are possibly of relevance to the study of humification processes [18], and is, therefore, avoided in this chapter.

C. Formation of Humic Substances: Current Concepts

The current concepts of the formation of humic substances can be divided into abiotic condensation and biopolymer degradation models. A review of both types of models is given by Hedges [20] and Hatcher and Spiker [21]. In the degradative schemes, recalcitrant plant and microbial polymers are viewed as the precursors of humic substances, which are formed progressively via the humin, humic acid, and, finally, the fulvic acid step. The biopolymer degradation model assumes that recalcitrant plant and microbial biomacromolecules are selectively preserved during biodegradation, whereas the labile components of plants and microorganisms are completely mineralized. The recalcitrant biomacromolecules, which form the humin fraction, are further oxidized to form humic and fulvic macromolecules.

The condensation models propose that humic substances evolve from the polymerization of low-molecular-weight organic precursors, generated during biodegradation of plant and microbial residues. In this type of model the formation pathway progresses from fulvic acids, via humic acids, to humin. The abiotic condensation models proposed for soils include the polyphenol model and the melanoidin or browning reaction. In the polyphenol model, various phenols of plant or microbial origin are assumed to be oxidized to quinones, which condense with each other or with amino acids and ammonia to form humic macromolecules [20]. Simple sugars and amino acids, which condense to form dark, N-rich humic-like polymers, are the initial monomers in the melanoidin model. The melanoidin hypothesis has been criticized for several reasons: the precursor molecules are present in soils in only very low quantities and the reaction proceeds very slowly under acid or neutral conditions and natural temperatures. Therefore the precursor molecules probably do not persist long enough for a reaction to occur. Also, the structural characteristics of synthetic melanoidins, as determined by ^{13}C NMR spectroscopy, are different from natural humic substances [20]. Both types of models assume an increase in molecular weight from fulvic acids to humic acids to humin. These concepts, which are based exclusively on the conventional classification of solubility in acids and bases, are not very compatible with the concept of a "soil organic matter continuum."

The litter input to forest soils is mineralized in a two-stage process. A rapid initial phase of plant litter decomposition and transformation is followed by a second, slower phase. The preferential decomposition of easily mineralizable materials results in the selective preservation of refractory plant or microbial components [21]. In the second phase, the microbial biomass and its metabolic products, which were synthesized during the initial phase, and recalcitrant, selectively preserved compounds are decomposed [3,5]. The biodegradation of plant or microbial constituents can be limited because of several mechanisms. A compound may possess an intrinsic recalcitrance as a result of chemical structure or can be protected from microbial attach by associated recalcitrant compounds, such as lignin or melanins. These selectively preserved materials can be incorporated directly in humic compounds or undergo further transformation. Upon reaching the mineral soil, the (partly) degraded organic matter can be stabilized by precipitation or sorption on the mineral soil matrix. Organic matter in aggregated soils can also be physically protected from microbial attack [7]. Transformation of organic matter to humic compounds may also be promoted by a catalytic effect of soil minerals [19,20]. Carbon from a variety of different plant and microbial sources contributes to the formation of humic matter. The formation of humus in forest soils is viewed here as the result of a combination of different processes, which can be summarized as *resynthesis by microorganisms, selective preservation,* and *direct transformation* (Figure 2). This chapter discusses the susceptibility of different classes of aliphatic and aromatic compounds classes of forest litter to these humification processes. However, the intensity of the individual processes can be different, depending on the soil environment. The sum of these individual humification processes is expressed in the morphology of different humus types in forest soils.

III. METHODOLOGICAL APPROACH TO THE STUDY OF HUMIFICATION PROCESSES

A. Experimental Approach

Various types of experiments have been used to study decomposition and humification processes in forest soils. They can be grouped into field litterbag experiments, laboratory decomposition experiments, and studies of humus profiles. Litterbag experiments are used mainly to determine the decomposition rates under field conditions. Laboratory decomposition experiments are used to study the decomposition rates of bulk material or individual components of plant litter as affected by variations in the individual environmental factors that control the decomposition rate. These experiments are generally restricted to months or years of duration, so the data obtained give information only on the biodegradation of litter. To study the humification processes, another approach is necessary. In undisturbed soils, such as most forest soils, the equilibrium between litter input and the decomposition and humification processes results in the accumulation of organic matter in distinct horizons in the forest floor and the associated mineral soil. The morphological changes observed with depth in forest soils are directly related to the degree of decomposition, ranging from fresh plant litter material to completely humified, i.e., macromorphologically unidentifiable, soil organic matter [11]. Thus, humus profiles provide a natural system for following the course of humification by evaluating changes in chemical structure with depth.

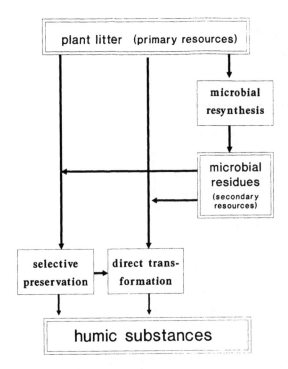

Figure 2   Schematic representation of the different humification processes operating in the transformation of litter to humic compounds.

B. Fractionation

The elucidation of humification processes can be approached by using different separation techniques, such as those used in classical humus fractionation, separation according to density or particle size, and selective chemical treatment (Figure 3). However, none of the fractions obtained is pure; they remain as mixtures, the compositions of which are highly dependent on the precursor material and the stage of decomposition. The most promising approach is to use a combination of the fractionation techniques mentioned above. However, most studies using these techniques lack a comparison of the results obtained for specific fractions with the results obtained for the bulk sample. It is important to determine not only the qualitative differences between fractions but also the quantitative changes, i.e., the changes in yields of individual fractions.

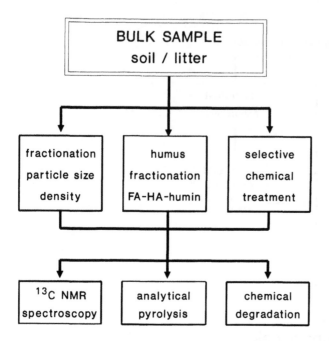

Figure 3   Schematic representation of the combination of the complementary analytical techniques of solid state ^{13}C NMR spectroscopy, chemical degradation, and thermal degradation, which can be used either on bulk samples or after fractionation of soil organic matter.

The classical method of fractionation, based on the solubility characteristics in acid and alkaline aqueous solutions, is the most widely used fractionation technique in studies on structural characterization of humic compounds. The organic material from soils that is insoluble in 0.5 M NaOH is defined as humin; the soluble material is further fractionated in humic acids, which precipitate as the pH is lowered to 2, and fulvic acids, which remain in solution. This classical fractionation procedure has been criticized, as it is difficult to correlate these fractions with the biochemical processes operating during biodegradation and humification [18]. The chemical composition of the humic acids extracted from soils by the classical fractionation procedures depends on the solubility characteristics of the various polymers present in the bulk soil sample. The majority of biopolymers originating from plant litter remain in the humin fraction [21—23]. However, several components of plants and microorganisms are soluble in alkaline solvents and, therefore, reside in the fulvic acid fraction (mainly polysaccharides) and humic acid fraction (polysaccharides, peptides, lignin) [23]. Significant amounts of fulvic and humic acids are obtained from plant material by the conventional alkaline extraction techniques [22]. As a chemical separation of specific classes of compounds and humic substances is not possible, any humic isolates extracted are a mixture of different compounds.

The physical fractionation techniques based on particle size or density can also be used to separate the humified soil organic matter from detritus. As pointed out by Oades [18], the light fraction ($<1.6-2.0$ mg m^{-3}) can be compared to the macroorganic particles (>250 μm), at least in mineral soils. Elliott and Cambardella [24] have provided an overview of the different techniques suitable for physical fractionation of soil organic matter.

The selective chemical treatment technique is used to remove fractions that are susceptible to a chemical treatment, such as acid or base hydrolysis, delignification, or extraction with organic solvents. The residue is supposed to be more homogeneous in composition, which allows easier structural characterization. The material removed by chemical treatment can also be studied in detail. Most widely used is hydrolysis with HCl or H_2SO_4, which removes proteinaceous materials and carbohydrates from the sample [e.g., 25].

C. Analytical Methods

Advances in instrumentation and techniques that permit characterization of macromolecules have provided new insights into the chemical composition of soil organic matter and humus [7,26−28]. For the characterization of complex organic mixtures, such as forest soil organic matter, analytical approaches using a combination of different destructive and nondestructive methods seem most promising (Figure 3). The combination of analytical methods described here can be applied to bulk soil samples as well as to soil organic matter fractions obtained by the fractionation procedures discussed above.

Chemical Degradative Methods

Numerous methods are used for the quantitative determination of various compound classes in plant litter and soil organic matter after chemical degradation. Hydrolysis, followed by chromatographic separation and quantification of the monomeric hydrolysis products, is used mainly for carbohydrates (neutral sugars, uronic acids, amino sugars) and amino acids. The amount and stage of oxidative decomposition of lignin are estimated using the system of lignin parameters obtained from oxidation with CuO [22,26].

Hydroxy fatty acids derived from cutin and suberin can be determined after saponification or transesterification. For some other plant constituents such as tannins and the highly aliphatic biomacromolecules found in plants and algae (cutan, suberan, alginan), suitable chemical degradative methods do not exist. All wet-chemical methods described are applicable to both bulk soil organic matter and humic isolates. For some classes of plant compounds (e.g., carbohydrates, amino acids), the chromatographic separation and determination of individual monomers can be replaced by colorimetric methods, e.g., the spectrophotometric procedure for carbohydrates described by Johnson and Sieburth [29].

Other chemical degradative methods that have been used in investigations of the structure of humic substances (e.g., permanganate oxidation, sodium amalgam reduction) do not enable unambiguous structural characterizations. The major problems associated with these methods are low yields of degradation products and the generation of artifacts [30]. Therefore, the significance of aromatic and phenolic structures as building blocks of humic substances has probably been overemphasized.

Analytical Pyrolysis

Analytical pyrolysis has found wide application in structural studies of plant litter and soil organic matter. A thermal energy pulse applied to a macromolecule causes fracture of weaker bonds and yields pyrolysis products characteristic of the original structure [31]. For studies of humic substances, a widely used technique is flash pyrolysis (Py) with gas chromatographic (GC) separation of the pyrolysis products before identification with electron impact (EI) mass spectrometry (MS; Py-GC-MS) [32]. In Py-MS, the products pass directly into the ion source of a mass spectrometer operating with low EI energies (10 to 15 eV) or with field ionization (Py-FIMS), so that mainly molecular ions are recorded [31,33]. Secondary reactions are avoided by conducting the pyrolysis in a rapid stream of inert gas or in a vacuum. Temperature-programmed pyrolysis in combination with time-resolved field ionization has been shown to be especially useful for the characterization of different compound classes in the organic matter of forest soils. Pyrolysis is also especially useful for compound classes that are difficult to analyze by chemical degradative methods (e.g., cutan, suberan, lignin) and may be applied to whole soils as well as to isolated organic fractions [28,32,34]. However, the results obtained from analytical pyrolysis are essentially qualitative and should be combined with other analytical techniques to enhance precision and avoid misleading conclusions [35].

CPMAS ^{13}C NMR Spectroscopy

Developments in ^{13}C nuclear magnetic resonance (NMR) have made it possible to obtain well-resolved and chemically informative spectra of solid samples. The technique of cross-polarization magic angle spinning (CPMAS) has resolved many of the problems associated with line broadening and low signal intensity in solid-state NMR. The technique, which is carried out nondestructively on dry, ground samples, provides an estimate of the chemical composition of bulk soil organic matter. The problems associated with the lack of resolution and low signal-to-noise ratios for CPMAS ^{13}C NMR analyses of mineral soil samples can be overcome by prior deashing with HCl/HF without major structural changes in the organic chemical composition of the carbon-enriched samples [36]. The content of paramagnetic species, such as Fe^{3+}, can be reduced by extraction with dithionite [27]. If the conditions for CPMAS experiments in relation to the relaxation behavior of the ^{13}C and ^{1}H nuclei are chosen carefully, quantification of the NMR data is possible [27]. However, cautious sample pretreatment is necessary to ensure that the data obtained can be used in a quantitative way [37]. The literature on NMR of forest soils is listed in a bibliography by Preston and Rusk [38].

The relaxation behavior of ^{13}C and ^{1}H nuclei can also be exploited for structural investigations with the dipolar dephasing technique. With this pulse sequence, the contributions of protonated and nonprotonated carbons and of mobile and rigid alkyl-carbon structures to the signal intensity of ^{13}C NMR spectra can be calculated [26,39]. The dipolar dephasing technique has been used successfully for the structural characterization of complex plant biomacromolecules, such as lignins and tannins [40], and microbial materials isolated from soils [41].

IV. COMPOSITION AND DISTRIBUTION OF THE INPUT TO HUMIFICATION

A. Primary Resources

The basis of structural investigations of the organic matter of forest soils is a detailed knowledge of the morphological and molecular composition of the parent litter material. Forest litter layers are composed almost exclusively of aboveground plant litter, including leaves or needles, woody debris (logs, branches, twigs), fruits, and buds. In deeper humus horizons and in the mineral soil, the contribution of root litter is significant. In some ecosystems, the belowground inputs from the turnover of fine roots may contribute more to the decomposition cycle of organic matter than aboveground litterfall [42]. According to McClaugherty et al. [43], fine root

litter input, which is especially significant in the Oh and mineral
soil horizons, is similar in magnitude to the production of foliar
litter. Vogt et al. [44] estimated the input of root litter to be 20
to 50% of the total C input to temperate forest soils. High amounts
of partially decomposed root residues were found in different forest soils after density fractionation and subsequent morphological
and structural characterization [45,46]. Woody debris is also an
important component of forest ecosystems, comprising 24 to 39% of
the total aboveground organic matter input [47].

Results from ^{13}C NMR spectroscopy show that forest litter layers consist of about 55% O-alkyl carbon, 20% aromatic carbon, 5%
carboxyl carbon, and 20% alkyl carbon (Figure 4). The overall
composition of litter from different tree species and different sites
is remarkably similar [48,49]. The signal intensity in the O-alkyl
region is due mainly to polysaccharides (cellulose and hemicelluloses) and lignin side chains. The aromatic signal intensity can
be attributed mainly to lignin and also to tannins. Lignin is a
complex three-dimensional polymer biosynthesized by dehydrogena-
tive polymerization of three phenylpropane monomers (coniferyl,

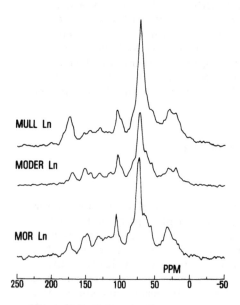

Figure 4   CPMAS ^{13}C NMR spectra of the bulk samples of three
different litter layers under ash (mull), beech (moder), and spruce
(mor). From Ref. 48.

sinapyl, and p-coumaryl alcohol). The monomers are linked together
by several different carbon-carbon and ether linkages, most of
which are not readily hydrolyzable. The relative proportions of each
monomeric unit, commonly referred to as guaiacyl/vanillyl, syringyl,
and p-hydroxyphenyl units, in the lignin of a particular plant species depend on its phylogenetic origin [50]. The lignin of hard-
woods, such as beech, consists of about equal proportions of guaiacyl and syringyl monomers, whereas softwood lignin is composed
mainly of guaiacyl units. Proteins and extractable lipids account
for 30 to 40% of the alkyl carbon. About 60% of the total carbon
signal intensity of the NMR spectra of the litter layers can be
identified. The carbon unaccounted for by wet-chemical methods
is found mainly in the alkyl carbon region and is most likely from
aliphatic biomacromolecules of the protective layers in higher plants
(cuticle and periderm), i.e., the polyesters, cutin and suberin,
and cutan and suberan, the highly aliphatic biomacromolecules discovered in plant cuticles [51]. Cutin consists of characteristic
long-chain (mainly C_{16} and C_{18}) hydroxy and epoxy fatty acids,
which are cross-linked to form a three-dimensional amorphous poly-
ester-type biomacromolecule. Suberin is composed of a mixture of
hydroxy fatty acids and phenolic constituents [52]. Woody debris
consists mainly of O-alkyl carbon from polysaccharides and aromatic
carbon from lignin [53]. The bark of woody debris is rich in suberin and tannins. Although considerable progress has been made
in recent years, data on the chemical composition of plant litter are
still lacking for many tree species and quantitative data on composition of roots are lacking almost completely.

B. Secondary Resources

Secondary resources are composed mainly of the remains of fungal
and bacterial cell walls. Animal residues are quantitatively of minor importance [5]. Microbial standing biomass contributes approximately 1 to 4% to the organic carbon in soils [54]. Microbial residues are even more complex in composition than are plant remains.
Bacterial cell walls are composed of peptidoglycans (murein), lipids,
and lipopolysaccharides, which contain a variety of unusual monomers [55]. Many bacteria produce extracellular polysaccharides
consisting of neutral or acidic sugar monomers [56]. The cell walls
of fungi contain proteins, chitin, chitosan, cellulose, and a variety
of noncrystalline polysaccharides, mainly mannans and glucans. Despite the heterogeneity of fungal cell walls with respect to the variety of macromolecules present, they can be subdivided into (1)
an inner layer of chitin, glucans, or cellulose, which form the skel-
etal wall components and are embedded in various matrix polymers
(mainly glucans), and (2) an outer layer composed of noncrystalline
polysaccharides [57]. The outer layer is soluble in dilute alkali,

leaving the inner layer as a residue [58]. Recent studies indicate
that microbial residues also contain macromolecular alkyl carbon
structures [51,59] in addition to extractable lipids. Melanins are
minor cell wall components in many fungi, but they have a significant role in protection against lysis by enzymes and irradiation [57,
60].

Information on the chemical composition of secondary resources
is mainly qualitative. Baldock et al. [41] isolated mixed bacterial
and fungal cultures from an agricultural soil and found that the
bacterial materials contained more alkyl and carboxyl carbon but
less O-alkyl and acetal carbon than the fungal materials. The microbial materials isolated from a forested Typic Dystrochrept were
similar in composition to the bacterial materials described by Baldock
et al. [41], suggesting that mainly bacterial material was isolated
[61]. Certainly, more information is needed on the chemical composition of secondary resources in forest soils.

V. CHEMICAL STRUCTURAL TRENDS
DURING HUMIFICATION

A. Bulk Samples

Several forest humus profiles have been investigated by a combination of wet-chemical degradative methods and ^{13}C NMR spectroscopy
[26,62–64]. Figure 5 shows an example of the conventional CPMAS
^{13}C NMR spectra of bulk samples from a moder humus profile developed under European beech. The signals in the NMR spectrum
of the litter layer can be assigned to several major plant compound
classes. The signal at 72 ppm, together with the signal at 105 ppm
and shoulders around 62 and 89 ppm, is characteristic for polysaccharides. Signals from lignin are found at 119, 130, 150, and 56
ppm and result from protonated, C-substituted, O-substituted aromatic C, and methoxyl C, respectively. The major signal in the
alkyl C region is found at 30 ppm and is assigned to C in long-
chain paraffinic structures of lipids, waxes, cutin, and other aliphatic biomacromolecules. A more detailed signal assignment for
the ^{13}C NMR spectra of plant litter is given elsewhere [48,49].

Changes in the overall chemical composition of forest soil organic matter can be observed with CPMAS ^{13}C NMR. Based on
area measurements of the indicated chemical shift ranges, general
changes in structural composition have been described for a number of forest humus profiles. A decrease of signal intensities in
the O-alkyl-C region (50 to 110 ppm) is found with increasing
depth as a result of the preferential degradation of carbohydrates.
The signal intensities in the chemical shift region of 110 to 160
ppm (aromatic C), mainly arising from lignin, remain constant with
depth. A relative increase in the intensity of the carboxyl-C

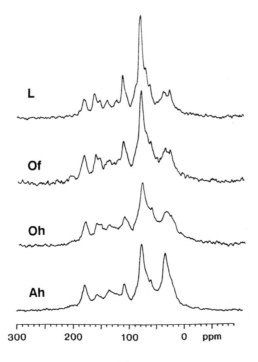

L

Of

Oh

Ah

300   200   100   0   ppm

Figure 5   CPMAS ^{13}C NMR spectra of the bulk samples from a moder humus profile under beech.   From Ref. 64.

signal (175 ppm) and of the alkyl-C signal (30 ppm) is observed. The chemical shift region of this alkyl C (0 to 50 ppm) represents about 15 to 20% of the total organic carbon in litter layers and up to 30 to 40% in the mineral soil horizons [63,64].   The increase in intensity of the alkyl-C signal during biodegradation of forest litter has also been confirmed in litterbag experiments with Norway spruce litter [65].   During decay of woody debris in forest soils (Figure 6), an increase in alkyl signal intensity could not be observed [53].

To compare the results from wet-chemical analyses and ^{13}C NMR spectroscopy, the contribution of carbon in specific compound classes (extractable lipids, amino acids, polysaccharides, and CuO-lignin) to the four chemical shift ranges of the NMR spectra can be calculated [63,64].   The highest percentage of identified structures is found in the chemical shift range 50 to 110 ppm (Table 1).   Polysaccharides and lignin side chains, as well as small amounts of protein, make up 74 to 78% of the total O-alkyl carbon in the litter layers and 34 to 48% in the humus layers.   Proteins and extractable

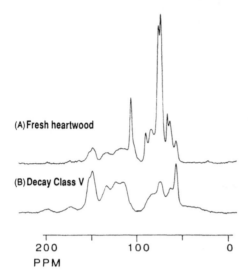

(A) Fresh heartwood

(B) Decay Class V

200   100   0
PPM

Figure 6   CPMAS ^{13}C NMR spectra of decayed Douglas fir logs: (A) fresh heartwood; (B) highly decomposed Douglas fir after 192 years residence time on the forest floor. From Ref. 46.

lipids account for 17 to 50% of the alkyl carbon (0 to 50 ppm). CuO-lignin accounts for 22 to 71% of the aromatic carbon (110 to 160 ppm), with the percentage identified decreasing within the profile. However, the relative amount of aromatic carbon as determined by ^{13}C NMR remains constant. This trend has been attributed to a relative enrichment with recalcitrant C-C—linked lignin structures, which are not attacked by CuO, as a function of depth [33,64]. Even in the mineral soil horizons, at least 40% of the carbon is identified as plant- or microbial-derived biomacromolecules. This value is considered a minimum, because of losses during the degradative determination of specific compound classes.

The organic structures not accounted for by wet-chemical methods constitute 35 to 45% of the carbon in the O horizons and 57 to 66% in the mineral soil horizons. The amount of nonidentified O-alkyl carbon is comparatively low at 20 to 44%, but this increases with depth. The increasing percentage of the unaccountable resonance in the chemical shift range 50 to 110 ppm is probably from noncarbohydrate O-alkyl carbon structures. Higher amounts of chemically unidentified carbon are found in the chemical shift range for aromatic carbon (29 to 78%), alkyl carbon (54 to 83%), and carboxyl carbon (73 to 88%). The percentage of unidentified aromatic

Table 1   Percentage of Carbon Contributed by Specific Compound Classes[a] to the Signal Intensity in the Chemical Shift Ranges of the NMR Spectra of Different Forest Humus Types[b]

Humus type and soil	Hori-zon	220–160 (carboxyl)	160–110 (aromatic)	110–50 (O-alkyl)	50–0 (alkyl C)	Total
Mull	L	19	26	80	31	55
Cumulic	LAh	23	22	66	23	39
Moder	L	27	41	73	45	59
Typic	Of	16	40	64	38	49
Dystrochrept	Ofh	19	43	64	17	44
	Ah	21	35	60	32	44
Mor	L	26	71	73	44	65
Lithic	Of	17	59	79	33	55
Udorthent	Oh	13	51	56	28	43
	Aeh	12	25	39	46	34

[a]Sum of extractable lipids, amino acids, cellulosic and noncellulosic carbohydrates, and CuO-oxidizable lignin.
[b]From Ref. 64.

carbon generally increases with depth, which can be partly explained by the decrease in yields of lignin phenols represented by the CuO oxidation products, as described above. A large percentage (54 to 83%) of the alkyl carbon is not identified by wet-chemical methods. Limitations in analytical techniques have resulted in an underestimation of the alkyl carbon content of soil humic substances [66].

B. Individual Plant and Microbial Constituents

Polysaccharides

Polysaccharides are the major component of forest litter. The input of polysaccharides into forest soil from plant litter consists of about 20 to 25% cellulose and 20 to 30% hemicelluloses [48]. The major structural difference is that cellulose is a crystalline polymer of glucose, whereas hemicelluloses are composed of various pentoses and hexoses. The amounts and types of monomers are different for different tree species [67]. Plant carbohydrates provide the major C and energy source for microorganisms. Most of the C of plant carbohydrates is, therefore, mineralized within several months or years [68]. As depth increases, a decrease of the overall carbohydrate content is observed. In soil, the plant-derived polysaccharides are decomposed preferentially to lignin [69]. This is reflected by the complete loss of cellulose, which is mainly of plant origin, with depth. The cellulose content in forest soils decreases from 20 to 25% in the litter layer to less than 3% in the A horizon [63,64]. Similar results are obtained in laboratory experiments using ^{14}C-labeled polysaccharides [69–71].

The major part of the polysaccharide C that remains in soils after prolonged periods of degradation is present as hydrolyzable carbohydrates [72]. These noncrystalline polysaccharides make up a high proportion of the O-alkyl carbon in the humified horizons (Oh, Ah). Microorganisms provide an input of noncellulosic carbohydrates during decomposition. The suggestion that in increasing contribution of microbial remains is observed as a function of increased degradation is corroborated by an increase of microbial-derived amino sugars as depth increases. This input is most pronounced for mull and moder profiles, but it is not readily observed in mor profiles (Table 2). An accumulation of microbial residues

Table 2   Accumulation of Microbial-Derived Lipids and Amino Sugars in Different Forest Humus Types[a]

Humus type and soil	Horizon	Lipid-P[b] (mg kg^{-1} OM)[d]	Amino sugars[c] (mg g^{-1} OM)
Mull	L	19.8	9.5
Cumulic	LAh	23.2	29.9
Hapludoll	Ah	19.3	45.7
Moder	L	16.2	6.5
Typic	Of	21.2	17.7
Dystrochrept	Ofh	17.4	25.1
	Ah	8.3	25.8
Mor	L	18.9	5.0
Lithic	Of	16.2	10.5
Udorthent	Oh	10.8	16.8
	Aeh	tr.[e]	12.2

[a]From Refs. 26 and 93.
[b]Content of P in polar lipid extract.
[c]Sum of glucosamine and galactosamine.
[d]OM = organic matter.
[e]tr. = traces.

was also found by Joergensen [73]. The content of hexosamines in decomposing beech litter (*Fagus silvatica*) was 10 to 20 times higher than the amounts present in the living microbial biomass. Microbially produced carbohydrates are presumably more resistant to decomposition than plant-derived polysaccharides. Fungal cell walls are very resistant in soils, due to the presence of substantial amounts of melanins in the walls [68]. There is a lack of data that provide evidence for the processes that result in a stabilization of these polysaccharides in forest soils. The concurrent increase in carbohydrates from microbial sources with depth at the expense of plant-derived carbohydrates suggests that *microbial resynthesis* is the major humification process [26,41,64].

Aromatic Compounds

CPMAS ^{13}C NMR spectroscopy shows that forest soil organic matter contains about 20 to 30% aromatic carbon [63,64]. The major aromatic components of plant litter are lignin and tannins. Evidence of the fate of lignin during decomposition in soils is obtained from chemical degradation and analysis of the monomers released, e.g., by CuO oxidation [22,74,75] or from analytical pyrolysis [63,76, 77]. Lignin biodegradation is attributed mainly to white-rot fungi, but bacteria, fungi imperfecti, and actinomycetes have also been shown to degrade lignin [68–71,78–80]. Studies of lignin breakdown in vitro have shown that the key reactions are oxidative cleavage of phenylpropanoid side chains, demethylation of methoxyl groups, hydroxylation of aromatic rings, and cleavage of aromatic rings while they are still in the macromolecule [70,81]. Lignin is attacked preferentially at the ether linkages. Carbon-carbon—linked lignin structural units, such as pinoresinol, phenylcourmaran, and biphenyl units, are more resistant to biodegradation than the ether-linked structures [82]. In most cases, woody angiosperm lignin is decomposed at a higher rate than coniferous lignin. Biodegradation of lignin in forest soils results in a modification of the remnant lignin polymer, which has a lower content of intact lignin moieties due to ring cleavage and a higher degree of side chain oxidation [83–85]. This is reflected by an increase in the acid/aldehyde ratio of the CuO oxidation products with depth in forest soils (Figure 7), which can be used as an indication of an increase in side-chain oxidation of the remnant lignin [22,64,74,75, 83]. Pyrolysis—field ionization—mass spectrometry shows [76,86] that the more recalcitrant C-C—linked moieties are selectively preserved compared to the ether-linked moieties in forest soils. This general pattern of lignin degradation is observed regardless of the type of parent litter material, soil pH, or climatic factors [85]. High biological activity in soils (bioturbation) in combination with the presence of clay results in the incorporation of only slightly

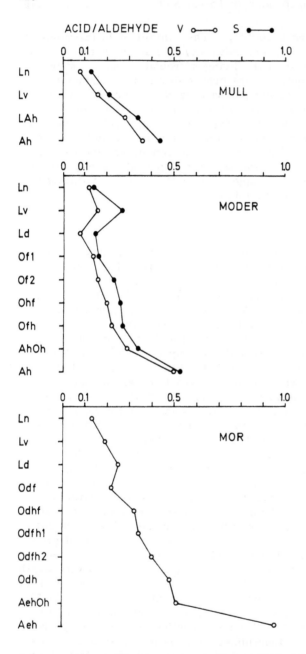

Figure 7   Acid/aldehyde ratio of the vanillyl (o) and syringyl units
(●) of a mull, moder, and mor humus profile under ash (mull),
beech (moder), and spruce (mor).   From Ref. 83.

Table 3 Acid/Aldehyde Ratios for the Vanillyl Unit of the CuO Oxidation Products in A Horizons Associated with Different Forest Humus Types[a]

Soil (location	Main vegetation	$(ac/al)_V$[b]
Mull		
Rendoll (Göttinger Wald)	European beech	0.36
	European ash	0.32
Cumulic Hapludoll (Allersdorf)	European ash	0.35
Moder		
Typic Haplorthod (Farrenleite)	European beech	0.51
Typic Dystrochrept (Wülfersreuth)	Norway spruce	0.52
Typic Dystrochrept (Eremitage)	European beech	0.50
Mor		
Typic Dystrochrept (Schweinsbacher Sattel)	Norway spruce	0.67
Lithic Udorthent (Schneeberg)	Norway spruce	0.95
Aeric Haplaquept (Häusellohe)	Norway spruce/ Scots pine	0.97

[a]From Refs. 83 and 84.
[b]$(ac/al)_V$ = ratio of vanillic acid to vanillin obtained from CuO oxidation.

modified plant materials into soil organic matter. This is represented by lower acid/aldehyde ratios in the A horizons of mull than in the A horizons of moder and mor humus types (Table 3).

During microbial degradation of lignin, chemically altered, polymeric, water-soluble lignin fragments are produced [87,88], which give low yields of CuO oxidation products and show high acid/aldehyde ratios [85]. These modified lignin degradation products contribute significantly to the aromatic carbon structures in mineral soil horizons. In the mineral horizons of some forest soils, lignin structural units characterized by a strong chemical alteration accumulate, as the result of the precipitation or sorption of highly altered water-soluble lignin fragments [85].

The changes in the aromatic C content and structures of forest soil humic acids at different stages of humification were studied using solid state ^{13}C NMR and dipolar dephasing, a technique that differentiates between protonated and nonprotonated C, in combination with molecular-level characterization of lignin-derived phenols by CuO oxidation. The humic acid fraction isolated from fresh litter of European beech and Norway spruce shows mainly peaks attributable to aromatic C derived from lignin and tannin structures, which are both partly extractable by alkaline solvents. The most prominent feature of the NMR spectra is the decrease in phenolic C and methoxyl C with increasing degree of humification and the simultaneous increase in the signal intensity at 130 ppm in the ^{13}C NMR spectra (Figure 8).

Detailed structural assignments for this signal can be obtained by measuring the percentage of signal intensity from protonated and nonprotonated C by dipolar dephasing ^{13}C NMR spectroscopy [89], which enables the percentage of phenolic C (aryl-O) and C-substituted aromatic C (aryl-C) to be calculated from the NMR data. The percentage of nonprotonated aromatic C remains constant and the C-substituted aromatic C fraction increases when humification proceeds (Table 4). These observations are in good agreement with the decreasing yields of lignin-derived CuO oxidation products and an increasing degree of oxidative decomposition (side chain oxidation) in the lignin-derived structures, as described above. The higher number of carboxyl groups caused by ring cleavage and side chain oxidation results in an increasing percentage of aromatic C extractable with alkaline solvents (Figure 9).

If it is assumed that lignin is the primary precursor of the aromatic C components of humic acids in forest soils [6,90], then the lignin structure is altered considerably during humification, resulting in lignin-derived aromatic structures with a high degree of C

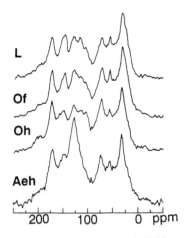

Figure 8   CPMAS ^{13}C NMR spectra of the humic acid fraction isolated from a mor humus profile (Lithic Udorthent) under Norway spruce. From Ref. 64.

Table 4   Fraction of Protonated ($f_a{}^{a,H}$) and Nonprotonated ($f_a{}^{a,N}$) Aromatic C and Aryl-O C of the Total Aromatic C Signal Intensity (ar-O/ar) of the Humic Acid Fraction[a]

Site (main vegetation/humus type)		$f_a{}^{a,H}$ [b]	$f_a{}^{a,N}$ [b]	ar-O/ar[c]
Eremitage	L	0.31	0.69	0.38
(beech/moder)	Of	_[d]	—	0.37
	Oh	—	—	0.31
	Ah	0.31	0.69	0.29
Farrenleite	Oh	0.27	0.73	0.34
(beech/moder)	OhAeh	0.25	0.75	0.32
	Bs	0.27	0.73	n.d.[e]
Schneeberg	L	0.39	0.61	0.34
(spruce/mor)	Of	—	—	0.37
	Oh	0.37	0.63	0.29
	Aeh	0.33	0.67	0.24

[a]From Ref. 89.
[b]Coefficient of variation is ±8%.
[c]Coefficient of variation is ±1%.
[d]— = not determined.
[e]n.d. = not detected.

substitution and carboxyl functionality. The compex picture obtained for the structural changes of lignin in forest soils leads to the conclusion that lignin undergoes *several humification processes.* The microbial degradation of lignin results in the mineralization of lignin with concomitant ring cleavage and side chain oxidation. According to Martin and Haider [68] and Haider [71], the processes of lignin mineralization are extracellular, and lignin does not provide a carbon or energy source for microorganisms. Therefore, it is assumed that microbial resynthesis of lignin carbon in microbial biomass carbon does not occur. The more refractory lignin components are *selectively preserved.* There is evidence for a *direct transformation* of lignin, resulting in a decrease of the relative percentage of aryl-O C and a relative increase of C-substituted aromatic structures with depth.

The leaves and barks of several tree species are high in tannins [40,91]. Tannins are heterogeneous compounds and provide problems for analysis by wet-chemical methods. Therefore, information on the structural changes that they undergo during

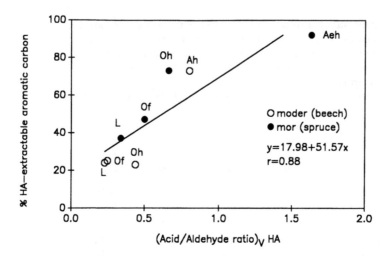

Figure 9 Amount of aromatic carbon extracted in the humic acid fraction as a function of the acid/aldehyde ratio of the humic acid fraction. From Ref. 64.

humification is scarce. Nonetheless, evidence is provided by dipolar dephasing ^{13}C NMR for the presence of tannins in the humic acid fractions extracted from forest soils. From these data, it can be estimated that tannins contribute up to about 20% to the total aromatic carbon content of forest soils [89]. Obviously, the fate of tannins during humification needs further attention.

Extractable Lipids

Extractable lipids (30% of the total alkyl carbon signal intensity) show an increasing contribution of microbial lipids with depth, especially in acid forest soils [92,93]. Table 2 shows an increase in phospholipids, which are mainly of microbial origin, in the mineral soils from mor and moder to mull humus types. The pattern of microbial and plant-derived lipids is, therefore, assumed to follow a trend that also suggests *microbial resynthesis* as the major humification process [72].

Cutin, Suberin, and Other Aliphatic Biomacromolecules

CPMAS ^{13}C NMR spectroscopy has shown that forest soil organic matter contains significant amounts (15 to 30% of alkyl carbon [26, 46,62–62,76]. The major alkyl-C components of forest soil organic matter, in addition to extractable and bound lipids [94], are the

plant polyesters, cutin and suberin, and other nonsaponifiable aliphatic components [51,52,61,95−98]. The nonlipid components constitute a major fraction of soil organic matter and associated humic
substances. Cutin is decomposed or transformed in forest soils and
does not accumulate in forest floor horizons [99]. This has been
confirmed in litterbag experiments in the field and in laboratory decomposition experiments [96,100]. Figure 10 shows the decomposition pattern of bulk litter and cutin in the laboratory. Cutin
is decomposed at a higher rate than bulk litter over a period of
465 days. Similar results were obtained by Goni and Hedges [101].
Cutin was found to be less stable than bulk organic matter during

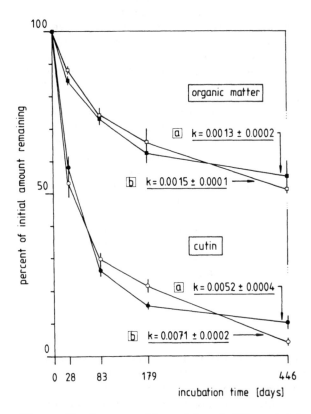

Figure 10   Decomposition of beech litter, total and the cutin fraction, under laboratory conditions for 465 days. (a) Beech litter;
(b) beech litter in the presence of artificial soil; k, decomposition
rate constant, d^{-1}. From Ref. 100.

diagenesis of buried needles in a coastal marine environment.* In mineral soil horizons, the input of root litter becomes significant, as demonstrated by the presence of suberin [34,99].

Analysis by solid-state CPMAS ^{13}C NMR and Curie point pyrolysis—gas chromatography—mass spectrometry indicated that the alkyl carbon moieties are altered significantly with increasing depth and decomposition in the soil profile. The NMR data show that the aliphatic structures in forest soil organic matter can be assigned to rigid carbon moieties and mobile carbon moieties [103]. Mobile carbons are found in compounds that are at temperatures near their melting point or in gel-like structures. They have also been found in algae and in microbial residues isolated from soils [41,51,61,104]. At depth in the soil profiles investigated, the mobile components are lost, but the rigid aliphatic components appear to be selectively preserved. It was hypothesized that the mobile and rigid carbon types are possibly associated with different types of macromolecules. Cutin and suberin from leaves, barks, and roots show a high proportion of mobile carbon structures, whereas the resistant nonsaponifiable aliphatic biomacromolecule, which has been identified recently in the cuticle of several plants [51,97,102,105], is composed almost exclusively of rigid carbon structures. This new nonsaponifiable aliphatic biomacromolecule produces a homologous series of alkane-alkene pairs upon flash pyrolysis. However, the aliphatic materials in the humified soil horizons bear no resemblance to the resistant aliphatic nonsaponifiable biomacromolecules in fresh leaf cuticles [61,96]. This lack of resemblance is probably due to the fact that selective preservation of resistant, nonsaponifiable plant macromolecules is not the dominant process leading to the accumulation of alkyl carbon moieties in forest soil organic matter. Alternatively, it seems possible that the structural differences observed between the alkyl carbon moieties in forest litter and humified soil horizons result from *direct transformation* of the material during humification, which results in increased cross-linking and, therefore, reduced mobility.

VI. CONCLUSIONS

The occurrence of one or another humification process in soils is modulated by the chemical composition of the primary and secondary resources and environmental factors. For carbohydrates and proteins, the major humification pathway seems to be microbial

---

*The preservation potential of cutin has to be regarded as low compared to cutan [102].

resynthesis, alshouth some of the polysaccharies may survive by being protected by associated recalcitrant molecules. It is clear that there is a transformation of lignin structural units such that the aromatic units lose oxygen functional groups as depth in the soil profile increases. These results are not in agreement with models of humification in soils [19–21], which would lead to an increase in aryl-O content during humification of lignin or high contents of aryl-O functionality in humic acids. The more recalcitrant lignin components are selectively preserved. Knowledge about the biodegradation and humification of tannins is scarce and certainly needs further attention. The same is true for the humification pathway for microbial melanins in soils. Furthermore, the humification of alkyl-C structures in forest soils needs to be investigated further. The process of transformation of cutin/suberin and other aliphatic biomacromolecules of plant and microbial origin also needs to be delineated. It is not clear whether the cross-linking occurs via O or C linkages. Other major problems which have not been addressed in this chapter are the humification pathways leading to the formation of N-, P-, and S-containing humic compounds. For many components of soil organic matter, the mechanisms of their stabilization against microbial degradation need to be investigated. Nonetheless, a picture is emerging of humic substances as a complex mixture of different plant- and microbial-derived compounds, which exist in soils in a continuum of degradative stages.

ACKNOWLEDGMENTS

The author thanks those who have contributed to the work reported in this chapter, in particular Pat Hatcher (The Pennsylvania State University, United States), Jan W. de Leeuw (Delft University of Technology, The Netherlands), Wolfgang Zech, Ludwig Haumaier, and Frank Ziegler (University of Bayreuth, Germany), and Markus Riederer (University of Karlsruhe, Germany). Part of the research reviewed in this chapter has been supported by the Deutsche Forschungsgemeinschaft (SBF 137, TP C1) and NATO grant 0799/87.

REFERENCES

1. Schlesinger, W. H. 1977. Carbon balance in terrestrial detritus. Annu. Rev. Ecol. Syst. 8:51–81.
2. Meentemeyer, V., E. O. Box, and R. Thompson. 1982. World paterns and amounts of terrestrial plant litter production. BioScience 32:125–128.
3. Jenkinson, D. S. 1988. Soil organic matter and its dynamics, p. 564–607. *In* A. Wild (ed.), Russell's soil conditions and plant growth. Longman, Harlow, U.K.

4. Schlesinger, W. H. 1990. Evidence from chronosequence studies for a low carbon-storage potential of soils. Nature 348:232—234.

5. Swift, M. J., O. W. Heal, and J. M. Anderson. 1979. Decomposition in terrestrial ecosystems. Blackwell, Oxford.

6. Ertel, J. R., P. Behmel, R. F. Christman, W. J. A. Flaig, K. M. Haider, G. R. Harvey, P. G. Hatcher, J. I. Hedges, J. P. Martin, F. K. Pfaender, and H.-R. Schulten. 1988. Genesis, p. 105—112. *In* F. H. Frimmel and R. F. Christman (eds.), Humic substances and their role in the environment. John Wiley & Sons, Chichester.

7. Oades, J. M. 1989. An introduction to organic matter in mineral soils, p. 89—159. *In* J. B. Dison and S. B. Weed (eds.), Minerals in soil environments. Soil Science Society of America, Madison, Wisconsin.

8. Gunnarsson, T., P. Sundin, and A. Tunlid. 1988. Importance of leaf litter fragmentation for bacterial growth. Oikos 52:303—308.

9. Eijsackers, H., and A. J. B. Zehnder. 1990. Litter decomposition: a Russian matriochka doll. Biogeochemistry 11:154—174.

10. Anderson, J. M. 1988. Invertebrate-mediated transport processes in soils. Agric. Ecosyst. Environ. 24:5—19.

11. Angehrn-Bettinazzi, C., P. Lüscher, and J. Hertz. 1988. Thermogravimetry as a method for distinguishing various degrees of mineralization in macromorphologically-defined humus horizons. Z. Pflanzenernaehr. Bodenk. 152:177—183.

12. Schaefer, M., and J. Schauermann. 1990. The soil fauna of beech forests: comparison between a mull and a moder soil. Pedobiologia 34:299—314.

13. Blume, H.-P. 1965. Die Charakterisierung von Humuskörpern durch Streu- und Humus-Stoffgruppenanalysen unter Berücksichtigung ihrer morphologischen Eigenschaften. Z. Pflanzenernaehr. Düng. Bodenkd. 111:95—114.

14. Babel, U. 1975. Micromorphology of soil organic matter, p. 369—473. *In* J. E. Gieseking (ed.), Soil components, Vol. 1, Organic compounds. Springer-Verlag, Berlin.

15. Ohta, S., A. Suzuki, and K. Kumada. 1986. Experimental studies on the behavior of fine organic particles and water-soluble organic matter in mineral soil horizons. Soil Sci. Plant Nutr. 32:15—26.

16. Waksman, S. A. 1938. Humus: origin, chemical composition and importance in nature. Baillière, Tindall and Cox, London.

17. Hayes, M. H. B., P. MacCarthy, R. L. Malcolm, and R. S. Swift. 1989. The search for structure: setting the scene, p. 3—31. *In* M. H. B. Hayes, P. MacCarthy, R. L. Malcolm, and R. S. Swift (eds.), Humic substances. II. John Wiley & Sons, Chichester.

18. Oades, J. M. 1988. The retention of organic matter in soils. Biogeochemistry 5:35–70.
19. Stevenson, F. 1992. Humus chemistry. John Wiley & Sons, New York.
20. Hedges, J. I. 1988. Polymerization of humic substances in natural environments, p. 45–58. *In* F. H. Frimmel and R. F. Christman (eds.), Humic substances and their role in the environment. John Wiley & Sons, Chichester.
21. Hatcher, P. G., and E. C. Spiker. 1988. Selective degradation of plant biomolecules, p. 59–74. *In* F. H. Frimmel and R. F. Christman (eds.), Humic substances and their role in the environment. John Wiley & Sons, New York.
22. Ertel, J. R., and J. I. Hedges. 1985. Sources of sedimentary humic substances: vascular plant debris. Geochim. Cosmochim. Acta 49:2097–2107.
23. Waksman, S. A., and K. R. Stevens. 1930. A critical study of the methods for determining the nature and abundance of soil organic matter. Soil Sci. 30:97–116.
24. Elliott, E. T., and A. A. Cambardella. 1991. Physical separation of soil organic matter. Agric. Ecosyst. Environ. 34:407–419.
25. Schnitzer, M., and C. M. Preston. 1983. Effects of acid hydrolysis on ^{13}C NMR spectra of humic substances. Plant Soil 75:201–211.
26. Kögel-Knabner, I., P. G. Hatcher, and W. Zech. 1990. Decomposition and humification processes in forest soils: implications from structural characterization of forest soil organic matter. Trans. 14th Int. Congress Soil Sci., Kyoto, Vol. V:218–223.
27. Wilson, M. A. 1987. NMR techniques and applications in geochemistry and soil chemistry. Pergamon Press, Oxford.
28. Schnitzer, M. 1990. Selected methods for the characterization of soil humic substances, p. 65–89. *In* P. MacCarthy, C. E. Clapp, R. L. Malcolm, and P. R. Bloom (eds.), Humic substances in soil and crop sciences: selected readings. ASA, SSSA, Madison, Wisconsin.
29. Johnson, K. M., and J. McN. Sieburth. 1977. Dissolved carbohydrates in seawater. I. A precise spectrophotometric analysis for monosaccharides. Mar. Chem. 5:1–13.
30. Norwood, D. L. 1988. Critical comparison of structural implications from degradative and nondegradative approaches, p. 133–148. *In* F. H. Frimmel and R. F. Christman (eds.), Humic substances and their role in the environment. John Wiley & Sons, Chichester.
31. Bracewell, J. M., K. Haider, S. R. Larter, and H.-R. Schulten. 1989. Thermal degradation relevant to structural studies of humic substances, p. 181–222. *In* M. H. B. Hayes,

P. MacCarthy, R. L. Malcolm, and R. S. Swift (eds.), Humic substances. II. John Wiley & Sons, Chichester.

32. Saiz-Jimenez, C., and J. W. de Leeuw. 1986. Chemical characterization of soil organic matter fractions by analytical pyrolysis—gas chromatography—mass spectrometry. J. Anal. Appl. Pyrol. 9:99—119.

33. Hempfling, R., W. Zech, and H.-R. Schulten. 1988. Chemical composition of the organic matter in forest soils. 2. Moder profile. Soil Sci. 146:262—276.

34. Hempfling, R., N. Simmleit, and H.-R. Schulten. 1991. Characterization and chemodynamics of plant constituents during maturation, senescence and humus genesis in spruce ecosystems. Biogeochemistry 13:27—60.

35. Nip, M., J. W. de Leeuw, P. J. Holloway, J. P. T. Jensen, J. C. M. Sprenkels, M. de Poorter, and J. J. M. Sleeckx. 1987. Comparison of flash pyrolysis, differential scanning calorimetry, ^{13}C NMR and IR spectroscopy in the analysis of a highly aliphatic biopolymer from plant cuticles. J. Anal. Appl. Pyrol. 11:287—295.

36. Preston, C. M., M. Schnitzer, and J. A. Ripmeester. 1989. A spectroscopic and chemical investigation on the de-ashing of humin. Soil Sci. Soc. Am. J. 53:1442—1447.

37. Hatcher, P. G., and M. A. Wilson. 1991. The effect of sample hydration on the ^{13}C CPMAS NMR spectra of fulvic acids. Org. Geochem. 17:293—299.

38. Preston, C. M., and A. C. M. Rusk. 1990. A bibliography of NMR application for forestry research. Pac. For. Centre Information Report BC-X-322, Forestry Canada, BC, Canada.

39. Wilson, M. A., R. J. Pugmire, and D. M. Grant. 1983. Nuclear magnetic resonance spectroscopy of soils and related materials. Relaxation of ^{13}C nuclei in cross polarization nuclear magnetic resonance experiments. Org. Geochem. 5:121—129.

40. Wilson, M. A., and P. G. Hatcher. 1988. Detection of tannin in modern and fossil barks and in plant residues by high-resolution solid-state ^{13}C nucler magnetic resonance. Org. Geochem. 12:539—546.

41. Baldock, J. A., J. M. Oades, A. M. Vassallo, and M. A. Wilson. 1990. Solid-state CP/MAS ^{13}C N.M.R. analysis of bacterial and fungal cultures isolated from a soil incubated with glucose. Aust. J. Soil Res. 28:213—225.

42. Raich, J. W., and K. J. Nadelhoffer. 1989. Belowground carbon allocation in forest ecosystems: global trends. Ecology 70:1346—1354.

43. McClaugherty, C. A., J. D. Aber, and J. M. Mellilo. 1984. Decomposition dynamics of fine roots in forested ecosystems. Oikos 42:378—386.

44. Vogt, K. A., C. C. Grier, and D. J. Vogt. 1986. Production, turnover, and nutrient dynamics of above- and belowground detritus of world forests. Adv. Ecol. Res. 15:303–377.

45. Beudert, G., I. Kögel-Knabner, and W. Zech. 1989. Micromorphological, wet-chemical and ^{13}C NMR spectroscopic characterization of density fractionated forest soils. Sci. Total Environ. 81/82:401–408.

46. Preston, C. M. 1992. The application of NMR to organic matter inputs and processes in forest ecosystems of the Pacific northwest. Sci. Total Environ. 113:107–120.

47. Harmon, M. E., J. F. Franklin, F. J. Swanson, P. Sollins, S. V. Gregory, J. D. Lattin, N. H. Anderson, S. P. Cline, N. G. Aumen, J. R. Sedell, G. W. Lienkaemper, K. Cromack, and K. W. Cummins. 1986. Ecology of coarse woody debris in temperate ecosystems. Adv. Ecol. Res. 15:133–301.

48. Kögel, I., R. Hempfling, W. Zech, P. G. Hatcher, and H.-R. Schulten. 1988. Chemical composition of the organic matter in forest soils. I. Forest litter. Soil Sci. 146:124–136.

49. Wilson, M. A., S. Heng, K. M. Goh, R. J. Pugmire, and D. M. Grant. 1983. Studies of litter and acid insoluble soil organic matter fractions using ^{13}C-cross polarization nuclear magnetic resonance spectroscopy with magic angle spinning. J. Soil Sci. 34:83–97.

50. Sarkanen, K. V., and C. H. Ludwig. 1971. Lignins. Wiley-Interscience, New York.

51. Tegelaar, E. W., J. W. de Leeuw, and C. Saiz-Jimenez. 1989. Possible origin of aliphatic moieties in humic substances. Sci. Total Environ. 81/82:1–17.

52. Kolattukudy, P. E. 1981. Structure, biosynthesis, and biodegradation of cutin and suberin. Annu. Rev. Plant Physiol. 32:539–567.

53. Preston, C. M., P. Sollins, and B. G. Sayer. 1990. Changes in organic components for fallen logs by ^{13}C nuclear magnetic resonance spectroscopy. Can. J. Forest Res. 20:1382–1391.

54. Kassim, G., J. P. Martin, and K. Haider. 1981. Incorporation of a wide variety of organic substrate carbons into soil biomass as estimated by the fumigation procedure. Soil Sci. Soc. Am. J. 45:1106–1112.

55. Rogers, H. J., H. R. Perkins, and J. B. Ward. 1980. Microbial cell walls and membranes. Chapman & Hall, London.

56. Hepper, C. M. 1975. Extracellular polysaccharides of soil bacteria, p. 93–110. *In* N. Walker (ed.), Soil microbiology. Butterworths, London.

57. Peberdy, J. F. 1990. Fungal cell walls—a review, p. 5–30. *In* P. J. Kuhn, A. P. J. Trinci, M. J. Jung, M. W. Goosey,

and L. G. Copping (eds.), Biochemistry of cell walls and membranes in fungi. Springer-Verlag, Berlin.

58. Wessels, J. G. H., and J. H. Sietsma. 1981. Fungal cell walls: a survey, p. 294–352. In W. Tanner and F. A. Loewus (eds.), Encyclopedia of plant physiology, Vol. 13B, Plant carbohydrates II. Springer-Verlag, Berlin.

59. Zelibor, J. L., L. Romankiw, P. G. Hatcher, and R. R. Colwell. 1988. Comparative analysis of the chemical composition of mixed and pure cultures of green algae and their decomposition residues by ^{13}C nuclear magnetic resonance spectroscopy. Appl. Environ. Microbiol. 54:1051–1060.

60. Bell, A. A., and M.-H. Wheeler. 1986. Biosynthesis and functions of fungal melanins. Annu. Rev. Phytopathol. 24: 411–451.

61. Kögel-Knabner, I., J. W. de Leeuw, and P. G. Hatcher. 1992. Nature and distribution of alkyl carbon in forest soil profiles: implications for the origin and humification of aliphatic biomacromolecules. Sci. Total Environ. 117/118:175–185.

62. Zech, W., I. Kögel, A. Zucker, and H. Alt. 1985. CP-MAS-^{13}C-NMR Spektren organischer Lagen einer Tangelrendzina. Z. Pflanzenernaehr. Bodenkd. 148:481–488.

63. Hempfling, R., F. Ziegler, W. Zech, and H.-R. Schulten. 1987. Litter decomposition and humification in acidic forest soils studied by chemical degredation, IR and NMR spectroscopy and pyrolysis field ionization mass spectrometry. Z. Pflanzenernaehr. Bodenkd. 150:179–186.

64. Kögel-Knabner, I., W. Zech, and P. G. Hatcher. 1988. Chemical composition of the organic matter in forest soils: III. The humus layer. Z. Pflanzenernaehr. Bodenkd. 151: 331–340.

65. Zech, W., M.-B. Johansson, L. Haumaier, and R. L. Malcolm. 1987. CPMAS ^{13}C NMR and IR spectra of spruce and pine litter and of the Klason lignin fraction at different stages of decomposition. Z. Pflanzenernaehr. Bodenkd. 150:262–265.

66. Hatcher, P. G., M. Schnitzer, L. W. Dennis, and G. E. Maciel, 1981. Aromaticity of humic substances in soils. Soil Sci. Soc. Am. J. 45:1089–1094.

67. Fengel, D., and G. Wegener. 1984. Wood: chemistry, ultrastructure, reactions. De Gruyter, Berlin.

68. Martin, J. P., and K. Haider. 1986. Influence of mineral colloids on turnover rates of soil organic carbon, p. 283–304. In P. M. Huang and M. Schnitzer (eds.), Interactions of soil minerals with natural organics and microbes. Soil Science Society of America, Madison, Wisconsin.

69. Haider, K. 1986. Changes in substrate composition during the incubation of plant residues in soil, p. 133–147. In V. Jensen, A. Kjoller, and L. H. Sorensen (eds.), Microbial communities in soil. Elsevier, London.

70. Haider, K. 1986. The synthesis and degradation of humic substances in soil. Trans. XIII Congress ISSS, p. 644–656.
71. Haider, K. 1991. Problems related to the humification processes in soils of the temperate climate, p. 55–94. *In* J.-M. Bollag and G. Stotzky (eds.), Soil biochemistry, Vol. 7. Marcel Dekker, New York.
72. Stott, D. E., and J. P. Martin. 1990. Synthesis and degradation of natural and synthetic humic material in soils, p. 37–63. *In* P. MacCarthy, C. E. Clapp, R. L. Malcolm, and P. R. Bloom (eds.), Humic substances in soil and crop sciences: selected readings. ASA, SSSA, Madison, Wisconsin.
73. Joergensen, R. G. 1991. Organic matter and nutrient dynamics of the litter layer on a forest Rendzina under beech. Biol. Fertil. Soils 11:163–169.
74. Ertel, J. R., and J. I. Hedges. 1984. The lignin component of humic substances: distribution among soil and sedimentary humic, fulvic, and base-insoluble fractions. Geochim. Cosmochim. Acta 48:2065–2074.
75. Hedges, J. I., R. A. Blanchette, K. Weliky, and A. H. Devol. 1988. Effects of fungal degradation on the CuO oxidation products of lignin: a controlled laboratory study. Geochim. Cosmochim. Acta 52:2717–2726.
76. Hempfling, R., and H.-R. Schulten. 1990. Chemical characterization of the organic matter in forest soils by Curie point pyrolysis–GC/MS and pyrolysis–field ionization mass spectrometry. Org. Geochem. 15:131–145.
77. Saiz-Jimenez, C., and J. W. de Leeuw. 1984. Pyrolysis gas chromatography mass spectrometry of isolated synthetic and degraded lignins. Org. Geochem. 56:417–442.
78. Rayner, A. D. M., and L. Boddy. 1988. Fungal decomposition of wood: its biology and ecology. John Wiley & Sons, Chichester.
79. Shoemaker, H. E., and M. S. A. Leisola. 1990. Degradation of lignin by *Phanerochaete chrysosporium*. J. Biotechnol. 13:101–109.
80. Zimmermann, W. 1990. Degradation of lignin by bacteria. J. Biotechnol. 13:119–130.
81. Kirk, T. K., and R. L. Farrell. 1987. Enzymatic "combustion": the microbial degradation of lignin. Annu. Rev. Microbiol. 41:465–505.
82. Haider, K., K. W. Kern, and L. Ernst. 1985. Intermediate steps of microbial lignin degradation as elucidated by ^{13}C NMR spectroscopy of specifically ^{13}C-enriched DHP-lignins. Holzforschung 39:23–32.
83. Kögel, I. 1986. Estimation and decomposition pattern of the lignin component in forest soils. Soil Biol. Biochem. 18:589–594.
84. Ziegler, F., I. Kögel, and W. Zech. 1986. Alteration of gymnosperm and angiosperm lignin during decomposition in

forest humus layers. Z. Pflanzenernaehr. Bodenkd. 149:323–331.

85. Kögel, I., F. Ziegler, and W. Zech. 1988. Lignin signature of subalpine Rendzinas (Tangel- and Moderrendzina) in the Bavarian Alps. Z. Pflanzenernaehr. Bodenkd. 151:15–20.

86. Hempfling, R., and H.-R. Schulten. 1989. Selective preservation of biomolecules during humification of forest litter studied by pyrolysis–field ionization mass spectrometry. Sci. Total Environ. 81/82:31–40.

87. Ellwardt, P.-C., K. Haider, and L. Ernst. 1981. Untersuchungen des mikrobiellen Ligninabbaus durch 13-C-NMR-Spektroskopie an spezifisch 13-C-angereichertem DHP-Lignin aus Coniferylalkohol. Holzforschung 35:103–109.

88. Seelenfreund, D., C. Lapierre, and R. Vicuna. 1990. Production of soluble lignin-rich fragments (APPL) from wheat lignocellulose by *Streptomyces viridosporus* and their partial metabolism by natural bacterial isolates. J. Biotechnol. 13:145–158.

89. Kögel-Knabner, I., P. G. Hatcher, and W. Zech. 1991. Chemical structural studies of forest soil humic acids: aromatic carbon fraction. Soil Sci. Soc. Am. J. 55:241–247.

90. Hayes, M. H. B., P. MacCarthy, R. L. Malcolm, and R. S. Swift. 1989. Humic substances: the emergence of structural forms, p. 689–733. *In* M. H. B. Hayes, P. MacCarthy, R. L. Malcolm, and R. S. Swift (eds.), Humic substances II. John Wiley & Sons, Chichester.

91. Benner, R., P. G. Hatcher, and J. I. Hedges. 1990. Early diagenesis of mangrove leaves in a tropical estuary: bulk chemical characterization using solid-state ^{13}C NMR and elemental analyses. Geochim. Cosmochim. Acta 54:2003–2013.

92. Jambu, P., E. Fustec, and R. Jacquesy. 1978. Les lipides des sols, nature, origine, évolution, propriétés. Sci. Sol 4:229–239.

93. Ziegler, F., and W. Zech. 1989. Distribution pattern of total lipids and lipid fractions in forest humus. Z. Pflanzenernaehr. Bodenkd. 152:287–290.

94. Ziegler, F. 1989. Changes of lipid content and lipid composition in forest humus layers derived from Norway spruce. Soil Biol. Biochem. 21:237–243.

95. Holloway, P. J. 1984. Cutins and suberins, the polymeric plant lipids, p. 321–345. *In* H. J. Mangold (ed.), CRC handbook of chromatography, lipids, Vol. 1. CRC Press, Boca Raton, Florida.

96. Kögel-Knabner, I., P. G. Hatcher, E. W. Tegelaar, and J. W. de Leeuw. 1992. Aliphatic components of forest soil organic matter as determined by solid state ^{13}C NMR and analytical pyrolysis. Sci. Total Environ. 113:89–106.

97. Nip, M., E. W. Tegelaar, J. W. de Leeuw, P. A. Schenck, and P. J. Holloway. 1986. A new non-saponifiable highly aliphatic and resistant biopolymer in plant cuticles. Evidence from pyrolysis and ^{13}C-NMR analysis of present-day and fossil plants. Naturwissenchaften 73:579—585.

98. Tegelaar, E. W., G. Hollman, P. van der Vegt, J. W. de Leeuw, and P. J. Holloway. 1992. Chemical characterization of the outer bark tissue of some angiosperm species: recognition of an insoluble, non-hydrolyzable highly aliphatic biopolymer (suberan). Org. Geochem., submitted.

99. Kögel-Knabner, I., F. Ziegler, M. Riederer, and W. Zech. 1989. Distribution and decomposition pattern of cutin and suberin in forest soils. Z. Pflanzenernaehr. Bodenkd. 152: 409—413.

100. Ziegler, F., and W. Zech. 1990. Decomposition of beech litter cutin under laboratory conditions. Z. Pflanzenernaehr. Bodenkd. 153:373—374.

101. Goni, M. A., and J. I. Hedges. 1990. The diagenetic behavior of cutin acids in buried conifer needles and sediments from a coastal marine environment. Geochim. Cosmochim. Acta 54:3083—3093.

102. Tegelaar, E. W., J. H. F. Kerp, H. Visscher, P. A. Schenk, and J. W. de Leeuw. 1991. Bias of the palaeobotanical record as a consequence of variations in the chemical composition of higher vascular plant cuticles. Palaeobiology 17:133—144.

103. Kögel-Knabner, I., and P. G. Hatcher. 1989. Characterization of alkyl carbon in forest soils by CPMAS ^{13}C NMR spectroscopy and dipolar dephasing. Sci. Total Environ. 81/82:169—177.

104. Wilson, M. A., B. D. Batts, and P. G. Hatcher. 1988. Molecular composition and mobility of torbanite precursors: implications for the structure of coal. Energy Fuels 2:668—672.

105. Tegelaar, E. W., J. Wattendorf, and J. W. de Leeuw. 1992. A comparative structural, chemical and histochemical study of extant *Symplocos paniculata* and Eocene *Symplocos hallensis* leaf cuticles. Rev. Paleobot. Paleonol., in press.

4

Pesticide Interactions with Cyanobacteria in Soil and Pure Culture

KADIYALA VENKATESWARLU *Sri Krishnadevaraya University, Anantapur, India*

I. INTRODUCTION

Cyanobacteria (blue-green algae) are morphologically the most diverse and complicated of the prokaryotes [1]. Classification of the cyanobacteria alongside the eubacteria in the kingdom Prokaryota is a truer representation of their phylogenetic position than classification with the other algal forms [2]. The cyanobacteria have a worldwide distribution and are common constituents of microalgal communities on temperate soils, particularly on moist neutral to alkaline soils [3], and in temperate habitats [4]. Thus, cyanobacteria together with microalgae probably account for between 4 and 27% of the total microbial biomass on dry weight basis in agricultural soils [5,6]. Also, cyanobacteria are ubiquitous in a variety of associated environments, such as irrigation and drainage systems, lakes, and ponds [2].

In addition to their potential for contributing to productivity in a wide array of agricultural and ecological situations, many genera of cyanobacteria are major diazotrophs capable of fixing free nitrogen (N_2) of the atmosphere and increasing markedly the nitrogen content of soil to an annual total up to 90 kg N ha^{-1} yr^{-1} [3]. Available evidence indicates that N_2-fixing cyanobacteria make a major contribution to the fertility of soils [7]. As N_2-fixing cyanobacteria are probably more widespread than their bacterial counterparts [8], the importance of diazotrophic cyanobacteria for use in "algalization" (introduction of mixed cultures of free-living species) in rice culture of tropical and subtropical soils has been well recognized [9]. Thus, in terms of ultimate N input,

particularly in rice paddies, algalization is about one-third of the
cost of chemical fertilizer [9]. Furthermore, these prokaryotes are
of importance in synthesizing substances that support plant growth
[10,11], in accumulating phosphorus as polyphosphates which are
released in an available form during autolysis [12], and in binding
soil particles together, thereby reducing or preventing erosion [2].
With their possible use also as cattle or human food substitutes, the
cyanobacteria may have a more realistic role to play in the immediate future as a source for extraction of high-value individual components, such as phycobilin proteins for coloring foodstuffs [7].

Intensive and extensive cultivation in several countries of new
high-yielding crop varieties with high nitrogen input has resulted
in serious outbreaks of insect pests and pathogens. As a consequence, in agricultural practice throughout the industrial world
and to a lesser extent in the developing countries, pesticides are
an important factor in maintaining high agricultural productivity.
(Pesticides mentioned in text are listed in Table 1.) In addition
to the application of a variety of conventional pesticides, synthetic
pyrethroids have been used for effective control of household and
agricultural insect pests [13].

Soil forms the respository or the largest reservoir for all kinds
of pesticides whether they are aimed at soil or foliage. Generally,
1% or often 0.1% of a chemical hits the target pest while the remainder drifts into the environment, contaminating soil, water, and biota
[14–17]. Furthermore, pesticides usually enter the soil environment
at or below the concentrations recommended for agricultural use,
mostly in the order of 0.5 to 5 kg ha^{-1} [18]. The entry of pesticides into soil might have far-reaching consequences, as they would
disturb the delicate equilibrium between a microorganism and its environment, both involved in an important biological process. Thus
interference of the heavily used pesticides with the normal activities
of soil populations of cyanobacteria may be expected to have potentially serious effects on the overall fertility of soil. A large body of
work discussing the response of microalgae and cyanobacteria to
pesticides has been reviewed by Cox [19], Andrews [20], Butler
[21], Wright [22], McCann and Cullimore [6], Metting [3], Lal [23],
and Padhy [24]. However, in view of the recent accumulation of
much more literature in this area of research, an attempt has been
made to provide updated information on the interactions of pesticides with cyanobacteria in soils and in pure culture.

II. EFFECTS ON INDIGENOUS
POPULATIONS IN SOIL

Studies that deal with the influence of pesticides on indigenous populations of cyanobacteria are limited in number. These investigations

established the toxic effects by applying pesticides either to soil samples in the laboratory or to fields under nonflooded or flooded conditions. Most reports indicate the toxicity of pesticides to the total population of both cyanobacteria and microalgae in soil.

A. Herbicides

The postemergence herbicide 2,4-D (2,4-dichlorophenoxyacetic acid) received the most attention with respect to its influence on cyanobacteria in soil. Stimulation of growth of *Tolypothrix tenuis* was observed in soils treated with 0.04 ppm 2,4-D [25]. Pillay and Tchan [26] reported no suppression of growth of *Nostoc* and *Anabaena* species when 2,4-D was applied to a field at the rates of 1 and 5 ppm. The growth of filamentous cyanobacteria, such as *Aulosira fertilissima, T. tenuis, Anabaena* sp., and *Anacystis nidulans*, was unaffected at 200 kg ha^{-1} 2,4-D in rice fields [27]. The primary productivity of mixed fresh water phytoplankton in fish ponds was inhibited by 2,4,5-T and silvex [28]. In an attempt to determine the influence of herbicides on N_2-fixing cyanobacteria in a laboratory study, Cullimore and McCann [29] treated soil cores from a grassland loam with 2,4-D, MCPA (2-methyl-4-chlorophenoxyacetic acid), trifluralin and TCA (trichloroacetate) at 1 and 100 ppm. An overall reduction in cell numbers was observed for potential nitrogen-fixing cyanobacteria, but species of *Lyngbya* and *Nostoc* were more resistant than those of *Scytonema*, exhibiting less than 50% sensitivity to the four herbicides. Lipnitskaya and Kruglov [30] reported that *Phormidium tenue* was resistant to atrazine, simazine, and prometryne in a heavy loam soil. Whereas atrazine, at 5 kg ha^{-1}, reduced the population of microalgae in a light and an organic soil, *Phormidium autumnale* was tolerant to this herbicide [31]. Growth of cyanobacteria in rice fields was enhanced as a consequence of the inhibitory effect of simetryn on green algae [32]. In a light loamy derno-podzolic soil, atrazine, simazine, propazine (all at 2 kg ha^{-1}), and prometryne (at 1 to 3 kg ha^{-1}) inhibited growth of cyanobacteria and microalgae for a longer period [33]. Soil applications of chlortoluron, terbutryne, methabenzthiazuron, chloridazon, and dinosebacetate at the recommended doses caused total suppression of growth and acetylene (C_2H_2) reduction activity of cyanobacteria for several weeks [34].

Monuron, diuron, and neburon at 1 and 5 ppm were nontoxic to species of *Anabaena* and *Nostoc* in field soils [26]. *Tolypothrix tenuis* was not adversely affected in soil by prometone at 2 kg ha^{-1} [25]. Applications of lenacil and pyrazon to an arable field soil reduced the population density and species diversity of cyanobacteria, green algae, and diatoms [35]. The nitrogen-fixing activity of *Anabaena variabilis* was reduced by lenacil and pyrazon.

Table 1   Chemical Designations of Pesticides Mentioned in the Text

Common or other name	Chemical name
Alachlor	2-Chloro-2',6-diethyl-*N*-(methoxymethyl)-acetanilide
Aldicarb	2-Methyl-2-(methylthio)propionaldehyde *O*-methylcarbamoyloxime
Aldrin	1,2,3,4,10,10-Hexachloro-1,4,4a,5,8,8a-hexahydro-1,4-*endo*,*exo*-5,8-dimethano-naphthalene
Aminotriazole (Amitrole)	3-Amino-1,2,4-triazole
Atrazine	2-Chloro-4-ethylamino-6-isopropylamino-s-triazine
Barban	4-Chlorobut-2-ynyl-3-chloroca banilate
Bavistin	2-(Methoxycarbonyl) benzimidazole
Benomyl	Methyl-1-(butylcarbamoyl)-2-benzimidazole carbamate
Benzthiocarb	*S*-(4-Chlorobenzyl)-*N*,*N*-diethylthiol car-bamate
BHC	1,2,3,4,5,6-Hexachlorocyclohexane
Blitox	50% copper as copper oxychloride
Brassicol	Pentachloronitrobenzene
Butachlor	2-Chloro-2',6'-diethyl-*N*-(butoxymethyl acetanilide)
Camphachlor	Chlorinated camphenes with 67 to 69% chlorine
Captafol	*N*-(1,1,2,2-tetrachloroethylthio)-3a,4,7,7a-tetrahydrophthalimide
Captan	*N*-trichloromethylmercapto-4-cyclohexene-1,2-dicarboximide
Carbaryl	1-Naphthyl *N*-methylcarbamate
Carbendazim	Methyl benzimidazol-2-yl carbamate
Carbofuran	2,3-Dihydro-2,2-dimethylbenzofuran-7-yl methylcarbamate

(continued)

Table 1 (Continued)

Common or other name	Chemical name
Carboxin	5,6-Dihydro-2-methyl-1,4-oxathiin-3-car-boxanilide
Ceresan	N-(Ethylmercury)-p-toluene sulfonilide
Chloridazone	5-Amino-4-chloro-2-phenylpyridazin-3-one
Chloropicrin	Trichloronitromethane
Chlorpropham	Isopropyl N-(3-chlorophenyl)carbamate
Chlorpyriphos	O,O-Diethyl O-(3,5,6-trichloro-2-pyridyl)-phosphorothioate
Chlortoluron	3-(3-Chloro-p-tolyl)-1,1-dimethylurea
Cypermethrin	α-Cyano-3-phenoxyphenyl-3-(2,2-dichloro-vinyl)-2,2-dimethylcyclopropane car-boxylate
2,4-D	2,4-Dichlorophenoxyacetic acid
Dalapan	Sodium α,α-dichloropropionate
DCMU	See Diuron
DDD	2,3,5-Trichloropyridin-4-ol
DDE	2,2-Bis(p-chlorophenyl)-1,1-dichloroethyl-ene
DDT	2,2-Bis(p-chlorophenyl)-1,1,1-trichloro-ethane
Diallate	S-2,3-Dichloroallyl diisopropylthiocarbam-ate
Diazinon	O,O-Diethyl O-(2-isopropyl-4-methyl-6-pyrimidinyl)phosphorothioate
Dichlone	2,3-Dichloro-1,4-naphthoquinone
Dieldrin	1,2,3,4,10,10-Hexachloro-exo-6,7-epoxy-1,4,4a,5,6,7,8,8a-octahydro-1,4-endo, exo-5,8-dimethanonaphthalene
Dinoseb	2-(1-Methyl-n-propyl)-4,6-dinitrophenol
DNP	2,4-Dinitrophenol

(continued)

Table 1   (Continued)

Common or other name	Chemical name
Diquat	6,7-Dihydrodipyrido (1,2-*a*:2',1')-*c*-pyrazinediium ion (9,10-dihydro-8a,10a-diazoniaphenanthrene-2A)
Diuron	3-(3,4-Dichlorophenyl)-1,1-dimethylurea
Dursban	*O,O*-Diethyl *O*-3,5,6-trichloro-2-pyridyl phosphorothioate
Endosulfan	6,7,8,9,10,10-Hexachloro-1,5,5a,6,9,9a-hexahydro-6,9-methano-2,4,3-benzodi-oxathiepin-3-oxide
Endrin	1,2,3,4,10,10-Hexachloro-6,7-epoxy,1,4,4a,5,6,7,8,8a-octahydro-1,4-*endo,exo*-5,8-dimethanonaphthalene
EPTC	*S*-Ethyl dipropylthiocarbamate
Ethrel	2-Chloroethylphosphonic acid
Fenitrothion	*O,O*-Dimethyl *O*-(3-methyl-4-nitrophenyl) phosphorothioate
Fenvalerate	Cyano(3-phenoxyphenyl)-methyl 4-chloro-(1-methylethyl)benzeneacetate
Fluchloralin	*N*-(2-Chloroethyl)-2,6-dinitro-*N*-propyl-4-(trifluromethyl)-aniline
Fluometuron	*N*'-(1,1-Dimethyl-3-(α, α, α-trifluro-*m*-tolyl))urea
Flurodifen	*p*-Nitrophenyl-α, α, α-trifluro-2-nitro-*p*-tolyl ether
Folpet	*N*-(Trichloromethylthio)phthalimide
Fytolan	50% copper oxychloride
Glyphosate	*N*-(Phosphonomethyl)glycine
HOE 2997	2-Dichloroacetamide-3-chloro-1,4-naphtho-quinone
Lenacil	3-Cyclohexyl-6,7-dihydro-1*H*-cyclopenta-pyrimidine-2,4-dione
Lindane	γ-1,2,3,4,5,6-Hexachlorocyclohexane

(continued)

Table 1   (Continued)

Common or other name	Chemical name
Linuron	3-(3,4-Dichlorophenyl)-1-methoxy-1-methylurea
Machete	See butachlor
Magnacide-H	2-Propenal
Malathion	*O,O*-Dimethyl *S*-1,2-di(ethoxycarbonyl) ethyl phosphorodithioate
Mancozeb	Zinc ion and manganese ethylene bis-dithiocarbamate
Maneb	Manganous ethylene bis-dithiocarbamate
MBC	Methylcarbamoyl benzimidazole
MCPA	2-Methyl-4-chlorophenoxyacetic acid
MCPB	4-(4-Chloro-2-methylphenoxy)butyric acid
Metamitron	4-Amino-4,5-dihydro-3-methyl-6-phenyl-1,2,4-triazin-5-one
Methabenzthiazuron	1-(Benzothiazol-2-yl)-1,3-dimethylurea
Metobromuron	3-(*p*-Bromophehyl)-1-methoxy-1-methyl-urea
Metribuzin	4-Amino-6-(1,1-dimethylethyl)-3-(methyl-thio)-1,2,4-triazin-5(4*H*)-one
Methylparathion	*O,O*-Dimethyl *O-p*-nitrophenyl phosphoro-thioate
Milstem	5-Butyl-2-ethylamino-6-methylpyrimidin-4-ol
MNP	*m*-Nitrophenol
Molinate	*S*-Ethyl hexahydro-1-*H*-azepine-1-carbo-thioate
Monocrotophos	Dimethyl(*E*)-1-methyl-2-metnyl carbamoyl-vinyl phosphate
Monuron	3-(*p*-Chlorophenyl)-1,1-dimethylurea tri-chloroacetate

(continued)

Table 1   (Continued)

Common or other name	Chemical name
MSMA	Methylarsonic acid
Nabam	Disodium ethylene bis-dithiocarbamate
Neburon	1-n-Butyl-3-(3,4-dichlorophenyl)-1-methyl-urea
Oxine	8-Hydroxyquinolinate
Panacide	2,2'-Dihydroxy-5,5'-dichlorophenylmethane
PAP	p-Aminophenol
Paraquat	1,1'-Dimethyl-4,4'-bipyridylium ion
Parathion	O,O-Diethyl O-p-nitrophenyl phosphorothioate
PCP	Pentachlorophenol
Permethrin	3-Phenoxybenzyl-3-(2,2-dichlorovinyl)-2,2-dimethyl cyclopropane carboxylate
Phorate	O,O-Diethyl-S-(ethylthiomethyl)phosphorodithioate
Picloram	4-Amino-3,5,6-trichloropicolinic acid
PNP	p-Nitrophenol
Prometone	2,4-Bis(isopropylamino)-6-methoxy-s-triazine
Prometryne	2,4-Bis(isopropylamino)-6-methylthio-s-triazine
Propanil	3,4-Dichloropropionanilide
Propazine	2-Chloro-4,6-bis(isopropylamino)-s-triazine
Propham	N-Phenyl isopropyl carbamate
Pyrazol	5-Amino-4-chloro-2-phenyl-3($2H$)-pyradazinone
Quinalphos	O,O-Diethyl-O-quinoxalin-2-yl phosphorothioate
Saturn	See thiobencarb
Silvex	2-(2,4,5-Trichlorophenoxy)propionic acid
Simazine	2-Chloro-4,6-bis(ethylamino)-s-triazine

(continued)

Table 1 (Continued)

Common or other name	Chemical name
Simetryne	2-Methylthio-4,6-bis(ethylamino)-s-triazine
Stam f-34	See propanil
2,4,5-T	2,4,5-Trichlorophenoxyacetic acid
TCA	Trichloroacetate
Terbutryne	2-*tert*-Butylamino-4-ethylamino-6-methyl-thio-1,3,5-triazine
Thimet	See phorate
Thiobencarb	S[4-(chlorophenyl)methyl] diethylcarbamothioate
Thiram	Tetramethylthiuram disulfide
Topsin-M	1,2-Bis(3-methoxycarbonyl-2-thioureido)-benzene(thiophanate-methyl)
Toxaphene	Chlorinated camphene
Trichlorfon	Dimethyl 2,2,2-trichloro-1-hydroxyethyl-phosphonate
Trifluralin	α,α,α-Trifluoro-2,6-dinitro-N,N-dipropyl-p-toluidine
Vapam	Sodium methyldithiocarbamate
Zectran	4-Dimethylamino-3,4-xylyl methylcarbamate
Zineb	Zinc ethylene bis-dithiocarbamate
Ziram	Zinc dimethyldithiocarbamate

Nitrogen fixation by *Nostoc muscorum* and *Anabaena oryzae* was not inhibited by treatment of a paddy soil with propanil, molinate, benzthiocarb, and flurodifen [36]. Paraquat totally inhibited the cyanobacteria in rice fields [37]. Investigating the impact of four herbicides on nitrogen fixation by *T. tenuis* and *Calothrix brevissima* in Hungarian soils, Ibrahim [38] obseved that 0.1 ppm EPTC (*S*-ethyl dipropylthiocarbamate) (0.1 ppm) was most toxic, propanil and molinate (both at 1 ppm) were innocuous, and trifluralin was stimulative. Dinoseb effectively reduced the population of cyanobacteria and

microalgae in a loamy sand by more than 90%, but trifluralin, linuron, and metribuzin were nontoxic [39]. Pife [40] demonstrated
that *Oscillatoria* sp., which together with *Hamtzschia* sp. formed
more than 90% of the soil algal population, was less sensitive to the
application of trifluralin at a rate of 2 mg kg^{-1} applied to the top
layer in field plots of a sandy loam soil. Picloram, at 50 ppm, suppressed the growth of *Cylindrospermum licheniforme* in soil [41].
Dalapon at 1 and 5 ppm inhibited growth of *Nostoc* and *Anabaena*
species in fields [26], and *A. variabilis* showed a high tolerance
to molinate in soil [42].

B. Fungicides

In spite of the extensive use of fungicides in modern agriculture,
few data are available on their effects on indigenous populations
of cyanobacteria in soil. Fitzerald and Skoog [43] reported the
total elimination of blooms of cyanobacteria in fields that received
dichlone at 10 to 100 ppb. In rice fields, *Anabaena fertilissima*
and *Tolypothrix tenuis* were not affected by high concentrations
of zineb [44]. HOE 2997 (2-dichloroacetamide-3-chloro-1,4-naph-
thoquinone), a fungicide with algicidal activity, significantly inhibited the growth of *Anabaena* sp. in a rice field [45]. Doneche
[46] observed a reduction in the numbers of cyanobacteria and
microalgae in soil after treatment with mancozeb. Wegener et al.
[34] found no detectable effect on populations of cyanobacteria or
on C_2H_2-reducing activity when the fungicide carbendazim was applied to soil at the recommended rates.

The effect of dichlone in clearing cyanobacteria from ponds was
studied, and the concentrations of dichlone required to clear
blooms of *Microcystis aeruginosa* and non-bloom-forming cyanobacteria were 30 and 55 ppm, respectively [47]. Also, selective
elimination of *Cylindrospermum licheniforme* and *M. aeruginosa*
from ponds was observed with 2 ppm dichlone [48].

C. Insecticides

An application of lindane (gamma-BHC) at only 50 kg ha^{-1} to soil
under flooded conditions, similar to those of rice fields, greatly
inhibited the nitrogen-fixing cyanobacteria [49]. Similarly, BHC
at less than 10 ppm (close to field application rate) stimulated the
growth of *Aulosira fertilissima* in fields [50]. Even higher rates
of BHC application (15 g of 10% BHC m^{-2}) to rice fields had no adverse effect on growth of *Tolypothrix tenuis* [51]. Field application rates of lindane ranging from 5 to 50 kg ha^{-1} significantly increased the indigenous populations of cyanobacteria in submerged
tropical soils, probably because of the total elimination of preda-
tory crustaceans [52]. Although lindane was innocuous to *A.*

fertilissima, endrin at 10 ppm inhibited this cyanobacterium in rice fields [50]. Dursban, at 2.4 ppb, stimulated growth of cyanobacteria in artificial ponds [53]. Watanabe [54] reported that 100 ppm PCP (pentachlorophenol) had no inhibitory effect on cyanobacteria in rice fields, but the green algae were suppressed. The total population of cyanobacteria and microalgae increased in a soil under non-flooded conditions [50% WHC (water-holding capacity)] when endosulfan was applied at levels close to the recommended rates [55].

The application of parathion at concentrations between 1 and 5 ppm to standing water in rice fields was not inhibitory to the growth of *T. tenuis* [56]. Growth of *Anabaena* sp. in a lake was enhanced, possibly as a consequence of the elimination of predators, by application of parathion [57]. Muralikrishna and Venkateswarlu [55] reported that parathion application, at 1 and 10 kg ha^{-1}, to a red laterite soil under flooded and nonflooded conditions greatly reduced the cyanobacteria, as well as microalgae; the inhibitory effect was more pronounced in nonflooded soil. However, a 0.5 kg ha^{-1} application of parathion, which is close to the field application rate, stimulated the cyanobacterial population under nonflooded conditions. In contrast, Naumann [58] observed no toxic effects on population of microalgae and cyanobacteria when methyl parathion was applied at rates up to 300 kg ha^{-1} to a loam soil with a pH of 7.2. Methylparathion, at 25 ppm, increased the growth of cyanobacteria in rice fields, probably as a result of the complete suppression of grazers [59]. Similarly, elimination of predators and an increase in growth of cyanobacteria were reported in rice fields receiving phorate at 10 g m^{-2} [60]. Sethunathan and MacRae [61] observed that diazinon stimulated populations of cyanobacteria in the standing water of rice fields, and chlorpyrifos enhanced cyanobacterial populations in rice fields [62].

Whereas single or two successive applications of monocrotophos to a black vertisol had no inhibitory effect on several species of cyanobacteria, two successive additions of quinalphos, even at levels close to the recommended rates (0.5 to 2 kg ha^{-1}), were significantly toxic to the soil population [63]. *Gloeocystis gigas, Synechococcus elongatus, Nostoc linckia, N. punctiforme,* and *Phormidium* sp. were among the most frequently occurring forms in soil samples treated once with these two insecticides, whereas species of *Nostoc* and *Phormidium* were more predominant in soil receiving two successive applications of both monocrotophos and quinalphos. In another study, Megharaj et al. [64] observed that two successive applications of monocrotophos and even a single application of quinalphos, both at their recommended doses, to flooded soil resulted in a significant decrease in the size and species composition of cyanobacteria and microalgae.

Sevin, the commercial formulation of carbaryl, when applied to rice fields at 10 ppm had no inhibitory effect on growth of *Aulosira*

fertilissima [50]. Carbaryl, at 0.5 kg ha^{-1}, stimulated the cyanobacterial population and was nontoxic even at 2.5 kg ha^{-1} in nonflooded soil [55]. An application of carbofuran, at concentrations
close to field doses, to soil under flooded and nonflooded conditions enhanced populations of both cyanobacteria and microalgae
[65]. Filamentous cyanobacteria, such as *Anabaena variabilis*,
Lyngbya gracilis, *Nostoc punctiforme*, and *Phormidium tenue*, pre-
dominated in soil that received carbofuran under both water reg-
imens.

Although pyrethroids are among the most potent insecticides
that have been widely used in agriculture and public health in
recent years [13], there has been only a single study [66] of the
impact of this new class of pesticides on native populations of cyanobacteria in soil. Single or two repeated applications of cypermethrin or a single application of fenvalerate to soil at 1 kg ha^{-1}
had no inhibitory effect on populations of cyanobacteria and microalgae, but two applications of fenvalerate at concentrations of 0.75
to 5 kg ha^{-1} resulted in a significant increase in the soil populations. Interestingly, there was no change in population size when
one of the organophosphate insecticides (monocrotophos or quinalphos) was applied to the soil samples between two additions of cypermethrin or fenvalerate, all added at 1 kg ha^{-1} [66].

The nitrophenolic compounds represent major soil pollutants by
virtue of their importance as pesticides and also as key intermediates in the breakdown of several aromatics by soil microorganisms
[67]. Thus, *o*-nitrophenol (ONP), *p*-nitrophenol (PNP), and 2,4-
dinitrophenol (DNP) are listed as priority pollutants in soil by the
U.S. Environmental Protection Agency [68]. The possible toxic effects of these nitrophenolics on populations of cyanobacteria and
microalgae were determined in a black vertisol from cotton fields
[69]. Increasing concentrations (0.5 to 50 µg ml^{-1}) of PNP, MNP,
and DNP resulted in a gradual increase in toxicity toward both
populations in soil.

III. EFFECTS ON PURE CULTURES

The studies concerning nontarget effects of pesticides on cyanobacteria dealt primarily with growth and related phenomena of the
pure cultures as toxicity criteria. Much of the information on the
effects of herbicides, in particular, on cyanobacteria has come indirectly from the study of their mode of action. For instance,
herbicides have been shown to influence chiefly the process of
photosynthesis by affecting electron transport as well as morphology [70]. Among the herbicides, triazine, phenyl ureas, and
bipyridyls are inhibitors of photosynthesis, while phenoxy compounds, phenylcarbamates, and acylanilides act as inhibitors not

primarily involving photosynthesis [6]. Fungicides are known to affect the microorganisms, in general, by acting directly on the cell membranes, cell organelles, or nucleus. However, there is no satisfactory explanation of the mechanism of action of insecticides on microorganisms.

A. Herbicides

Many investigations have assessed the influence of herbicides on growth and nitrogen-fixing activity of pure cultures of cyanobacteria. Phenoxycarboxylic herbicides, the inhibitors not primarily involving photosynthesis, have been extensively examined in in vitro studies using microalgae, but cyanobacteria have received little attention in studies with these herbicides. The application of 250 ppm 2,4-D was ineffective in controlling the growth of bloom-forming *Microcystis aeruginosa* [57]. Gamble et al. [71] observed that 0.5 ppm 2,4-D inhibited the growth of *Tolypothrix tenuis*, and *Anabaena flos-aquae* tolerated 60 ppm 2,4-D [71]. Dry matter, chlorophyll content, and nitrogen-fixing activity of *T. tenuis* decreased at concentrations of 2,4-D ranging from 0.045 to 4.5 ppm [73]. Low doses (10^{-4} M) of 2,4-D stimulated the growth of *Nostoc muscorum, N. punctiforme*, and *Cylindrospermum* sp., but concentrations corresponding to field application rates (1.1 × 10^{-2} M) inhibited both growth and nitrogen fixation [74]. Lundqvist [75] observed inhibition of nitrogen-fixing activity of *N. muscorum, N. punctiforme*, and *Cylindrospermum* sp. at 1.1 × 10^{-2} M 2,4-D. Even at 440 ppm 2,4-D had no adverse effect on the growth of *Cylindrospermum licheniforme* in agar plate culture [41]. The maximum level of 2,4-D that permitted growth of 27 strains of cyanobacteria ranged from 5 to 500 ppm [27,44]. Concentrations of 2,4-D between 90 and 100 ppm were inhibitory to the growth of *Anacystis nidulans*, whereas 50 ppm enhanced growth [76]. However, levels of 2,4-D ranging from 100 to 150 ppm, 600 to 800 ppm, and 1000 ppm, were nontoxic, toxic, and lethal, respectively, to the growth of a *Cylindrospermum* sp. [77]. Chlorophyll production in *Oscillatoria lutea* was also unaffected by 100 ppm 2,4-D [78]. Das and Singh [79] reported weak mutagenic action of 2,4-D in *Anabaenopsis raciborskii, Anabaena aphanizomenoides, A. spiroides*, and *Microcystis flos-aquae*. The LD_{50} value of 2,4-D was found to be 400 ppm for *M. flos-aquae*, but the highest limit of tolerance was 1200 ppm (B. Das, Ph.D. thesis, Utkal University, India, 1978). The growth and respiration of *Anabaena variabilis* were not affected by 10 μM 2,4-D [80]. Incorporation of 2,4-D in liquid cultures at 10 ppm, 10 to 600 ppm, and 800 to 1000 ppm resulted in stimulation, toxicity, and lethality, respectively, toward *A. raciborskii*; for reasons that were not clear, the tolerance limit and

lethal dose in agar medium were 90 and 100 ppm, respectively [81].
N. linckia was completely eliminated by 2000 ppm 2,4-D [82].

Mycrocystis aeruginosa and Cylindrospermum licheniforme were
not affected by 2,4,5-T at 2 ppm [48]. Another phenoxy herbicide, MCPA, at low doses accelerated the growth of N. muscorum,
N. punctiforme, and Cylindrospermum sp. [74]. MCPA, at 1.9×10^{-5} M, stimulated cyanobacterial nitrogen fixation, but 1.9×10^{-2}
M (close to the field application rate) inhibited this process [75].
DaSilva et al. [83] observed initial inhibition by 20 ppm MCPA followed by partial recovery of nitrogen-fixing activity in several cyanobacteria. MCPA and diallate, both at 0.01 to 1 ppm, enhanced
the growth of a Nostoc sp. isolated from uncultivated silt loam; 100
ppm MCPA and 10 ppm diallate reduced its growth [84].

Trichome migration in Phormidium sp. was inhibited by several
triazines at levels ranging from 0.3 to 15 ppm [85]. Significant
inhibition of photosynthesis in Anabaena cylindrica was observed
with atrazine even at 0.2 ppm, but nitrogenase activity, oxygen
production, and growth were inhibited at concentrations of atrazine
above 1 ppm [86]. Concentrations of atrazine above 100 ppm inhibited C_2H_2 reduction in species of Anabaena [87,88], and EC_{50}
values ranged from 0.1 to 0.5 ppm atrazine for growth and 40 to
70 ppm for C_2H_2-reducing activity in three species of Anabaena
[89]. Of the degradation products of atrazine, deethylated atrazine was the most toxic to Anabaena sp.

Anabaena flos-aquae was sensitive to 2 ppm simazine [72], whereas Tolypothrix tenuis exhibited tolerance to this concentration
[25]. Electron microscopic observations of the simazine-treated
cells of Anacystis nidulans revealed modification of the thylakoids
besides changes in photosynthetic pigments [90]. Arvick et al.
[91] observed inhibition of growth of Anabaena cultures treated
with 1 ppm metribuzin, the herbicide known to interrupt the Hill
reaction and thereby interfere with electron flow in photosynthesis [92], and growth and photosynthesis in A. nidulans was completely inhibited within 10 h by 1 ppm metribuzin [93]. Although
2 kg ha^{-1} propazine was lethal to Aulosira fertilissima and 200 kg
ha^{-1} was toxic to almost all strains of Anabaena and Nostoc, even
concentrations of 1000 and 2000 kg ha^{-1} were not lethal to T.
tenuis and A. nidulans, respectively [27]. Prometryne was nontoxic to T. tenuis [25], but at 0.25 to 12 µg ml^{-1} prometryne decreased the frequency of heterocysts in Anabaena sp., and the
cultures did not recover even after washing [94].

Monuron was less toxic to cyanobacteria than to microalgae [95].
Shilo [96] demonstrated the inhibitory effects of monuron and diuron toward Anabaena variabilis, A. spiroides, Nostoc sp., and
Plectonema boryanum. Anabaena flos-aquae was sensitive to 2
ppm monuron [72]. Aulosira fertilissima did not tolerate 2 kg
ha^{-1} diuron, linuron, and fluometuron, and Tolypothrix tenuis

was sensitive to 2 kg ha^{-1} diuron. However, most strains of *Nostoc* and *Anabaena* and *Anacystis nidulans* were not inhibited at higher levels of these herbicides [27,44]. Monuron and linuron caused initial inhibition of nitrogen fixation by cyanobacteria, but the inhibition was followed by subsequent recovery and/or stimulation [83]. Hawxby et al. [80] reported suppression of growth of *Lyngbya birgei* and *A. variabilis* with 10 μM fluometuron. Furthermore, fluometuron caused cellular disintegration, disruption of photosynthetic processes, reorientation and redistribution of phycobiliproteins, and lowered RNA contents within 12 h, followed by total lethality after 60 h of exposure to the herbicide [97]. A spontaneous mutant of *Nostoc muscorum* was shown to be capable of tolerating 0.44 mM monuron [98].

Anabaena cylindrica and *Tolypothrix tenuis* were less sensitive to propham and chlorpropham than were *Anacystis nidulans* and *Gloeocapsa alpicola*, but lower concentrations of propham, chlorpropham, or barban stimulated the growth of both *A. cylindrica* and *T. tenuis* [22]. Concentrations of propham between 3 and 4 ppm changed the pigment color from brown to blue-green in *T. tenuis*, but this change was readily reversed with removal of the herbicide [99]. Chlorpropham was consistently more toxic to green algae than to cyanobacteria, and it was more inhibitory to green algae and cyanobacteria than its metabolite, 3-chloraniline [100]. Inhibition of growth of cyanobacteria by chlorpropham was attributed to its adverse effects on photosynthesis, respiration, and synthesis of RNA, protein, and lipid [101].

Anabaena doliolum was intolerant of four different herbicides, even at low concentrations, when grown without an added nitrogen source (nitrogen-fixing conditions) [102]. Although picloram was nontoxic, dinoseb, prometryne, and fluometuron inhibited growth of *Lyngbya birgei* [103]. Propanil, at 1.8 ppm, inhibited the synthesis of chlorophyll and, at 10 ppm, totally suppressed the growth of *Tolypothrix tenuis* [73]. Growth and nitrogen fixation in *T. tenuis* and *Calothrix brevissima* were stimulated by 0.01 ppm propanil, but 0.1 ppm was inhibitory [38]. Propanil was also reported to cause initial inhibition, followed by recovery, of photosynthesis in *Anabaena cylindrica*; the toxicity was less when the culture was grown in vitro under nitrogen-fixing conditions than when supplied with nitrate [104]. Moreover, either 0.2 ppm propanil or 70 ppm DCA (3,4-dichloraniline), the primary degradation product of propanil, caused 50% growth inhibition in *A. cylindrica*, *A. variabilis*, *Nostoc entophytum*, *N. tenuis*, *N. muscorum*, and *Gloeocapsa alpicola*. An increase in pH (up to 9) of the culture medium stimulated the growth of *Nostoc calcicola* and reduced the toxicity of Stam f-34, the commercial formulation of propanil [105]. Addition of glucose, acetate, and amino acids, such as arginine, aspartate, serine, threonine, and glutamine, also effectively reversed

the inhibitory action of propanil, even when the herbicide was present at a lethal dose of 30 µg ml^{-1}. Propanil inhibited heterocyst
differentiation and nitrogen fixation and caused a decrease in the
content of total protein, nucleic acids, carbohydrate, and chlorophyll *a* in *N. calcicola* [106]. Growth, heterocyst differentiation,
and N$_2$ fixation of *Anabaena doliolum* were inhibited by 30 µM propanil [107], but the toxicity could be reversed by addition of
glucose, sodium acetate, and amino acids, such as arginine, aspartate, serine, threonine, and glutamine.

Diquat, the bipyridyl herbicide, was toxic and also lytic to
Nostoc muscorum [108]. Both diquat and paraquat, in the presence of light, reduced the levels of chlorophyll, carotenoid, and
phycocyanin pigments and resulted in total lethality of *Coccoch-
loris penyocystis*, *Microcolum vaginatus*, *Anabaena variabilis*, *A.
spiroides*, *Nostoc* sp., and *Plectonema boryanum*, all isolated
from fish ponds [96]. Application of paraquat at 0.6 and 1 ppm
to a fish reservoir for weed control had significant toxic effects
on the populations of cyanobacteria as well as green algae [109].
DaSilva et al. [83] reported that paraquat and diquat were lethal
to *Chlorogloea fritschii* at 25 ppm and to *Anacystis nidulans* at
100 ppm and that 20 ppm diquat and paraquat inhibited the nitrogen-fixing activity of species of *Anabaena*, *Aulosira*, *Calothrix*,
Chlorogloea, *Cylindrospermum*, *Tolypothrix*, and *Westiellopsis*.
Vaishampayan [110] reported a powerful mutagenic effect of paraquat in *N. muscorum*, causing both forward mutation (streptomycin and methylamine resistance) and backward mutation (*nif*$^-$
het$^-$ to *nif*$^+$ *het*$^+$). *Anabaena variabilis* was sensitive to 12 derivatives of 1,3-diphenyl-3-(2-pyridyl) ureas and thioureas at 100
ppm [111].

Maule and Wright [100] demonstrated, with a replica dish technique, that diuron, propanil, and atrazine were the most inhibitory, chlorpropham was moderately toxic, and MCPA and glyphosate were least toxic toward numerous cyanobacteria as well as
green algae. *Nostoc muscorum* and another cyanobacterial strain,
G4, showed a high degree of tolerance to DCMU (diuron), metamitron, and metribuzin [112]. Low concentrations (0.01 to 5 ppm)
of diuron, atrazine, and paraquat inhibited growth of *Aphanocapsa*
strains 6308 and 6714, *Anabaena variabilis*, and *Nostoc* strain MAC,
while higher amounts (>100 ppm) of MCPA, MCPB [4-(4-chloro-2-
methylphenoxy)butyric acid], 2,4-D, milstem, and ethrel caused
growth inhibition of these cyanobacteria [113]. In a simulated
stream periphyton, *Phormidium minnesotense* was severely inhibited by 1 and 10 mg kg^{-1} atrazine, whereas MSMA (methylarsonic
acid) and paraquat caused only a slight reduction of growth in
species of *Calothrix* and *Phormidium*, respectively [114].

Growth and pigment levels in *Anacystis nidulans* were reduced
by 50 ppm aminotriazole, but the inhibition was reversible [115].

The phosphorylase activity of *Oscillatoria princeps* was inhibited by aminotriazole [116], and trifluralin, up to 25 ppm, when added to the cultures 10 days after inoculation, reduced the growth of *Tolypothrix tenuis* [73]. Nitrogen fixation by *T. tenuis* was also shown to be affected by molinate, trifluralin, and Stam. Cyanobacteria, as well as green algae, were adversely affected by the application of chloropicrin [49]. Ibrahim [38] noticed inhibition in growth of *T. tenuis* even at 1 ppm trifluralin and 0.1 ppm EPTC (*S*-ethyl dipropylthiocarbamate). *Oscillatoria terrebriformis* and *O. agardhi* were highly sensitive to phenylmercury acetate at 50 to 500 ppm [117]. DaSilva et al. [83] observed initial inhibition, followed by recovery and stimulation, of cyanobacterial nitrogen fixation by 20 ppm amitrole. Growth and nitrogen fixation in species of *Nostoc* and *Anabaena* were inhibited by molinate and thiobencarb [42]. However, molinate enhanced, whereas thiobencarb inhibited, the growth, nitrogen fixation, and chlorophyll *a* content of both *Nostoc muscorum* and *Anabaena oryzae* [118]. Pandey [119] demonstrated that the toxicity to cyanobacteria of a herbicide applied to rice, Saturn (thiobencarb), was effectively reduced by raising the pH of the medium from 7 to 10. Thiobencarb, at 10 ppm, and 500 ppm 2,4-D reduced growth and chlorophyll content of *Anabaena variabilis* [120]. Concentrations of thiobencarb up to 50 ppm and of butachlor up to 100 ppm significantly reduced dry mass, contents of nitrogen and chlorophyll, and heterocyst formation in species of *Anabaena*, *Nostoc*, and *Oscillatoria* [121].

Although alachlor and butachlor were innocuous to heterocyst differentiation in *Nostoc muscorum* and its mutants, growth was inhibited after 20 min of exposure to the toxicants [122,123]. Also, mutagenicities of alachlor and butachlor were strong on back mutation of the *nif⁻ het⁻* mutant to the wild type in *N. muscorum*. Machete (butachlor) inhibited heterocyst differentiation and nitrogen fixation in *Anabaena doliolum* [124]. Photosynthesis by *Anabaena* sp. was totally inhibited by 2.5 ppm magnacide-H (2-propenel), which also caused cell disintegration [125]. Stam f-34 and DCMU inhibited growth and heterocyst differentiation in strains of *N. muscorum*, both with and without nitrogen-fixing activity, but the inhibitory effect was lost when the medium was supplemented with citrulline, methionine, and valine [126]. *Gloeocapsa* sp. showed considerable resistance to butachlor, fluchloralin, and propanil, whereas *N. muscorum* was moderately susceptible to these herbicides [127]. Butachlor, fluchloraline, and propanil also inhibited photosynthetic oxygen evolution, but stimulated respiratory oxygen uptake in *N. muscorum* and *Gloeocapsa* sp. [128]. Nitrogenase, nitrate reductase, and glutamine synthetase in *N. muscorum* were inhibited by the three herbicides, whereas butachlor and fluchloralin stimulated nitrogenase and glutamine synthetase in *Gloeocapsa* sp. Contents of phycocyanin and chlorophyll *a* and nitrogenase activity in species of *Nostoc, Anabaena,* and *Calothrix* were depressed under the impact of thiocarbazil, molinate, prestil-achlor, and dimethametryn [129]. N_2 fixation in *A. doliolum* and *N. muscorum* was adversely affected at 20 to 80 µg/ml alachlor or 2 to 8 µg/ml butachlor [130].

B. Fungicides

A bloom-forming strain of *Microcystis aeruginosa* was totally inhib-ited by 5 ppm oxine [47]. The sensitivity to concentrations of ziram ranging from 0.25 to 2 ppm followed the order diatoms > cy-anobacteria > green algae [131]. Most strains of *Anabaena* and *Nostoc* tested were insensitive to higher levels (e.g., 100 kg ha^{-1}) of zineb [27]. In contrast, zineb at 1 and 30 ppm was toxic to the growth of *Cylindrospermum* sp. and *Nostoc muscorum,* and concentrations of ziram up to 0.5 ppm inhibited growth of these cyanobacteria [24]. Moore [132] reported the suppression of growth of *N. muscorum* by diethyl dithiocarbamate, and copper dimethyl dithiocarbamate was highly toxic to two species of *Oscil-latoria* [117]. In *N. muscorum,* 1 ppm nabam caused inhibition of the photoassimilation of acetate, 1 ppm captan did not suppress growth, and 1 ppm vapam was nontoxic [132]. Similarly, 1 ppm captan inhibited the growth of *Anacystis nidulans* [133], whereas even 500 ppm captan only slightly inhibited the growth of *Westiel-lopsis prolifica, Anabaena fertilissima, Nostoc* sp., *Tolypothrix tenuis,* and *Calothrix* sp. [134].

The growth of *Anacystis nidulans* was suppressed by 140 ppb dichlone [135], whereas that of *Anabaena* sp. was suppressed by concentrations as low as 10 ppb [136]. The formation of hetero-cysts in *Nostoc* sp. was reduced by captan, zineb, blitone, blitox, and Rovral, all at 500 ppm [137]. Twenty-seven strains of cyano-bacteria were not affected by 5 to 50 ppm mancozeb [44]. Ziram, zineb, and mancozeb did not affect the frequency of heterocysts in *Cylindrospermum,* but the nitrogenase activity was adversely affected [24]. Maneb was toxic to two species of *Oscillatoria* [117]. *Aulosira fertilissima,* species of *Anabaena,* and *Nostoc* sp. were more sensitive to ceresan than *Tolypothrix tenuis* and *A. nidulans* [27]. *Anabaena* was sensitive to 10 ppm captafol and folpet [136]. Gupta and Saxena [138] observed significant toxicity of panacide to *A. fertilissima* and *Nostoc* sp. Butachlor was significantly more toxic to species of *Oscillatoria* and *Cyl-indrospermum* than was HOE 2997 [139].

Species of *Calothrix, Westiellopsis, Aulosira, Tolypothrix,* and *Nostoc* exhibited differential sensitivity to Topsin-M [1,2-bis(3-methoxycarbonyl-2-thioureido)benzene(thiophanate-methyl)] at concentrations from 50 to 1000 ppm [140]. Rovral was relatively less toxic to species of *Nostoc* and *Tolypothrix* than to those of *Westiellopsis, Aulosira,* and *Calothrix,* but thiram was inhibitory to the growth of the first two genera. Thimet, at 100, 500, 500, and 300 ppm had harmful effects on species of *Westiellopsis*, *Aulosira*, *Tolypothrix*, and *Calothrix*, respectively, whereas species of *Nostoc* did not tolerate even 1 ppm [141]. Higher concentrations (500 to 1000 ppm) of brassicol than of bavistin and fytolan were tolerated by the genera mentioned above [142]. Exposure to 45.5 ppm blitox produced blitox-resistant mutants of *Nostoc linckia* and *N. muscorum*, which were defective in H_2 and NO_3 metabolism [143]. Benomyl, captan, carboxin, and thiram, at concentrations ranging from 10^{-3} to 5×10^{-1} M, in a medium with or without nitrate completely inhibited growth of species of *Nostoc*, *Anabaena*, *Lyngbya*, and *Mastigocladus laminosus* [144]. Species of *Aulosira* and *Tolypothrix* were more resistant to MBC (methylcarbamoyl benzimidazole), the metabolite of benomyl, than were species of *Nostoc* and *Calothrix* [134].

C. Insecticides

DDT [2,2-*bis*(p-chlorophenyl)-1,1,1-trichloroethane] is a potent agent that alters the structural integrity of chloroplasts, in addition to changing cell permeability in microorganisms [70]. Gregory et al. [145] reported that *Anacystis nidulans* was not adversely affected by 1 ppm DDT although the accumulation of DDT was higher (99 to 848 times) from the culture medium. However, 800 ppb DDT in the presence of 1% NaCl inhibited the growth of *A. nidulans*, but the inhibition could be reversed by increasing the concentration of calcium in the medium [146]. Even 20 ppm DDT had no effect on the growth of *S. elongatus* [147]. *Anabaena cylindrica* was more sensitive to 1000 ppb DDT than was *Microcystis aeruginosa* [148]. Growth of *A. nidulans* was markedly inhibited by DDT and its metabolites, DDD and DDE; DDD was most toxic, followed by DDE and DDT [149]. DDT, at 10 ppm, also inhibited growth of *M. aeruginosa* [150]. Both growth and cell morphology of *Synechococcus elongatus* were unaffected by 90 ppb DDT [151]. Growth of two strains of *Aphanocapsa*, as well as of *Anabaena variabilis* and a *Nostoc* strain, was hindered by DDT at concentrations up to 20 ppm [113]. Similarly, initial inhibition of chlorophyll content and size reduction of cells of *A. variabilis* were observed with 1 ppm DDT [152].

A 16% solution of BHC caused a 50% decrease in growth of *Microcystis aeruginosa* [47]. In contrast, several isomers of BHC were nontoxic to this cyanobacterium [48]. BHC, at 1 ppm, caused stimulation, no effect, and 50% inhibition of nitrogenase activity in *Westiellopsis prolifica*, *Hapalosiphon welwitschii*, and *H. fontinalis*, respectively [153]. Singh [154] reported that 5 ppm commercial lindane (gamma-BHC) was nontoxic to *Cylindrospermum* sp., *Aulosira fertilissima*, and *Plectonema boryanum*, but 200 ppm was lethal. However, the maximum concentration of BHC that permitted growth of these three cyanobacteria ranged from 10 to 55 ppm. Although

50 ppm lindane was lethal to *Anabaena aphanizomenoides* and *Anabaenopsis raciborskii*, concentrations of 80 and 100 ppm did not prevent growth of *Anabaena spiroides* and *Microcystis flos-aquae*, respectively [155]. Lindane, at levels as low as 2 to 4 ppm, inhibited growth of *Nostoc muscorum* and *Wollea bharadwajae* [156], and 5 ppm inhibited growth of *Anacystis nidulans* [157]. The dry weight, chlorophyll content, and photosynthetic activity of *A. nidulans* were reduced by lindane at concentrations ranging from 0.1 to 100 ppm [158]. The reduction in total nitrogen content of lindane-adapted strains of *A. fertilissima*, *Anabaena doliolum*, and *Nostoc* sp. was ascribed to inhibition of dinitrogen fixation, protein formation, or both [159]. However, Subramanian [160] reported stimulation of growth in *Anabaena* sp. at levels up to 100 ppm of lindane.

Aldrin, dieldrin, and endrin, all at 1 ppm, were nontoxic to microalgae and to the cyanobacteria *Microcystis aeruginosa* and *Anabaena cylindrica* [148]. In contrast, dieldrin inhibited growth of *Anacystis nidulans* [157], and 1 ppm aldrin and dieldrin inhibited the growth of *A. cylindrica*, *A. nidulans*, and *Nostoc muscorum* [161]. Clegg and Koevenig [162] observed prevention of growth of *M. aeruginosa* with less than 5 ppm aldrin, dieldrin, or endrin. Furthermore, 950 ppm of aldrin and photoaldrin was required to suppress the growth of *A. nidulans* and a marine isolate, *Agmenellum quadruplicatum* [163]. Thus, a critical examination of the effects of aldrin on a particular cyanobacterium (e.g., *A. nidulans*) shows apparently conflicting results. Although growth of *A. nidulans* was inhibited by 950 ppm endrin, *A. quadruplicatum*, in addition to being sensitive to all concentrations of endrin tested, was stimulated by ketoendrin [163]. Singh [154] reported tolerance of *Cylindrospermum* sp., *Aulosira fertilissima*, and *Plectonema boryanum* to endrin up to 600 ppm. Dieldrin and its analogs were also found to inhibit the growth of *A. nidulans* [149].

Endosulfan and its metabolites, α-endosulfan, β-endosulfan, endosulfanidiol, endosulfan ether, endosulfan sulfate, hydroxyendosulfan ether, and endosulfan lactone, all at 1.25 mg kg^{-1}, were reported to have no significant effects on growth of a *Phormidium* sp. during a 14-day exposure [164]. However, Sardeshpande and Goyal [165] observed inhibition of growth of *Anabaena iyengarii*, *Hapalosiphon intricatus*, *Calothrix membranacea*, and *C. bharadwajae* by endosulfan at concentrations ranging between 1 and 50 ppm. Endosulfan above 1 ppm was lethal to *Anabaena flos-aquae*, and even the growth-promoting activity of fernoxone (sodium salt of 2,4-D) was reversed by 1 ppm endosulfan [166]. The growth of *Anabaena* sp. and *Aulosira fertilissima* was also adversely affected at 1 μg ml^{-1} endosulfan, and the cultures were completely bleached by 50 μg ml^{-1} [167]. Endosulfan, however, lowered ^{14}C uptake and nitrogenase activity considerably in *A. fertilissima* but not in an *Anabaena* sp. Palmer and Maloney [48] reported that *Microcystis aeruginosa* was more sensitive than *Cylindrospermum licheniforme* to 2 ppm toxaphene. Growth of *C. licheniforme* and *M. aeruginosa* was prevented by 2 ppm camphachlor. Addition of pentachlorophenol to the culture medium even 10 days after inoculation suppressed the growth of two species of *Oscillatoria* [117].

In contrast to the organochlorine insecticides, less attention has been paid to determining the toxic effects of organophosphorus insecticides toward cyanobacteria [168], although this group of insecticides is widely used in agriculture. Cultures of *Anacystis nidulans*, *Euglena gracilis*, and *Scenedesmus obliquus* were not adversely affected by 1 ppm parathion despite the accumulation of parathion at levels up to 72 ppm [145]. Ahmed and Venkataraman [50] reported that 10 ppm diazinon was nontoxic to *Aulosira fertilissima*, and *A. fertilissima*, *Cylindrospermum* sp., and *Plectonema boryanum* were observed to tolerate a commercial formulation of diazinon at concentrations as high as 400 ppm [154]. Growth of *Chlorogloea* was completely inhibited by 200 ppm malathion, and nitrogen fixation by *Nostoc muscorum*, *Tolypothrix tenuis*, and *Westiellopsis muscicola* was markedly decreased by 100 ppm [83]. Torres and O'Flaherty [78] reported that malathion at 1 ppm reduced the chlorophyll content in *Oscillatoria* sp.; low concentrations (0.01 to 0.55 ppb) stimulated chlorophyll production. Growth of *N. muscorum* was drastically reduced by 500 ppm malathion [169], and *A. fertilissima* was more sensitive to malathion than *Anabaena* sp. [167]. Phorate concentrations of 1000, 500, and 300 ppm were nontoxic to species of *Westiellopsis*, *Aulosira*, and *Tolypothrix*, respectively, but even 1 ppm was highly toxic to *Nostoc* sp. [141]. Phorate at 1 ppm increased nitrogen fixation in *Calothrix bharadwajae* and enhanced growth and nitrogen fixation in *Anabaena iyengarii* and *Calothrix membranacea* [165]. About 40% inhibition of nitrogenase activity in *Hapalosiphon welwitschii* var. *vaginatus* and 50% inhibition in *H. fontinalis* were observed with 1 ppm phorate; *Westiellopsis prolifica*, however, exhibited pronounced enhancement of nitrogenase activity up to 10 ppm phorate [153]. Application of phorate at concentrations up to 250 μg g^{-1} was highly toxic to nitrogen fixation and chlorophyll content of *N. muscorum* and *N. piscinale* in soil cultures but not in liquid cultures [170]. *Anabaena* (Arm 310) and *A. fertilissima* were quite sensitive to fenitrothion, chlorpyriphos, and DDT, but the toxic effects were dependent on the type and nature of the insecticide, the organisms, and the experimental conditions [171].

Synechococcus elongatus, *Nostoc linckia*, and *Phormidium tenue* responded differentially to monocrotophos and quinalphos [172]; monocrotophos was toxic to *P. tenue* at 50 and 100 μg ml^{-1} and to *N. linckia* at 100 μg ml^{-1}, while the growth of *S. elongatus* was enhanced at even 100 μg ml^{-1}. Although significant enhancement in growth of *P. tenue* was observed with 100 µg ml^{-1} quinalphos, concentrations above 5 or 10 µg ml^{-1} exerted inhibitory effects on *S. elongatus* and *N. linckia*. Both monocrotophos and quinalphos inhibited nitrogen fixation by *N. linckia* [173]. Trichlorfon decreased the content of nitrogen compounds, increased carbohydrate fraction, and altered cell division and morphology of *Anabaena* PCC 7119 [174]. The contents of phycobiliproteins, chlorophylls, and total proteins decreased in trichlorfon-treated culture of *Anabaena* PCC 7119 [175].

N. *linckia* was highly sensitive to PNP, MNP, DNP, and catechol compared to the green alga *Chlorella vulgaris* [69]. Even a 0.5% glucose amendment of the culture medium did not reverse the toxicity of PNP and MNP toward *N. linckia* and *Synechococcus elongatus* [176]. PNP, MNP, and DNP, even at 5 µg ml^{-1}, inhibited the nitrogen-fixing activity of *N. linckia* [173]. Concentrations ranging from 5 to 50 µg ml^{-1} of PAP (*p*-aminophenol), the product of nitro group reduction of PNP, inhibited cell number, chlorophyll *a*, total carbohydrates, $^{14}CO_2$ uptake, and nitrate reductase and nitrogenase activity of *N. linckia* and *N. muscorum* [177]. ONP, MNP, and PNP, but not phenol, at concentrations above 20 µg ml^{-1}, significantly inhibited the cell constituents (such as chlorophyll *a*, protein, and carbohydrate), $^{14}CO_2$ uptake, and activities of nitrate reductase, nitrogenase, and glutamine synthetase in *N. linckia* [178]. Addition of 10 µM ATP (adenosine triphosphate) to MNP treatments reversed the toxicity. Transmission electron microscopy (TEM) revealed secretion of mucus around the filament and induction of spore formation in the culture subjected to nitrophenol toxicity.

Carbaryl, at 10 ppm, did not inhibit the growth of *Aulosira fertilissima* [50]. This concentration of carbaryl, however, enhanced the growth of *Cylindrospermum* sp. [24]. *Nostoc muscorum*, grown on solid minimal (i.e., nitrogen-free) medium, was highly sensitive to carbaryl [179]. *Synechococcus elongatus* was highly sensitive to even 5 µg ml^{-1} carbaryl when compared to *N. linckia* [180]. Enhancement of growth, survival, and nitrogen fixation of *N. muscorum* was observed when carbofuran was added at 25 ppm to the culture medium [181]. Growth and nitrogen fixation of *Anabaena iyengarii* and *Calothrix membranacea* were enhanced by 1 ppm carbofuran [165]. Carbofuran up to 10 ppm produced marked stimulation of nitrogen fixation in two strains of *Westiellopsis prolifica*, ARM 365 and ARM 366, *Hapalosiphon welwitschii* var. *vaginatus* ARM 364, and *Calothrix braunii* ARM 367, but a 50% reduction in nitrogenase activity in *Hapalosiphon fontinalis* was observed with 1 ppm [153]. Carbofuran was less toxic to *S. elongatus* and *N. linckia* than carbaryl [180]. Carbofuran, carbaryl, and dimethoate, but not endosulfan, all at 10 ppm, increased growth and N_2 fixation of *W. prolifica*, while higher concentrations (>100 ppm) of the insecticides suppressed photosynthetic oxygen evolution [182]. Snyder and Sheridan [183] observed that motility of *Oscillatoria terrebriformis* was arrested by zectran below 0.5 ppm, and 10 ppm inhibited growth of both *O. terrebriformis* and *Synechococcus lividus*. Aldicarb inhibited the growth of *Plectonema boryanum* [184].

The EC_{50} of permethrin toward growth of *Anabaena cylindrica* and *A. variabilis* was less than 10 ppm, but the EC_{50} values of various phenoxybenzyl degradation products of permethrin ranged from 1.4 to 8 ppm [185]. While determining the interaction between the solvent and pesticide in toxicity toward *A. cylindrica*, *A. variabilis*, and *A. inaequalis*, Stratton et al. [186] observed permethrin to be 20 times more toxic when it was incorporated with acetone at a final concentration of 1% (v/v) than with the admissible level of acetone (0.1%, v/v) in the bioassay system. Cypermethrin, at 10 to 50 µg ml^{-1}, stimulated or only slightly inhibited the growth of *Synechococcus elongatus*, and the corresponding levels of fenvalerate stimulated growth [187]. However, growth response, in terms of chlorophyll *a* concentration, was significantly enhanced in *Nostoc linckia*, but not in *Phormidium tenue*, by cypermethrin and fenvalerate at 10 to 50 µg ml^{-1}. In addition, cypermethrin and fenvalerate up to 20 µg ml^{-1} stimulated nitrogen fixation by *N. linckia* [173]. The toxic effect of PNP toward *N. linckia* was alleviated in the presence of 2 to 5 µg ml^{-1} PAP, the product of nitro group reduction of PNP [177].

D. Effects of Pesticide Combinations

It is a common practice in modern agriculture to apply different groups of pesticides, either simultaneously or in succession, for effective control of a variety of pests [185,188–190]. For instance, DuRant [190] reported excellent control of both cotton bollworm and boll weevil when insecticide combinations involving fenvalerate plus monocrotophos and permethrin plus chlorpyriphos were applied to the crop. Moreover, it has been highlighted in several reviews that the major, or frequently the only, means of degradation of numerous pesticides in the environment is microbiological [191–194]. Consequently, significant amounts of metabolites accumulate and are present with their parent compounds in the environment. It is, therefore, highly likely that the degradation products, if equally toxic, interact with their parent compounds to alter their individual toxicity patterns toward nontarget microorganisms and thus modify their respective environmental impacts [89,104]. The information on the interactions of pesticides, in combination or with their respective intermediate products of degradation, is sparse, although this is a real and essential situation that must be considered in pesticide microbiology.

In general, the response to the interactions may be additive, when two or more toxicants do not interact to any measurable degree, or synergistic or antagonistic, when the uptake or toxicity of the mixture is enhanced or reduced, respectively, over that of the individual component compounds [89]. For example, the herbicides atrazine and metobromuron interacted synergistically toward the growth of a green alga, *Chlamydomonas reinhardtii*, with the effect being more pronounced at a relatively low concentration of each [195]. Sometimes, combination of two pesticides at varying concentrations may result in more than one interaction effect. Thus, the herbicide atrazine and the pyrethroid insecticide permethrin interacted additively and antagonistically for growth, photosynthesis, and nitrogenase activity of *Anabaena inaequalis* [189]. However, mercury and atrazine or permethrin, when added simultaneously at day zero, interacted antagonistically on growth of *A. inaequalis* [196]. Likewise, combinations at differential concentrations of an organophosphate (monocrotophos or quinalphos) and a pyrethroid (cypermethrin or fenvalerate) yielded additive, antagonistic, and synergistic interaction effects on cell numbers of *Synechococcus elongatus* and *Scenedesmus bijugatus* [197].

Permethrin and its degradation products interacted to produce synergistic and antagonistic responses with *Anabaena cylindrica*, *A. variabilis*, *Chlorella pyrenoidosa*, and *Scenedesmus quadricauda* [185]. Depending on the actual test system employed (growth, photosynthesis, or C_2H_2-reducing activity), combinations of atrazine and its monodealkylated products elicited synergistic, antagonistic, and additive interaction responses in *Anabaena inaequalis* [89]. Carbaryl and its major hydrolysis product, 1-naphthol, interacted to elicit either additive or synergistic responses in *Synechococcus elongatus*, *Nostoc linckia*, and *Chlorella vulgaris* [198].

IV. ACCUMULATION AND METABOLISM OF PESTICIDES

The pesticide residues available in microbial environments are rapidly picked up by the microorganisms. In general, the amount of pesticide residue accumulated by an organism is quantitatively expressed in terms of percentage accumulation (the amount of substance accumulated as a percentage of the initial amount added) and bioaccumulation or bioconcentration factor (the ratio of the concentration of substance in the organism, in ppm, to its concentration in water, in ppm). The few reports available indicate that, in addition to being inhibited or stimulated, many cyanobacteria absorb and/or degrade pesticides, and the uptake is strongly influenced by temperature and pH [199].

Ware et al. [200] reported that *Oscillatoria* sp. accumulated more DDT and could serve as an indicator of DDT contamination in surface irrigation water. The bioaccumulation factor for 1 ppm DDT from culture solution was 849 times for *A. nidulans* during a 7-day incubation period; the corresponding value for 1 ppm parathion ranged between 50 and 116 [145]. DDT bioaccumulation factors, when DDT was incorporated at 1 ppm in the medium, for *Anabaena cylindrica* and *Microcystis aeruginosa* were 268 and 280, respectively [148]. Without any harmful effect on growth or cell morphology, about 84% of *p,p'*-DDT added at 99 ppb was accumulated and subsequently retained by *Synechococcus elongatus* [151]. A progressive increase, up to 500 h, in the amount of DDT accumulation was reported for *Anabaena variabilis* [152]. Furthermore, a linear relationship between cell number and the amount of DDT accumulated was evident during this incubation period. From 20 ppm aldrin, dieldrin, and endrin, the accumulation factors for *A. cylindrica* and *M. aeruginosa* ranged from 140 to 222 [148]. Dieldrin was accumulated by *A. nidulans* and *M. aeruginosa* (W. C. Dierksheide and R. M. Pfister, Abstr. Annu. Meet. Am. Soc. Microbiol., 1973, p. 32). A gradual reduction in the toxicity of BHC toward *Anabaena aphanizomenoides* and *Anabaena raciborskii* after repeated reinoculations of these species in the medium was ascribed to the rapid accumulation of the insecticide by the cultures [201]. *Anabaena* sp. and *Aulosira fertilissima* markedly accumulated DDT, fenitrothion, and chlorpyriphos [202]. Although there was only a small amount of uptake and accumulation of dieldrin and dimethoate by *Anabaena* sp. and *A. fertilissima*, the bioaccumulation factor for permethrin was very high [203]. The uptake and bioconcentration of aldrin and phorate in *Anabaena* sp. and *A. fertilissima* were directly related to the concentration of insecticide in the medium and inversely related to their water solubility [204].

Cyanobacteria, as well as microalgae, in addition to being potentially useful microorganisms for studies related to pesticide accumulation, are capable of degrading some pesticides [22,148]. However, the information on metabolism of pesticides by cyanobacteria is far from complete. During a 96-h incubation period, *Nostoc entophytum* formed 1.8 ppm DCA and *T. tenuis* formed 1 ppm DCA from 5 ppm propanil [104]. *Anacystis nidulans* converted propham and chlorpropham to their corresponding anilines [205]. *A. nidulans* was able to metabolize DDT [133,157]. Worthen [151] reported the conversion of DDT to DDE by *Synechococcus elongatus*. DDT was metabolized to DDD and DDE by *Anabaena* sp. and *Aulosira fertilissima* [202]. Nearly 88% of lindane incorporated at concentrations ranging from 0.5 to 2.5 ppm was metabolized by *A. nidulans* [133]. Aldrin was metabolized to dieldrin by both *Anabaena* sp. and *A. fertilissima* [204]. Two microalgae, *Chlorella vulgaris* and

Scenedesmus bijugatus, and three cyanobacteria, *Synechococcus elongatus*, *Phormidium tenue*, and *Nostoc linckia*, metabolized mono-crotophos and quinalphos fairly rapidly, yielding three or four unidentified metabolites [206]

V. HERBICIDE RESISTANCE

Their unique ability to fix nitrogen in aerobic habitats and carbon by the oxygenic eukaryotic plant mechanism makes the cyanobac-teria difficult yet important subjects for molecular biology [207]. However, there is rapidly developing interest in the molecular biol-ogy of cyanobacteria since mutant strains, molecular vectors, and gene libraries have become more common. Significant advances have been made, including transformation and use of *Anacystis* for photosynthesis and herbicide resistance studies [208,209] and the successful expression of an *Escherichia coli* gene on a plasmid vector in *Agmanellum* (= *Synechocystis*) [210].

The QB protein of photosystem II (PS II) is essential for oxy-genic photosynthetic electron transport and is the target of sev-eral herbicides which act by binding directly to the photosynthetic apparatus. The genome of the cyanobacterium *Anacystis nidulans* R_2 contains three genes for the QB protein [211]. The *psbA* gene of a mutant strain of *A. nidulans* R_2 was shown to code for the 32,000-dalton thylakoid membrane protein, which confers resistance to the herbicide diuron [209]. A cloned DNA fragment from within the coding region of the *psbA* gene transformed wild-type cells to herbicide resistance, proving that mutation within *psbA* is respon-sible for herbicide resistance. Furthermore, the mutation was shown to consist of a single nucleotide change that replaces serine at position 264 of the wild-type protein with alanine at that of the diuron-resistant mutant. A diuron-resistant transformant of *A. nidulans* R_2 contained a diuron-sensitive *psbAII* gene and a di-uron-resistant *psbAIII* gene, both of which were transcribed in the presence of diuron [212]. DCMU prevented binding of plasto-quinone in the QB site in a wild-type strain of *A. nidulans* R_2 containing functional *psbAI*, *psbAII*, and *psbAIII* genes [213]. All the PS II reaction centers were resistant to DCMU in the TaqI strain with a transformed *psbAI* gene. Also, strain KΔ1Taq3 with *psbAII* and transformed *psbAIII* was phenotypically DCMU resis-tant. Buzby et al. [214] cloned DNAs from a mutant of *Synecho-coccus* PCC 7002 that carry markers conveying resistance to atra-zine, diuron, and paraquat into the shuttle vector pAQE19 and transformed *E. coli*. Recombinant plasmids were recovered from atrazine- and diuron-resistant transformants. The DNA fragments containing atrazine and diuron resistance markers hybridized to a gene *psbA* probe.

The genes of a diazotrophic *Gloeocapsa* strain that conferred resistance to the rice field herbicides Machete and Basalin, when transferred to *Nostoc muscorum*, were expressed and maintained [215]. DNA from mutants of *Synechococcus* 6803, resistant to inhibitors of photosynthesis such as diuron, dinoseb, and pentachlormercuricbenzoate, transformed cells of the wild strain to an inhibitor-resistant state [216].

VI. CONCLUSIONS

Cyanobacteria contribute significantly to the important cycles of carbon, nitrogen, and possibly phosphorus in soil. The extensive and intensive application of a wide array of pesticides in agriculture clearly warrants study of the responses by cyanobacteria to these xenobiotics. However, studies concerning the interactions of pesticides with cyanobacteria have been limited to a selected group of these nontarget organisms. Available reports clearly indicate that the growth and metabolic activities of cyanobacteria are adversely affected by several pesticides used in agriculture. Thus, the well-established inhibitory effect of certain pesticides on biological N_2 fixation by cyanobacteria poses a major threat to soil fertility. Cyanobacteria have been utilized as biological assay organisms for anticipating crop response to both pesticides and fertilizers because of their biochemical similarity to higher plants and their short generation time. When predicting the environmental impact of the pesticides, it is, however, difficult to extrapolate data from in vitro studies to the natural conditions of the soil habitat, because of the possible intervention of various biotic and abiotic factors in ecological niches. Moreover, the available evidence suggests that a single pesticide can exert differential effects on different members of the cyanobacteria, thus raising potential problems in arriving at generalizations when evaluating the environmental safety of pesticides. Studies involving the potential environmental effects of extensively used pesticides in agriculture should, therefore, include a wide range of cyanobacteria isolated from the same field soils. The impact of metabolites of the pesticides, formed in the soil habitat, along with their parent compounds on cyanobacteria must be clearly understood. Emphasis should be directed to toxicity studies related to combinations of pesticides, because such applications may result in modification of toxicity patterns of the individual pesticides.

By comparison, the ability of cyanobacteria to biomagnify pesticides through the food chain has received little attention. Although microbial degradation is by far the most powerful means of detoxifying many pesticides in the environment, very little information is available on the degradation of pesticides by cyanobacteria. Because cyanobacteria are prokaryotic, they can be subjected to rDNA genetic engineering techniques to yield strains effective in detoxifying pesticides in agricultural and industrial areas. To make a significant contribution to agricultural productivity, development and introduction of mutants of heterocystous species, resistant to pesticides, must also be coupled with the traditional use of cyanobacterial biofertilizers.

ACKNOWLEDGMENTS

Thanks are due to Dr. B. Metting, Battelle, Pacific Northwest Laboratories, Richland, Washington, for inspiration and encouragement, to Dr. M. J. Daft, Department of Biological Sciences, University of Dundee, Dundee, for facilities and helpful discussions, and to Dr. M. Megharaj, Department of Genetics and Microbiology, University of Liverpool, Liverpool, for assistance in the preparation of the manuscript.

REFERENCES

1. Stanier, R. Y., and G. Cohen-Baziere. 1977. Phototrophic prokaryotes: the cyanobacteria. Annu. Rev. Microbiol. 31: 225–274.
2. Fogg, G. E., W. D. P. Stewart, P. Fay, and A. E. Walsby. 1973. The blue-green algae. Academic Press, New York.
3. Metting, B. 1981. The systematics and ecology of soil algae. Bot. Rev. 47:195–312.
4. Stewart, W. D. P. 1974. Blue-green algae, p. 202–237. *In* A. Quispel (ed.), The biology of nitrogen fixation. American Elsevier, New York.
5. Round, F. E. 1965. The biology of algae. Edward Arnold, London.
6. McCann, A. E., and D. R. Cullimore. 1979. Influence of pesticides on the soil algal flora. Residue Rev. 72:1–32.
7. Whitton, B. A., and H. G. Carr. 1982. Cyanobacteria: current perspectives, p. 1–8. *In* N. G. Carr and B. A. Whitton (eds.), The biology of cyanobacteria. Blackwell Scientific, Oxford.
8. Burns, R. C., and R. W. F. Hardy. 1975. Nitrogen fixation in bacteria and higher plants. Springer-Verlag, Berlin.
9. Roger, P. A., and S. A. Kulasooriya. 1990. Blue-green algae and rice. International Rice Research Institute, Los Banos, Manila, Philippines.
10. Shields, L. M., and L. W. Durrell. 1964. Algae in relation to soil fertility. Bot. Rev. 30:92–128.

11. Venkataraman, G. S. 1969. The cultivation of algae. ICAR, New Delhi.

12. Kaushik, B. D. 1985. Effect of native algal flora in nutritional and physico-chemical properties of sodic soil. Acta Bot. Indica 13:143−147.

13. Smith, T. M., and G. W. Stratton. 1986. Effects of synthetic pyrethroid insecticides on nontarget organisms. Residue Rev. 97:93−120.

14. Matsumura, F. 1972. Current pesticide situation in the United States, p. 33−60. *In* F. Matsumura, G. M. Boush and T. Misato (eds.), Environmental toxicity of pesticides. Academic Press, New York.

15. Tu, C. M., and J. R. W. Miles. 1976. Interaction between insecticides and soil microbes. Residue Rev. 64:17−65.

16. Jackson, R. B. 1963. Pesticide residues in soils, p. 825−842. *In* Soils: an Australian viewpoint. CSIRO, Academic Press, Melbourne.

17. Pimental, D., and L. Levitan. 1986. Pesticides: amounts applied and amounts reaching pests. Bioscience 36:86−91.

18. Hill, I. R., and S. J. L. Wright. 1978. The behavior and fate of pesticides in microbial environments, p. 79−136. I. R. Hill and S. J. L. Wright (eds.), Pesticide microbiology. Academic Press, London.

19. Cox, J. L. 1972. DDT residues in marine phytoplankton. Residue Rev. 44:23−28.

20. Andrews, J. H. 1976. The pathology of marine algae. Biol. Rev. 51:211−253.

21. Butler, G. L. 1977. Algae and pesticides. Residue Rev. 66: 19−62.

22. Wright, S. J. L. 1978. Interactions of pesticides with micro-algae, p. 535−602. *In* I. R. Hill and S. J. L. Wright (eds.), Pesticide microbiology. Academic Press, London.

23. Lal, S. 1984. Effects of insecticides on algae, p. 203−236. *In* R. Lal (ed.), Insecticide microbiology, Springer-Verlag, Berlin.

24. Padhy, R. N. 1985. Cyanobacteria and pesticides. Residue Rev. 95:1−44.

25. Platnova, V. P. 1967. The effect of 2,4-D, simazine, prometryne on soil algae. Trudy Kirov. sel'Khoz Inst. 20:215−221.

26. Pillay, A. R., and Y. T. Tchan. 1972. Study of soil algae. VII. Adsorption of herbicides in soil and prediction of their rate of application by algal methods. Plant Soil 36:571−594.

27. Venkataraman, G. S., and B. Rajyalakshmi. 1971. Tolerance of blue-green algae to pesticides. Curr. Sci. 40:143−144.

28. Butler, P. A. 1965. Effects of herbicides on estuarine fauna. Proc. 5th Weed Control Conf. 18:576−580.

29. Cullimore, D. R., and A. E. McCann. 1977. Influence of four herbicides on the algal flora of a prairie soil. Plant Soil 46: 499—510.

30. Lipnitskaya, G. P., and Y. V. Kruglov. 1967. The effect of triazine herbicides on soil algae. Works Inter-Univ. Conf. Kirowsk 20:222—225.

31. Mikhailova, E. I., and Y. V. Kruglov. 1973. Effect of some herbicides on soil algoflora. Pocvovedenie 8:81—85.

32. Yamagishi, A., and A. Hashizume. 1974. Ecology of green algae in paddy fields and their control with chemicals. Zasso Kenkyu 18:39—43.

33. Lebedeva, G. F., N. I. Chernova, and L. V. Epishina. 1978. The persistence of 1,3,5-triazines in a derno-podzolic soil and their influence on microflora. Vestn. Mosk. Univ. 16:44—46.

34. Wegener, K. E., R. Aldag, and B. Meyer. 1985. Soil algae: effects of herbicides on growth and C_2H_2 reduction (nitrogenase) activity. Soil Biol. Biochem. 17:641—644.

35. Zurek, L. 1981. The influence of herbicides lenacil and pyrazon on the soil algae. Ekol. Pol. 29:327—342.

36. El-Sawy, M., S. A. Z. Mahmoud, M. E. El-Haddad, W. A. Mashhour, and K. G. Salem. 1984. Effect of different herbicides on nitrogen fixation by blue-green algae in paddy soil, p. 297—306. In J. Szegi (ed.), Soil biology and conservation of the biosphere, Vol. 1.

37. Patel, N. P. 1972. Rice research in Fiji 1960—1970. Part III: Weed control. Fiji Agric. J. 34:27—34.

38. Ibrahim, A. N. 1972. Effect of certain herbicides on growth of nitrogen-fixing algae and rice plants. Symp. Biol. Hungary 11:445—448.

39. Lewis, J. A., G. C. Papavizas, and T. S. Hora. 1978. Effect of some herbicides on microbial activity in soil. Soil Biol. Biochem. 10:137—141.

40. Pife, A. E. 1967. The influence of the herbicide trifluralin, alone and in the presence of simulated acid rain, on the algae and cyanobacteria of a sandy loam soil. Proc. 1987 British Crop Protect. Conf., Weeds, Vol. 2:507—514.

41. Arvick, J. H., D. L. Wilson, and L. C. Darlington. 1971. Response of soil algae to picloram—2,4-D mixtures. Weed Sci. 19:276—278.

42. Zaitseva, I. I. 1979. The influence of soil herbicides on the nitrogen-fixing activity of blue-green algae. Byull. Vses. Nauchno-Issled. Inst. Skh. Mikrobiol. 32:60—62.

43. Fitzerald, G. P., and F. Skoog. 1954. Control of blue-green algal blooms with 2,3-dichloronaphthoquinone. Sewage Ind. Wastes 26:1136—1140.

44. Venkataraman, G. S., and B. Rajyalakshmi. 1972. Relative tolerance of nitrogen-fixing blue-green algae to pesticides. Indian J. Agric. Sci. 42:119–121.
45. Hartz, P., H. Rochling, and F. Mariouw-Smith. 1972. 2-Di-chloro-acetamido-3-chloro-1,4-naphthoquinone, a new algicide for application in rice and other cultures. Meded. Fac. Landbouwwet. Rijksuniv. Gent. 37:699–704.
46. Doneche, B. 1974. Effects du Mancozebe sur la microflora des sols du vignoble bordelais. Premiers résultats. C. R. Acad. Sci. Paris 278:3011–3014.
47. Fitzerald, G. P., G. C. Gerloff, and F. Skoog. 1952. Stud-ides on chemicals with selective toxicity to blue-green algae. Sewage Ind. Wastes 24:888–896.
48. Palmer, C. M., and T. E. Maloney. 1955. Preliminary screen-ing for algicides. Ohio J. Sci. 55:1–8.
49. Ishizawa, S., and T. Matsuguchi. 1966. Effects of pesticides and herbicides upon microorganisms in soil and water under waterlogged conditions. Bull. Nat. Inst. Agric. Sci. Tokyo B16:1–90.
50. Ahmed, M. H., and G. S. Venkataraman. 1973. Tolerance of *Aulosira fertilissima* to pesticides. Curr. Sci. 42:108.
51. Srinivasan, S., and N. Emayavaramban. 1977. Some observa-tions on blue-green algae. Aduthurai Rev. 1:98–102.
52. Raghu, K., and I. C. MacRae. 1967. The effect of the gamma-isomer of benzene hexachloride upon the microflora of submerged rice soils. I. Effect upon algae. Can. J. Microbiol. 13:173–180.
53. Hurlbert, S. H., M. S. Mulla, and H. R. Wilson. 1972. Ef-fects of an organophosphorus insecticide on the phytoplankton, zooplankton and insect populations of fresh water poinds. Ecol. Monogr. 42:269–299.
54. Watanabe, A. 1967. The blue-green algae as the nitrogen fix-ators, p. 77–85. Symp. IX Int. Congress Microbiol. (Moscow), 1966.
55. Muralikrishna, P. V. G., and K. Venkataswarlu. 1984. Effect of insecticides on soil algal population. Bull. Environ. Contam. Toxicol. 33:241–245.
56. Hirang, T., K. Shiraishi, and K. Nakano. 1955. Studies on the blue-green algae in low land paddy soil. Part I. On some conditions for the growth of BGA in paddy soil and its effect on growth of paddy rice plant. Shikoku Nogyo Shikenjo Ho-koku 2:121–137.
57. Cook, S. F., and J. D. Conners. 1963. The short-term side effects of the insecticidal treatment of clear lake, Lake County, California in 1962. Ann. Entomol. Soc. 99:70–73.

58. Naumann, K. 1970. Zur Dynamik der Bodenmikroflora nach
    Anwendung von Pflanzenschutzmittein II. Die Reaktion ver-
    schiedener physiologischer Gruppen von Bodenbakterien auf
    den Einsatz von Parathion-methyl in Freiland. Zbl. Bakt. Abt.
    II 124:755–765.
59. Anonymous. 1977. Studies on biological nitrogen fixation, p.
    43. *In* Nuclear Institute for Agriculture and Biology (Five
    years of NIAB).
60. Srinivasan, S., and V. Ponnuswamy. 1978. Influence of
    weedicides on blue-green algae. Aduthurai Rev. 2:136–139.
61. Sethunathan, N., and I. C. MacRae. 1969. Some effects of
    diazinon on the microflora of submerged soils. Plant Soil 30:
    109–112.
62. Sivasithamparam, K. 1970. Some effects of an insecticide
    (Dursban) and a weed killer (linuron) on the microflora of
    a submerged soil. Riso 19:339–346.
63. Megharaj, M., K. Venkateswarlu, and A. S. Rao. 1986. Ef-
    fect of monocrotophos and quinalphos on soil algae. Environ.
    Pollut. A40:121–126.
64. Megharaj, M., K. Venkateswarlu, and A. S. Rao. 1988. Mi-
    crobial degradation and algal toxicity of monocrotophos and
    quinalphos in flooded soil. Chemosphere 17:1033–1039.
65. Magharaj, M., K. Venkateswarlu, and A. S. Rao. 1988. Tol-
    erance of algal population in rice soil to carbofuran applica-
    tion. Curr. Sci. 57:100–102.
66. Megharaj, M., K. Venkateswarlu, and A. S. Rao. 1986. In-
    fluence of cypermethrin and fenvalerate on natural soil algal
    population. Ecotoxicol. Environ. Saf. 12:141–145.
67. Dagley, S. 1967. The microbial metabolism of phenolics, p.
    287–317. *In* A. D. McLaren and G. W. Peterson (eds.), Soil
    biochemistry. Marcel Dekker, New York.
68. Keith, L. H., and W. A. Telliard. 1979. Priority pollutants.
    I. A perspective view. Environ. Sci. Technol. 13:416–423.
69. Megharaj, M., K. Venkateswarlu, and A. S. Rao. 1986. The
    toxicity of phenolic compounds to soil algal population and
    Chlorella vulgaris and *Nostoc linckia*. Plant Soil 96:197–203.
70. Tripathi, A. K. 1988. The cytology and biochemistry of
    pesticide microbiology. CRC Crit. Rev. Microbiol. 15:223–
    246.
71. Gamble, S. J. R., C. J. Mayhew, and W. E. Chappell. 1952.
    Respiration rates and plate counts for determining effect of
    herbicides on heterotrophic soil microorganisms. Soil Sci. 74:
    347–350.
72. Schluter, M. 1965. The effect of several fungicides and
    herbicides on fresh water algae under laboratory conditions.
    Z. Fisch. 314:303–315.

73. Hamdi, Y. A., A. S. El-Nawawy, and M. S. Tewfik. 1970. Effect of herbicides on growth and nitrogen fixation of the alga, *Tolypothrix tenuis*. Acta Microbiol. Pol. Ser. B 2:53—56.

74. Inger, L. 1970. Effect of two herbicides on nitrogen fixation by blue-green algae. Sud. Bot. Tidskr. 64:460—461.

75. Lundqvist, I. 1970. Effect of two herbicides on nitrogen fixation by blue-green algae. Svensk. Bot. Tidskr. 64:460—461.

76. Voight, R. A., and D. L. Lynch. 1974. Effects of 2,4-D and DMSO on prokaryotic and eukaryotic cells. Bull. Environ. Contam. Toxicol. 12:400—405.

77. Singh, P. K. 1974. Algicidal effect of 2,4-dichlorophenoxy-acetic acid on blue-green alga *Cylindrospermum* sp. Arch. Mikrobiol. 97:69—72.

78. Torres, A. M. R., and L. M. O'Flaherty. 1976. Influence of pesticides on *Chlorella, Chlorococcum, Stigeoclonium* (Chlorophyceae), *Tribonema, Vaucheria* (Xanthophyceae) and *Oscillatoria* (Cyanophyceae). Phycologia 15:25—36.

79. Das, B., and P. K. Singh. 1977. Mutagenicity of pesticides in blue-green algae. Microbiol. Lett. 5:103—106.

80. Hawxby, K., B. Tubea, J. Ownby, and E. Basler. 1977. Effect of various herbicides on four species of algae. Pestic. Biochem. Physiol. 7:203—209.

81. Das, B., and P. K. Singh. 1977. The effect of 2,4-dichlorophenoxy acetic acid on growth and nitrogen fixation of blue-green alga *Anabaenopsis raciborskii*. Arch. Environ. Contam. Toxicol. 5:437—445.

82. Tiwari, D. N., A. K. Pandey, and A. K. Mishra. 1981. Action of 2,4-dichlorophenoxyacetic acid and rifampicin on heterocyst differentiation in the blue-green alga, *Nostoc linckia*. J. Biosci. 3:33—39.

83. DaSilva, E. J., L. E. Henriksson, and E. Henriksson. 1975. Effect of pesticides on blue-green algae and nitrogen fixation. Arch. Environ. Contam. Toxicol. 3:193—204.

84. Metting, B., and W. Rayburn. 1979. The effects of the pre-emergence herbicide di-allate and the post-emergence herbicide MCPA on the growth of some soil algae. Phycologia 18: 269—272.

85. Noll, M., and U. Bauer. 1973. Rapid sensitive herbicide bioassay by inhibition of trichome-migration of blue-green algae. Zbl. Bakt. Hyg. 1 Abt. Orig. B157:178—181.

86. Rohwer, F., and F. Fluckiger. 1979. Effect of atrazine on growth, nitrogen fixation and photosynthetic rate of *Anabaena cylindrica*. Angew. Bot. 53:59—64.

87. Kallio, S., and R. E. Wilkinson. 1977. The effect of some herbicides on nitrogenase activity and carbon fixation in two sub-arctic lichens. Bot. Gaz. (Chicago) 138:468—473.

88. Holst, R. W., J. H. Yopp, and G. Kapusta. 1982. Effect of several pesticides on the growth and nitrogen assimilation of the *Azolla-Anabaena* symbiosis. Weed Sci. 30:54–58.
89. Stratton, G. W. 1984. Effects of the herbicide atrazine and its degradation products, alone and in combination, on phototrophic microorganisms. Arch. Environ. Contam. Toxicol. 13:35–42.
90. Mehta, R. S., and K. Hawxby. 1979. Effects of simazine on the blue-green alga *Anacystis nidulans*. Bull. Environ. Contam. Toxicol. 23:319–326.
91. Arvick, J. H., D. L. Hyzak, and R. L. Zimdahl. 1973. Effect of metribuzin and two analogues on five species of algae. Weed Sci. 21:173–175.
92. Trebst, A., and H. Wicloska. 1975. Inhibition of photosynthetic electron transport in chloroplasts by metribuzin. Z. Naturforsch. 30:499–503.
93. Eley, J. H., J. F. McConnell, and R. H. Catlett. 1983. Inhibition of metribuzin on growth and photosynthesis of the blue-green alga *Anacystis nidulans*. Environ. Exp. Bot. 23:365–368.
94. Yee, D., P. Weinberger, D. A. Johnson, and C. DeChacin. 1985. *In vitro* effects of the s-triazine herbicide, prometryne, on the growth of terrestrial and aquatic microflora. Arch. Environ. Contam. Toxicol. 14:25–31.
95. Maloney, T. E. 1958. Control of algae with chlorophenyl dimethyl-urea. J. Am. Water Works Assoc. 50:417–422.
96. Shilo, M. 1965. Study on the isolation and control of blue-green algae from fish ponds. Bamidgeh 17:83–93.
97. Mehta, R. S., and K. Hawxby. 1977. Action of herbicide on blue-green algae—ultrastructural observations on photosynthetic lamellae. Misc. Ser. Bot. Soc. Am. 154, p. 51, Langston University, Oklahoma.
98. Vaishampayan, A. 1984. Biological effects of a herbicide on a nitrogen-fixing cyanobacterium (blue-green alga): an attempt for introducing herbicide resistance. New Phytol. 96:7–11.
99. Maule, A., and S. J. L. Wright. 1983. Physiological effects of chlorpropham and 3-chloroaniline on some cyanobacteria and a green alga. Pestic. Biochem. Physiol. 19:196–202.
100. Maule, A., and S. J. L. Wright. 1984. Herbicide effects on the population growth of some green algae and cyanobacteria. J. Appl. Bacteriol. 57:369–379.
101. Ashton, F. M., O. T. DeVilliers, R. K. Glenn, and U. B. Duke. 1977. Localization of metabolic sites of action of herbicides. Pestic. Biochem. Physiol. 7:122–141.

102. Kapoor, K., and V. K. Sharma 1980. Effect of certain herbicides on survival, growth, and nitrogen fixation of the blue-green alga *Anabaena doliolum* Bharadwaja. Z. Allg. Mikrobiol. 20:465–469.

103. Tubea, B., K. Hawxby, and R. Mehta. 1981. The effects of nutrient, pH and herbicide levels on algal growth. Hydrobiologia 79:221–228.

104. Wright, S. J. L., A. F. Stainthrope, and J. D. Downs. 1977. Interactions of the herbicide propanil and a metabolite, 3,4-dichloroaniline, with blue-green algae. Acta Phytopathol. Hung. 12:51–60.

105. Pandey, A. K., V. Srivastava, and D. N. Tiwari. 1984. Toxicity of the herbicide Stam f-34 (propanil) on *Nostoc calcicola*. Z. Allg. Mikrobiol. 24:369–376.

106. Pandey, A. K. 1985. Effect of propanil on growth and cell constituents of *Nostoc linckia*. Pestic. Biochem. Physiol. 23:157–162.

107. Singh, S. 1990. Biological effects of rice-field herbicide Stam f-34 on the nitrogen-fixing cyanobacterium *Anabaena doliolum*. Proc. Indian Natl. Sci. Acad. B56:229–234.

108. Vaishampayan, A. 1984. Mutagenicity of a bipyridylium herbicide in a N_2-fixing cyanobacterium *Nostoc muscorum*. Environ. Int. 10:285–290.

109. Brooker, M. P., and R. W. Edwards. 1973. Effects of the herbicide paraquat on the ecology of a reservoir. I. Botanical and chemical aspects. Freshwater Biol. 3:157–175.

110. Vaishampayan, A. 1984. Powerful mutagenicity of a bipyridium herbicide in a nitrogen-fixing blue-green alga *Nostoc muscorum*. Mutat. Res. 138:39–42.

111. Goulding, K. H., N. V. Badami, and R. J. W. Cremlyn. 1978. The algicidal activity of some substituted 1,3-diphenyl ureas, 1-phenyl-3-(2-pyridyl) ureas and the corresponding thioureas. Pestic. Sci. 9:213–218.

112. Gadkari, D. 1988. Assessment of the effects of the photosynthesis-inhibiting herbicides diuron, DCMU, metamitron and metribuzin on growth and nitrogenase activity of *Nostoc muscorum* and a new cyanobacterial isolate, strain G4. Biol. Fert. Soils 6:50–54.

113. Hutber, G. N., L. J. Rogers, and A. J. Smith. 1979. Influence of pesticides on the growth of cyanobacteria. Z. Allg. Mikrobiol. 19:397–402.

114. Kosinski, R. J. 1984. The effect of terrestrial herbicides on the community structure of stream periphyton. Environ. Pollut. A36:165–189.

115. Kumar, H. D. 1963. Inhibition of growth and pigment production of a blue-green alga by 3-amino-1,2,4-triazole. Indian J. Plant Physiol. 6:150–155.

116. Frederick, J. F., and A. C. Gentile. 1960. The effect of 3-amino-1,2,4-triazole on phosphorylase of *Oscillatoria princeps*. Arch. Biochem. Biophys. 86:30–33.
117. Paraschiv, M., E. Serbanescu, G. H. Popovici, and C. Djendov. 1972. The algacide action of some chemical substances on blue-green algae. Rev. Roum. Biol. Ser. Bot. 17:195–201.
118. El-Haddad, M. E., S. A. Z. Mahmoud, M. El-Sawy, W. A. Mashhour, and K. G. Salem. 1984. Changes in metabolic activities of *Nostoc muscorum* and *Anabaena oryzae* in response to herbicide application. I. Effect of molinate and benzthiocarb, p. 277–285. *In* J. Szegi (ed.), Soil biology and conservation of the biosphere, Vol. 1.
119. Pandey, A. K. 1987. pH dependent Saturn toxicity and nitrogen fixation of *Nostoc calcicola*. Acta Bot. Indica 15: 231–235.
120. Ahluwalia, A. S. 1988. Influence of saturn and knockweed on the growth and heterocyst formation in the nitrogen-fixing blue-green alga. Pesticides 22:43–44.
121. Zargar, M. Y., and G. H. Dar. 1990. Effect of benthiocarb and butachlor on growth and nitrogen fixation by cyanobacteria. Bull. Environ. Contam. Toxicol. 45:232–234.
122. Singh, H. N., and A. Vaishampayan. 1978. Biological effects of the rice field herbicide 'machete' on various strains of the nitrogen-fixing blue-green alga *Nostoc muscorum*. Environ. Exp. Bot. 18:87–94.
123. Singh, H. N., H. R. Singh, and A. Vaishampayan. 1979. Toxic and mutagenic action of herbicide Alachlor (Lasso) on various strains of the nitrogen-fixing blue-green alga *Nostoc muscorum* and characterization of the herbicide-induced mutants resistant to methylamine and L-methionine-disulfoximine. Environ. Exp. Bot. 19:5–12.
124. Kashyap, A. K., and K. D. Pandey. 1982. Inhibitory effects of rice field herbicide Machete on *Anabaena doliolum* Bharadwaja and protection by nitrogen sources. Z. Pflanzenphysiol. 107:339–345.
125. Fritz-Sheridan, R. P. 1982. Impact of the herbicide magnacide-H (2-propenal) on algae. Bull. Environ. Contam. Toxicol. 28:245–249.
126. Vaishampayan, A., H. R. Singh, and H. N. Singh. 1978. Biological effects of rice-field herbicide Stam f-34 on various strains of the nitrogen-fixing blue-green alga *Nostoc muscorum*. Biochem. Physiol. Pflanzen 173:410–419.
127. Singh, L. J., and D. N. Tiwari. 1988. Some important parameters in the evaluation of herbicide toxicity in diazotrophic cyanobacteria. J. Appl. Bacteriol. 64:365–376.

128. Singh, L. J., and D. N. Tiwari. 1988. Effects of selected rice field herbicides on photosynthesis, respiration, and nitrogen assimilatory enzyme systems of paddy soil diazotrophic cyanobacteria. Pestic. Biochem. Physiol. 31:120—128.

129. Higazy, A., and M. Fayez. 1989. Effect of some herbicides on growth and diazotrophy of cyanobacteria isolated from Egyptian rice field. Ann. Agric. Sci. (Cairo) 34: 765—779.

130. Singh, V. P. 1989. *In vitro* effects of herbicides on rhizobia and cyanobacteria. Proc. Natl. Acad. Sci. India 59B: 349—356.

131. Maloney, T. E., and C. M. Palmer. 1956. Toxicity of six chemical compounds to thirty cultures of algae. Water Sewage Works 103:509—513.

132. Moore, R. B. 1967. Algae as biological indicators of pesticides. Abstr. J. Phycol. (Suppl.) 3:4.

133. Moore, R. B., and D. A. Dorward. 1968. Accumulation and metabolism of pesticides by algae. Abstr. J. Phycol. (Suppl.) 4:7.

134. Gangawane, L. V., and R. S. Saler. 1979. Tolerance of certain fungicides by nitrogen-fixing blue-green algae. Curr. Sci. 48:306—308.

135. Whitton, B. A., and K. MacArthur. 1967. The action of two toxic quinones on *Anacystis nidulans*. Arch. Mikrobiol. 57:147—154.

136. Bisiach, M. 1970. Algal infestations in Italian soils. Riso 19:129—134.

137. Gangawane, L. V., R. S. Saler, and L. Kulkarni. 1980. Effect of pesticides on growth and heterocyst formation in *Nostoc* sp. Marathwada Univ. J. Sci. 19:3—7.

138. Gupta, G. S., and P. N. Saxena. 1974. Effect of panacide on some green and blue-green algae. Curr. Sci. 43: 492—493.

139. Ferrante, G. M., and A. Battino-Viterbo. 1974. Agar plate technique for potential algacide screening. Riso 23:13—16.

140. Gangawane, L. V., and L. Kulkarni. 1979. Tolerance of certain fungicides by nitrogen-fixing blue-green algae and their side effects on rice cultures. Pesticides 13:37—39.

141. Gangawane, L. V. 1979. Tolerance of thimet by nitrogen-fixing blue-green algae. Pesticides 13:33—34.

142. Gangawane, L. V. 1980. Tolerance of N_2-fixing blue-green algae to brassicol, bavistin and fytolan. J. Indian Bot. Sci. 59:157—160.

143. Vaishampayan, A., and A. B. Prasad. 1982. Blitox-resistant mutants of the N_2-fixing blue-green algae *Nostoc linckia* and *Nostoc muscorum*. Environ. Exp. Bot. 22:427—435.

144. Cameron, H. J., and G. R. Julian. 1984. The effects of four commonly used fungicides on the growth of cyanobacteria. Plant Soil 78:409–415.

145. Gregory, W. W., J. K. Reed, and L. E. Priester. 1969. Accumulation of parathion and DDT by some algae and protozoa. J. Protozool. 16:69–71.

146. Batterton, J. C., G. M. Boush, and F. Matsumura. 1972. DDT: Inhibition of sodium chloride tolerance by the blue-green alga *Anacystis nidulans*. Science 176:1141–1143.

147. Morgan, J. R. 1972. Effects of Arochlor 1242R (a poly-chlorinated biphenyl) and DDT on cultures of an alga, protozoan, daphnid, ostrocod, and guppy. Bull. Environ. Contam. Toxicol. 8:129–137.

148. Vance, B. D., and W. Drummond. 1969. Biological concentration of pesticides by algae. J. Am. Water Works Assoc. 61:360–362.

149. Boush, G. M., and F. Matsumura. 1975. Pesticide degradation by marine algae. Natl. Tech. Inf. Serv. AD-A008, 75, 23.

150. VonWitsch, H., A. Munj, and C. Sherbe. 1975. Lichtabhangige DDT-wirkung bei mikroalgen. Ber. Dtsch. Bot. Ges. 88:191–196.

151. Worthen, L. R. 1973. Interception and degradation of pesticides by aquatic algae. Rpt. PB-223/506, Office of Water Resources Research. Distributed by National Technical Information Service, Springfield, Virginia.

152. Goulding, K. H., and S. Ellis. 1981. The interaction of DDT with two species of freshwater algae. Environ. Pollut. A25:271–290.

153. Kaushik, B. D., and G. S. Venkataraman. 1983. Response of cyanobacterial nitrogen fixation to insecticides. Curr. Sci. 52:321–323.

154. Singh, P. K. 1973. Effect of pesticides on blue-green algae. Arch. Mikrobiol. 89:317–320.

155. Das, B., and P. K. Singh. 1978. Pesticide (hexachloro-cyclohexane) inhibition of growth and nitrogen fixation in blue-green algae *Anabaenopsis raciborskii* and *Anabaena aphanizomenoides*. Z. Allg. Mikrobiol. 18:161–167.

156. Kar, S., and P. K. Singh. 1979. Detoxification of pesticides carbofuran and hexachlorocyclohexane by blue-green algae *Nostoc muscorum* and *Wollea bharadwajae*. Microbios Lett. 10:111–114.

157. Lazaroff, M., and R. B. Moore. 1966. Selective effects of chlorinated insecticides on algal population. Abstr. J. Phycol. (Suppl.) 2:7.

158. Kopeck, K., F. Fueller, W. Ratzman, and W. Simonis. 1976. Lichtabhangige Insektizidwirkungen und einzellige Algen. Ber. Dtsch. Bot. Ges. 88:269–281.

159. Sharma, V. K., and Y. S. Gaur. 1981. Nitrogen fixation by pesticide-adapted strains of paddy field cyanophytes. Int. J. Ecol. Environ. Sci. 7:117–122.

160. Subramanian, B. R. 1982. The effects of pesticides on nitrogen fixation and ammonium excretion by *Anabaena*, p. 569–586. Proc. Natl. Symp. Biol. Nitrogen Fixation, IARI, New Delhi.

161. Schauberger, C. W., and R. B. Wildman. 1977. Accumulation of aldrin and dieldrin by blue-green alga and related effects on photosynthetic pigments. Bull. Environ. Contam. Toxicol. 17:534–541.

162. Clegg, T. J., and J. L. Koevenig. 1974. The effect of four chlorinated hydrocarbon pesticides and one organophosphate pesticide on ATP levels in three species of photosynthetic freshwater algae. Bot. Gaz. 135:368–372.

163. Batterton, J. C., G. M. Boush, and F. Matsumura. 1971. Growth response of blue-green algae to aldrin, dieldrin, and their metabolites. Bull. Environ. Contam. Toxicol. 6:589–594.

164. Goebel, H., S. Gorbach, W. Kanuf, R. H. Rimpan, and H. Huttenback. 1982. Properties, effects of residues and analytics of the insecticide endosulfan. Residue Rev. 83:1–165.

165. Sardeshpande, J. S., and S. K. Goyal. 1982. Effect of insectidies on growth and nitrogen fixation by blue-green algae, p. 588–605. Proc. Natl. Symp. Biol. Nitrogen Fixation, IARI, New Delhi.

166. Chinnaswamy, R., and R. J. Patel. 1983. Effect of pesticide mixtures on the blue-green alga, *Anabaena flos-aquae*. Microbios Lett. 24:141–143.

167. Tandon, R. S., R. Lal, and V. V. S. N. Rao. 1988. Interaction of endosulfan and malathion with blue-green algae *Anabaena* and *Aulosira fertilissima*. Environ. Pollut. 52:1–9.

168. Anderson, J. R. 1978. Pesticide effects on nontarget soil microorganisms, p. 313–533. *In* I. R. Hill and S. J. L. Wright (eds.), Pesticide microbiology. Academic Press, London.

169. Tiwari, D. N., A. K. Pandey, and A. K. Mishra. 1979. Toxicity of malathion (*S*-1,2-di(ethoxycarbonyl) ethyldimethyl phosphorothiothionate), on growth and nitrogen fixation of cyanobacterium *Nostoc calcicola*. J. Sci. Res. Banaras Hindu Univ. 30:92–96.

170. Saha, K. C., S. Sannigrahi, S. K. Bandyopadhyay, and L. N. Mandal. 1984. Effect of phorate on nitrogen fixation by blue-green algae. J. Indian Soc. Soil Sci. 32:79–83.

171. Lal, S., D. M. Saxena, and R. Lal. 1987. Effects of DDT,
      fenitrothion and chlorpyrifos on growth, photosynthesis and
      nitrogen fixation in Anabaena (Arm 310) and Aulosira fertil-
      issima. Agric. Ecosyst. Environ. 19:197–209.
172. Megharaj, M., K. Venkateswarlu, and A. S. Rao. 1986.
      Growth response of four species of soil algae to monocro-
      tophos and quinalphos. Environ. Pollut. A42:15–22.
173. Megharaj, M., K. Venkateswarlu, and A. S. Rao. 1988.
      Effect of insecticides and phenolics on nitrogen fixation by
      Nostoc linckia. Bull. Environ. Contam. Toxicol. 41:277–281.
174. Marco, E., F. Martinez, and M. I. Orus. 1990. Physiologi-
      cal alterations induced by the organophosphorus insecticide
      trichlorfon in Anabaena PCC 7119. Environ. Exp. Bot. 30:
      119–126.
175. Orus, M. I., E. Marco, and F. Martinez. 1990. Effect of
      trichlorfon on N_2-fixing cyanobacterium Anabaena PCC 7119.
      Arch. Environ. Contam. Toxicol. 19:297–301.
176. Megharaj, M., K. Venkateswarlu, and A. S. Rao. 1989. Ef-
      fect of glucose addition on the toxicity of nitrophenols to
      Synechococcus elongatus and Nostoc linckia. Phykos 28:146–
      150.
177. Megharaj, M., H. W. Pearson, and K. Venkateswarlu. 1991.
      Toxicity of p-aminophenol and p-nitrophenol to Chlorella vul-
      garis and two species of Nostoc isolated from soil. Pestic.
      Biochem. Physiol. 40:266–273.
178. Megharaj, M., H. W. Pearson, and K. Venkateswarlu. 1991.
      Toxicity of phenol and three nitrophenols towards growth and
      metabolic activities of Nostoc linckia, isolated from soil. Arch.
      Environ. Contam. Toxicol. 21:578–584.
179. Vaishampayan, A. 1985. Mutagenic activity of alachlor,
      butachlor and carbaryl to a N_2-fixing cyanobacterium Nostoc
      muscorum. J. Agric. Sci. UK 104:571–576.
180. Megharaj, M., K. Venkateswarlu, and A. S. Rao. 1989. Ef-
      fect of carbofuran and carbaryl on growth of a green alga
      and two cyanobacteria isolated from soil. Agric. Ecosyst.
      Environ. 26:329–336.
181. Kar, S., and P. K. Singh. 1978. Toxicity of carbofuran to
      blue-green alga Nostoc muscorum. Bull. Environ. Contam.
      Toxicol. 20:707–714.
182. Adhikary, S. P. 1989. Effect of pesticides on the growth,
      photosynthetic oxygen evolution and nitrogen fixation of
      Westiellopsis prolifica. J. Gen. Appl. Microbiol. 35:319–325.
183. Snyder, C. E., and R. P. Sheridan. 1974. Toxicity of the
      pesticide zectran on photosynthesis, respiration and growth
      in four algae. J. Phycol. 10:137–139.

184. Mallison, S. M., and R. E. Cannon. 1984. Effects of pesticides on cyanobacterium *Plectonema boryanum* and cyanophage LPP-1. Appl. Environ. Microbiol. 95:910–914.
185. Stratton, G. W., and C. T. Corke. 1982. Toxicity of the insecticide permethrin and some degradation products towards algae and cyanobacteria. Environ. Pollut. A29:71–80.
186. Stratton, G. W., R. E. Burrell, M. L. Kurp, and C. T. Corke. 1980. Interactions between the solvent acetone and the pyrethroid insecticide permethrin on activities of the blue-green alga *Anabaena*. Bull. Environ. Contam. Toxicol. 24: 562–569.
187. Megharaj, M., K. Venkateswarlu, and A. S. Rao. 1987. Influence of cypermethrin and fenvalerate on a green alga and three cyanobacteria isolated from soil. Ecotoxicol. Environ. Saf. 14:142–147.
188. Sethunathan, N., T. K. Adhya, and K. Raghu. 1982. Microbial degradation of pesticides in tropical soils, p. 91–115. *In* F. Matsumura and C. R. Krishna Murti (eds.), Biodegradation of pesticides. Plenum Publishing, New York.
189. Stratton, G. W. 1983. Interaction effects of permethrin and atrazine combinations toward several nontarget microorganisms. Bull. Environ. Contam. Toxicol. 31:297–303.
190. DuRant, J. A. 1984. Cotton insect pests: field evaluation of selected insecticide treatments. J. Agric. Entomol. 1:201–211.
191. Helling, C. S., P. C. Kearney, and M. Alexander. 1971. Behavior of pesticides in soil. Adv. Agron. 23:147–240.
192. Bollag, J. M. 1974. Microbial transformation of pesticides. Adv. Appl. Microbiol. 18:75–130.
193. Alexander, M. 1981. Biodegradation of chemicals of environmental concern. Science 211:132–138.
194. Bollag, J. M. 1982. Microbial metabolism of pesticides, p. 126–167. *In* J. P. Rosazza (ed.), Microbial transformation of inactive compounds, Vol. 2. CRC Press, Boca Raton, Florida.
195. Loeppky, C., and B. G. Tweedy. 1969. Effects of selected herbicides upon growth of soil algae. Weed Sci. 17:110–113.
196. Stratton, G. W. 1985. Interaction effects of mercury-pesticide combinations towards a cyanobacterium. Bull. Environ. Contam. Toxicol. 34:676–683.
197. Megharaj, M., K. Venkateswarlu, and A. S. Rao. 1989. Interaction effects of insecticide combinations towards the growth of *Scendesmus bijugatus* and *Synechococcus elongatus*. Plant Soil 114:159–163.

198. Magharaj, M., K. Venkateswarlu, and A. S. Rao. 1990. In-
      teraction effects of carbaryl and its hydrolysis product, 1-
      naphthol, towards three isolates of microalgae from soil.
      Agric. Ecosyst. Environ. 31:293–300.
199. Fletcher, W. W., R. C. Kirkwood, and D. Smith. 1970. In-
      vestigations on the effect of certain herbicides on the growth
      of selected species of microalgae. Meded. Fac. Landb.
      Rijksuniv. Gent 35:855–867.
200. Ware, G. W., M. K. Dee, and W. P. Cahill. 1968. Water
      florae as indicator of irrigated water contamianted by DDT.
      Bull. Environ. Contam. Toxicol. 3:333–338.
201. Das, B., and P. K. Singh. 1977. Detoxification of the pes-
      ticide benzenehexachloride by blue-green algae. Microbios
      Lett. 4:99–102.
202. Lal, S., R. Lal, and D. M. Saxena. 1987. Bioconcentration
      and metabolism of DDT, fenitrothion and chlorpyriphos by the
      blue-green algae *Anabaena* sp. and *Aulosira fertilissima*. En-
      viron. Pollut. 46:187–196.
203. Kumar, S., R. Lal, and P. Bhatnagar. 1988. Uptake of
      dieldrin, dimethioate and permethrin by cyanobacteria,
      Anabaena sp. and *Aulosira fertilissima*. Environ. Pollut.
      54:55–61.
204. Dhanaraj, P. S., S. Kumar, and R. Lal. 1989. Bioconcen-
      tration and metabolism of aldrin and phorate by the blue-
      green algae *Anabaena* (ARM 310) and *Aulosira fertilissima*
      (ARM 68). Agric. Ecosyst. Environ. 25:187–193.
205. Wright, S. J. L., and A. Maule. 1982. Transformation of
      the herbicides propanil and chlorpropham by micro-algae.
      Pestic. Sci. 13:253–256.
206. Megharaj, M., K. Venkateswarlu, and A. S. Rao. 1987.
      Metabolism of monocrotophos and quinalphos by algae iso-
      lated from soil. Bull. Environ. Contam. Toxicol. 39:251–256.
207. Haselkorn, R. 1986. Organization of the genes for nitrogen
      fixation in photosynthetic bacteria and cyanobacteria. Annu.
      Rev. Microbiol. 40:525–547.
208. Golden, S., and L. A. Sherman. 1984. Optimal conditions
      for genetic transformation of the cyanobacterium *Anacystis
      nidulans* R$_2$. J. Bacteriol. 158:36–42.
209. Golden, S., and R. Haselkorn. 1985. Mutation to herbicide
      resistance maps within the *psbA* gene of *Anacystis nidulans*
      R$_2$. Science 229:1104–1107.
210. Buzby, J. S., R. D. Porter, and S. E. Stevans, Jr. 1985.
      Expression of the *Escherichia coli lacZ* gene on a plasmid
      vector in a cyanobacterium. Science 230:805–807.
211. Golden, S., J. Brusslan, and R. Haselkorn. 1986. Expres-
      sion of a family of *psbA* genes encoding a photosystem II

polypeptide in the cyanobacterium *Anacystis nidulans* R_2. EMBO J. 5:2789–2798.

212. Brusslan, J., S. Golden, and R. Haselkorn. 1987. Diuron resistance in the *psbA* multigen family of *Anacystis nidulans* R_2, p. 821–824. *In* J. Biggins and N. Nijhoff (eds.), Prog. Photosynth. Res. Proc. Int. Congr. Photosynth., 7th, 1986.

213. Robinson, H., S. Golden, J. Brusslan, and R. Haselkorn. 1987. Functioning of photosystem II in mutant strains of the cyanobacterium *Anacystis nidulans* R_2, p. 825–828. *In* J. Biggins and N. Nijhoff (eds.), Prog. Photosynth. Res. Proc. Int. Congr. Photosynth., 7th, 1986.

214. Buzby, J. S., R. O. Mumma, D. A. Bryant, J. Gingrich, R. H. Hawlton, R. D. Poster, C. A. Mullin, and S. E. Stevans, Jr. 1987. Genes with mutations causing herbicide resistance from the cyanobacterium *Snyechococcus* PCC 7002, p. 757–760. *In* J. Biggins and N. Nijhoff (eds.), Prog. Photosynth. Res. Proc. Int. Congr. Photosynth., 7th, 1986.

215. Singh, D. T., K. Nirmala, D. R. Modi, S. Katiyar, and H. N. Singh. 1987. Genetic transfer of herbicide resistance genes from *Gloeocapsa* sp. to *Nostoc muscorum*. Mol. Gen. Genet. 208:436–438.

216. Koksharova, O. A., and S. V. Shestakov. 1990. Mutants of the cyanobacterium *Synechococcus* 6803 resistant to the inhibitors of photosynthesis. Vestin. Mosk. Univ. Ser. XVI Biol. 1:42–47.

5

Inhibition of Nitrification in Soil by Allelochemicals Derived from Plants and Plant Residues

JOHN M. BREMNER *Iowa State University, Ames, Iowa*

GREGORY W. McCARTY *U.S. Department of Agriculture, Beltsville, Maryland*

I. INTRODUCTION

Recent literature on allelopathy reflects wide interest in hypotheses that plants and plant residues release allelochemicals that inhibit nitrification in soil [1–4]. The most cited hypothesis is that of Rice and Pancholy [5–7], who postulated that vegetation in climax ecosystems inhibits nitrification in soil by releasing phenolic compounds that retard oxidation of ammonium (NH_4^+) by nitrifying microorganisms. The most recent hypothesis is that of White [8,9], who proposed that vegetation in ponderosa pine ecosystems inhibits nitrification by releasing volatile terpenoids.

Although there is substantial support in the literature for hypotheses concerning inhibition of nitrification by allelochemicals, the basic concept that nitrification in various ecosystems is largely controlled by allelopathic interactions remains controversial, and several studies have cast serious doubt on the validity of this concept. The purpose of this chapter is to review evidence for and against current hypotheses concerning inhibition of nitrification by allelochemicals. Before discussing this evidence, it is necessary to discuss briefly the general concept of allelopathy and the methodology commonly used to study allelopathic interactions among plants and microorganisms.

II. ALLELOPATHY

A. Definition of Allelopathy

The word allelopathy was coined by Molisch [10] in 1937 to describe chemical interactions among plants and microorganisms. He intended that the term be used to cover stimulatory as well as inhibitory interactions, but some authors have used the term to describe only the harmful effects of one higher plant on another, presumably because allelopathy translates literally to "mutual suffering." Rice [11] adopted this more restricted definition in the first (1974) edition of his valuable monograph on allelopathy and defined allelopathy as "any direct or indirect harmful effect by one plant (including microorganisms) on another through production of chemical compounds that escape into the environment." But he modified this definition in the second (1984) edition of his monograph to include stimulatory as well as harmful effects [2]. He also emphasized that allelopathy involves the addition of a chemical compound to the environment and is thus separated from *competition*, which involves the removal or reduction of some factor from the environment that is required by some other plant sharing the habitat (e.g., water, nutrients, light). He further noted that confusion has arisen in the literature because some biologists consider allelopathy to be part of competition and that this confusion could be lessened by using the term *interference* to refer to the overall influence of one plant on another and to encompass both allelopathy and competition. Further use of the term *allelopathy* in this chapter will be as recommended by Rice [2] in the second edition of his monograph.

B. Methodological Problems in Studies of Allelopathy

Although there is an extensive literature on allelopathy [2], very few of the reports of allelopathic interactions in different ecosystems have contained satisfactory evidence that the interactions observed were truly allelopathic. Putnam and Tang [12] have commented on this problem in evaluating literature on allelopathic interactions and have suggested that the following sequence of studies be performed to obtain convincing proof of allelopathy:

1. Demonstrate interference using suitable controls, describe the symptomology, and quantitate the growth reduction.
2. Isolate, characterize, and assay the chemicals against species that were previously affected. Identification of chemicals that are not artifacts is a key step in proof of allelopathy.
3. Obtain toxicity with similar symptomology when chemical(s) are added back to the system.

4.   Monitor release of chemicals from the donor plant and detect them in the environment (soil, air, etc.) around the recipient and, ideally, in the recipient.

Few attempts have been made, however, to perform such a sequence of studies, and most reports of allelopathic interactions have not been accompanied by convincing proof of allelopathy. Following is a discussion of the methods most commonly used to study allelopathy and the defects of these methods.

Bioassays Used to Study Allelopathy

Bioassays are commonly used to obtain evidence for allelopathic interactions among plants [13]. These bioassays usually involve measurement of the effects of postulated allelochemicals in plant or soil extracts on early stages of plant development, such as seed germination or seedling growth, or on microbial processes, such as nitrification or nitrogen (N) fixation, in media treated with these extracts. They usually employ media such as germination paper, washed sand, agar, or liquid cultures [13], and they rarely involve use of the medium in which postulated allelopathic interactions must occur in terrestrial ecosystems, namely soil. The results of such bioassays must be treated with considerable caution because they cannot be safely extrapolated to the conditions found in natural ecosystems. For example, Stowe [14] applied nine commonly used bioassays to the seven most abundant plant species in an old-field in Illinois and found that the distribution patterns of species in this field were not significantly related to the results of these bioassays. He concluded that "the types of allelopathy which were tested by these bioassays were not demonstrably effective under field conditions, that perhaps any species can be shown to have allelopathic properties in bioassays, and that bioassays may, for many communities, have no ecological meaning."
   Bioassays are frequently performed using extracts of plant tissues or plant residues. Putnam and Tang [12] and Qasem and Hill [15] concluded that bioassays involving the use of plant extracts must be interpreted with caution because such extracts are likely to contain substances that are not released into the environment surrounding the plant and should not, therefore, be considered allelopathic. Moreover, the solute concentrations of such extracts are generally much higher than those of exudates or leachates from plant tissues or residues in natural ecosystems. The use of plant extracts with high solute concentrations can give misleading results because it is difficult, if not impossible, to account fully for possible effects of osmotic potential (water potential) that may be confused with allelopathic effects. It is noteworthy in this connection that the

hypothesis that the adverse effects of plant residues on crop yields are due to allelochemicals derived from these residues is based largely on studies showing that, when compared with distilled water, aqueous extracts of plant residues have an adverse effect on seed germination and seedling growth [16—21]. Because seed germination and seedling growth can be reduced by a delay in germination resulting from slow uptake of water by seeds, Krogmeier and Bremner [22] studied the possibility that the adverse effects of aqueous extracts of plant residues on seed germination and seedling growth might be at least partly due to water uptake by seeds being retarded by water-soluble constituents of these extracts. They found that the rates of water uptake and germination of seeds of corn (*Zea mays* L.), soybean (*Glycine max* L.), and wheat (*Triticum aestivum* L.) treated with aqueous extracts of residues of corn, sorghum (*Sorghum bicolor*), and wheat plants were appreciably slower than the corresponding rates for seeds treated with distilled water. They suggested that this might have resulted from the water potentials of these extracts because when seeds of corn, sorghum, and wheat were treated with a solution of polyethylene glycol 8000 having a water potential similar to that of the extracts of the plant residues tested, the rates of water uptake and seed germination were also slower than the corresponding rates for seeds treated with distilled water. These observations suggest that the adverse effects of aqueous extracts of plant residues on seed germination and seedling growth when compared with distilled water may be partly due to constituents of these extracts inducing water potential effects that reduce water uptake by germinating seeds.

To summarize, the bioassays that have been used to obtain evidence for allelopathic interactions among plants are open to a variety of criticisms, and the results of such assays must be interpreted with caution.

Identification and Quantification of Postulated Allelochemicals in Plants and Soils

The problems encountered in attempting to identify and quantify postulated allelochemicals in plants and soils can be illustrated by discussing the problems that have been encountered in studies to obtain support for the common assumption that phenolic compounds are important allelochemicals. Many studies of possible allelopathic effects of phenolic compounds have been reported, and there has been strong support for the hypothesis that phenolic compounds derived from plants or plant residues are allelopathic inhibitors of nitrification in soil and are responsible for the adverse effects of plant residues on crop growth [2]. These hypotheses have stimulated studies of the identity and concentration of the phenolic substances in soils and the release of phenolic compounds from plants and plant residues (see reviews by Rice [2] and Hartley and Whitehead [23]), but it is now recognized that the results of such studies depend greatly on the method used for extraction of phenolic compounds from soil [23]. The procedures used to extract phenolic compounds have often involved the use of strongly alkaline extractants, such as 2 M NaOH, but it is now generally accepted that such extractants solubilize phenolic substances that are physically or chemically bound on clay surfaces or organic matter and extract much larger amounts of phenolic substances than do neutral extractants such as water [24,25]. For example, Whitehead et al. [25] found that the amounts of phenolic acids extracted by 2 M NaOH from soils under permanent pasture were up to 2000 times greater than the amounts extracted by water. They also found that the amounts of phenolic acids extracted by water from a soil under permanent pasture were very small, being equivalent to concentrations in the soil solution ranging from 1.4 μM p-hydroxybenzoic acid to <10 nM ferulic acid. Kaminsky and Muller [24,26] cautioned against the use of alkaline reagents for extraction of phenolic compounds from soil because they found that use of such extractants can lead to production of phenolic artifacts and does not provide meaningful information concerning the nature and availability of the phenolic compounds in soil.

It has been commonly assumed that the phenolic compounds detected in soils are derived from vegetation and that the phenolic acid content of a soil is related to that of its vegetation. There is no basis for these assumptions, however, and there is evidence that they are invalid. For example, Whitehead et al. [25] were unable to detect any correlation between the phenolic contents of various plants and the corresponding contents of the soils in which the plants were grown, and they suggested that the phenolic compounds in these soils may have been produced by soil microorganisms. Whitehead et al. [25,27] also observed that large differences in the phenolic composition of different plants were often not reflected in the phenolic composition of soils under these plants, which suggests that the phenolic content of a soil may be determined largely by its microbial activity rather than by its vegetation. This seems a reasonable conclusion because it is well established that soil microorganisms can both produce and decompose phenolic compounds [23,28].

C. Proof of Allelopathy in Terrestrial Ecosystems

Proof of allelopathic interactions in various ecosystems ultimately requires demonstration that the postulated allelochemicals have an allelopathic effect under the conditions imposed by the ecosystem under study. For terrestrial ecosystems, the usual environment into which allelochemicals are released and must function is soil.

It is reasonable to expect, therefore, that the proof of allelopathic interactions in terrestrial ecosystems should include studies demonstrating the ability of postulated allelochemicals to exert their allelopathic effects in soils associated with these ecosystems, but few attempts have been made to perform such studies. For example, although there has been substantial support for hypotheses that phenolic compounds are important allelochemicals and are responsible for the adverse effects of plant residues on crop growth and for the slow rates of nitrification in some ecosystems (see Sections II.B and IV), the evidence cited to support these hypotheses has not included experiments showing that phenolic acids cause these allelopathic effects when added to soils. For adequate proof of allelopathic interactions, it is necessary to demonstrate that the postulated allelochemicals occur in soils associated with the ecosystems under study and that they exert allelopathic effects when they are added to these soils at concentrations at which they have been detected.

III. ALLELOPATHIC INHIBITION OF NITRIFICATION IN SOIL

Rice [2] outlined a theoretical basis for allelopathic inhibition of nitrification by late-succession or climax ecosystems. Briefly, he hypothesized that during the succession of vegetation in such ecosystems, plants that inhibit nitrification of ammonium in soil have a competitive advantage over other plants because oxidation of ammonium (NH_4^+) to nitrate (NO_3^-) by the nitrifying microorganisms in soils leads to conversion of a nonleachable form of nitrogen (NH_4^+) to a leachable form (NO_3^-), and plants cannot utilize NO_3^- without expending energy to reduce it to NH_4^+. Rice concluded that because inhibition of nitrification results in conservation of both energy and nitrogen, vegetation in late-succession or climax ecosystems contains plants that release allelochemicals that inhibit nitrification in soil.

A. Nitrification in Grassland Soils

Many workers have observed that the ratio of NH_4^+ to NO_3^- tends to be higher in grassland soils than in similar soils under cultivation [29—31], and various hypotheses have been proposed to explain this observation [32,33]. The most cited hypothesis is that grass roots release substances that inhibit nitrification in soil [30, 32]. A number of early reports [30,34—38] supported Theron's [30] hypothesis that nitrification in African grassland soils was retarded by unspecified substances released from the roots of grasses (notably *Hyparrhenia filipendula*), but the validity of this hypothesis has been questioned. For example, Harmsen and van Schreven [39] vigorously criticized the evidence cited by Theron [30] in support of the hypothesis, and Russell [40] doubted if root exudates could persist long enough in soils to inhibit nitrification. Robinson [41] concluded that nitrification in grassland soil was limited to the availability of NH_4^+ and found no evidence for inhibition of nitrification by toxins released by grass roots. Similarly, Purchase [42] found no convincing evidence for inhibition of nitrification by root secretions from *Hyparrhenia* sp. and concluded that nitrification in grassland soils was restricted by the availability of NH_4^+.

Support for the hypothesis that grasses inhibit nitrification in soil has been based largely on bioassays demonstrating that extracts of grass tissues can inhibit microbial growth and/or NH_4^+ oxidation by cultures of autotrophic nitrifying microorganisms grown on defined media [37,38,43–47]. As previously noted, however, the use of such bioassays is open to criticism (see Section II,B) because extracts of plant tissue contain substances that are not normally excreted by plants into soils and because bioassays indicating that plant extracts inhibit oxidation of NH_4^+ by nitrifying microorganisms in artificial media are not reliable tests of the ability of such extracts to inhibit nitrification in soil. More convincing evidence for the hypothesis that grasses may inhibit nitrification in soil was provided by Moore and Waid [48], who reported that root washings of ryegrass (*Secale cereale* L.) and wheat (*Triticum aestivum* L.) inhibited nitrification of NH_4^+ in columns of a clay loam soil. They found that the rate of NH_4^+ oxidation in these columns was not limited by the supply of NH_4^+ or other nutrients, and they discounted the possibility that immobilization or denitrification may have occurred in the columns. They concluded that nitrification in these columns was inhibited by the root washings, but they did not attempt to identify the substances causing this inhibition.

It has been commonly assumed in studies of the N cycle that nitrifying bacteria are poor competitors for NH_4^+ compared with plants and heterotrophic microoganisms and that microbial immobilization of NO_3^- is insignificant [49–51]. Davidson et al. [52] recently reported ^{15}N tracer studies of microbial production and consumption of NO_3^- in an annual grassland in California that provided data suggesting that these assumptions were not valid for this grassland system. They found that gross nitrification rates (calculated by dilution of the ^{15}N pool) ranged from 12 to 46% of gross mineralization rates during the growing season of the annual grasses studied and that pools of NH_4^+-N and NO_3^--N remained below 7 and 4 µg g^{-1} soil, respectively, but turned over about once a day. They also found that microbial assimilation of NO_3^- occurred at rates similar to previous estimates of plant uptake. They concluded from these observations that the two assumptions studied could not be

valid and that NO_3^- had an important role in the internal N cycle
of the grassland ecosystem investigated. This work casts doubt on
the validity of studies utilizing absolute or relative sizes of NH_4^+
and NO_3^- pools as indicators of the rate of nitrification in grassland ecosystems because it demonstrated that the net production of
NO_3^- is not related to the gross rate of nitrification in such ecosystems (see also Schimel et al. [53]). As previously noted, the
hypothesis that grasses inhibit nitrification in soil was based in
large part on observations that the ratio of NH_4^+ to NO_3^- tends
to be high in grassland soils [29–31].

To summarize, there is no satisfactory evidence to support the
hypothesis that grasses inhibit nitrification in soils by exuding
substances that retard oxidation of NH_4^+ by nitrifying microorganisms.

B. Nitrification in Forest Soils

Studies of the N status of forest ecosystems have shown that many
coniferous forest soils contain very little available N and that most
of this N is in the form of NH_4^+ with little or no NO_3^- present
[54]. Lodhi [55] hypothesized that different forest tree species
cause different degrees of allelopathic inhibition of nitrification by
soil microorganisms and that this inhibition has important ecological
implications because it maintains N as NH_4^+ in the ecosystem and
thereby conserves energy. Support for this hypothesis has been
provided by observations that clear-cutting of forests can lead to
proliferation of autotrophic nitrifiers [56] and to extensive loss of
N as NO_3^- [57]. As pointed out by Killham [58], however, stimulation of nitrification on clear-cutting cannot be assuredly related
to removal of allelopathic inhibition because clear-cutting leads to
many changes in soils (e.g., both the water potential and the temperature of a soil can change dramatically as a result of clear-cutting). Killham [58] reviewed the literature on factors that affect
nitrification in coniferous forest soils and concluded that if there
is allelopathic inhibition of nitrification in such soils, it is probably due to tannins and other polyphenols. He also concluded,
however, that "it will always remain doubtful as to what extent
allelochemicals may inhibit nitrification in coniferous forest systems (or any other soil system) because single allelopathic effects
cannot be factored out in real systems, and results of in vitro experiments where single potential allelopathic agents are tested cannot be extrapolated to field situations." He also argued that fungi
constitute by far the largest part of the nitrifier community in the
coniferous forest floor and that, because of the metabolic diversity
of nitrifying fungi, a variety of organic and inorganic nitrification
pathways may exist in soils of coniferous forests.

C. Nitrification in Agricultural Soils

Most of the research related to allelopathic inhibition of nitrification in soil has been performed with grassland and forest soils, and comparatively little attention has been given to agricultural systems since Russell's [59] observation that several crop plants appeared to inhibit nitrification. Moore and Waid [48] tested the effects of root exudates of five crops on nitrification in a clay loam soil and found that all of the exudates reduced the rate of nitrification in this soil. The effects of root exudates from rape (*Brassica napus* L.) and lettuce (*Lactuca sativa* L.) were temporary, but exudates from ryegrass (*Secale cereale* L.), wheat (*Triticum aestivum* L.), and onion (*Allium cepa* L.) had pronounced and persistent effects, with the effect of the ryegrass exudate being the most pronounced and persistent. These effects were attributed to inhibitors of nitrification in the exudates because they did not appear to be due to microbial immobilization of NH_4^+ or denitrification of NO_3^-. Other studies have also suggested that crop plants may have the ability to inhibit nitrification. For example, several cultivars of sorghum (*Sorghum bicolor*) and sunflower (*Helianthus annuus*) seem to have the ability to inhibit nitrification in soil [60, 61].

IV. EFFECTS OF PHENOLIC COMPOUNDS ON NITRIFICATION

Rice and Pancholy [5–7] hypothesized that climax vegetation inhibits nitrification in soils and that this is due to the production in such vegetation of phenolic compounds that inhibit oxidation of NH_4^+ by nitrifying microorganisms. This hypothesis has received substantial support [55,62–65], and it is cited extensively in the literature relating to allelopathy [1,2,4,11,66]. It is based on the following observations: (1) populations of *Nitrosomonas* and *Nitrobacter* tend to be lower, and the ratio of NH_4^+ to NO_3^- tends to be higher, in soil under climax vegetation than in soil under nonclimax vegetation [5,6]; (2) tannins and phenolic acids occur in climax vegetation and in soils under such vegetation, and very small amounts of these phenolic compounds inhibited NO_2^- production in an aqueous suspension of soil treated with $(NH_4)_2SO_4$ and a nutrient solution suitable for growth of *Nitrosomonas* [6,7].

McCarty and Bremner [67] evaluated the Rice-Pancholy hypothesis by studying the effects of different amounts of eight phenolic acids [p-hydroxybenzoic acid, p-coumaric acid, vanillic acid, ferulic acid, caffeic acid, ellagic acid, gallic acid, and chlorogenic acid (Figure 1)] and five tannins [from mangrove (*Rhizophoro mangle*),

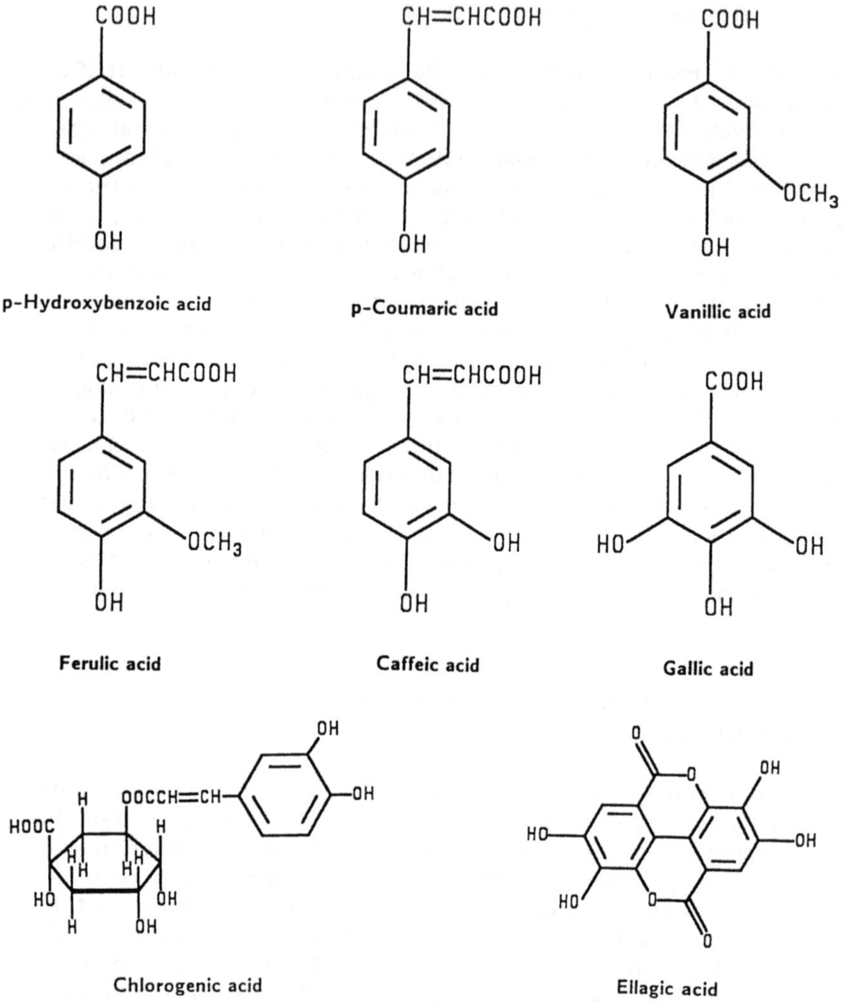

Figure 1   Structures of phenolic acid studied.

quebracho (*Quebrachia lorentzii*), mimosa (*Albizia julibrissin*), chest-
nut (*Castanea dentata*), and sumac (*Rhus coriaria*)] on nitrification
in soils treated with $(NH_4)_2SO_4$.   Table 1 shows the results they obtained in a comparison of the effects of different amounts of the
eight phenolic acids and two commercial nitrification inhibitors (nitrapyrin and etridiazole) on nitrification of NH_4^+ in soils incubated
at 30°C after treatment with $(NH_4)_2SO_4$.   The data reported show

that whereas the two commercial nitrification inhibitors markedly in-hibited nitrification when applied at concentrations as low as 1 or 5 μg g^{-1} soil, none of the phenolic acids had an appreciable effect on nitrification when applied at concentrations as high as 100 or 250 μg g^{-1} soil.

Rice and Pancholy [6,7] reported that very small amounts of caf-feic acid, ferulic acid, ellagic acid, gallic acid, and chlorogenic acid strongly inhibited nitrification of NH_4^+ by soil microorganisms added (as soil) to a nutrient solution suitable for growth of *Nitrosomonas*. They also reported that ferulic acid is a very potent inhibitor of nitrification and caused complete inhibition of nitrification when added at rates as low as 1.9 ng ml^{-1} to a nutrient solution inocu-lated with soil [7]. In contrast, McCarty and Bremner [67] found that ferulic acid had very little effect on nitrification of NH_4^+ in soil, even when it was added at concentrations as high as 500 μg g^{-1} soil (Table 1). To explain this divergence in results, it should be noted that although Rice and Pancholy [6,7] concluded that phe-nolic acids are potent inhibitors of nitrification in soil, they did not actually study the effects of such acids on nitrification of NH_4^+ in soil. Instead, they studied the effects of phenolic acids on the production of nitrite (NO_2^-) in an NH_4^+-containing medium inocu-lated with soil (0.02 g soil ml^{-1} medium). Their work did not per-mit any reliable conclusion concerning the effects of phenolic acids on nitrification of NH_4^+ because neither disappearance of NH_4^+ nor production of NO_3^- was measured. Production of NO_2^- is clearly not a reliable indicator of nitrification in a nutrient medium inocu-lated with soil because oxidation of NO_2^- to NO_3^- by soil micro-organisms is usually more rapid than oxidation of NH_4^+ to NO_2^-.

It is well established that phenolic compounds occur in many plant materials and are released into soil from decaying plant res-idues [2,23,68−72]. Many of these compounds are rapidly decom-posed by soil microorganisms [23,73−76], but some appear to be stabilized against microbiological degradation through sorption on organic matter, clay minerals, and hydroxy-Al or Fe compounds in soil [2,23,25,75−77]. Several phenolic compounds have been de-tected in soils [6,7,23,78−80], but the amounts detected have been much smaller than the amounts added to soils in the work reported in Table 1 (up to 250 or 500 μg g^{-1} soil). For example, Whitehead [78] found that the amounts of ferulic, p-coumaric, vanillic, and p-hydroxybenzoic acid in four soils ranged from 0.07 to 3.1 μg g^{-1} soil, and Guenzi and McCalla [79] found that the amounts of these acids in a tilled soil ranged from 1.2 to 14.4 μg g^{-1} soil. More-over, Wang et al. [80] found that surface (0 to 25 cm) samples of eight soils contained only 0 to 0.49 μg g^{-1} soil of these phenolic acids, and Rice and Pancholy [6] were unable to detect gallic or ellagic acid in surface (0 to 15 cm) samples of soil under a climax stand or to detect more than 3 μg g^{-1} soil of these compounds in

Table 1 Comparison of Effects of Phenolic Acids and Commercial
Nitrification Inhibitors on Nitrification of NH_4^+ in Soils[a]

Compound	Amount added ($\mu g\ g^{-1}$ soil)	% Inhibition of nitrification by compound (4 soils)[b]	
		Range	Average
Phenolic acids			
p-Hydroxybenzoic acid	10	0	0
	100	0–2	1
	250	1–7	3
p-Coumaric acid	10	0–1	1
	100	0–2	1
	250	0–5	3
Vanillic acid	10	0	0
	100	0–2	1
	250	0–4	2
Ferulic acid	10	0	0
	100	0–7	2
	250	0–12	5
	500	5–28	12
Caffeic acid	10	0	0
	100	0	0
	250	0–10	3
Gallic acid	10	0	0
	100	0–3	1
	250	0–5	2
Chlorogenic acid	10	0	0
	100	0–2	1
	250	0–4	2
Ellagic acid	10	0	0
	100	0–3	1
	250	1–16	5
Commercial nitrification inhibitors			
Nitrapyrin	0.1	0–7	4
	1.0	0–98	42
	5.0	47–99	81
	10.0	55–99	84

(continued)

Table 1   (Continued)

Compound	Amount added (μg g^{-1} soil)	% Inhibition of nitrification by compound (4 soils)[b]	
		Range	Average
Etridiazole	0.1	0—25	12
	1.0	29—99	68
	5.0	80—99	90
	10.0	81—99	92

[a]10-g samples of soil were incubated (30°C; 4 ml water) for 14 d after treatment with 2 mg N as $(NH_4)_2SO_4$ and the amount of compound specified.
[b]Soils used were Harps (Typic Calciaquoll), Canisteo (Typic Haplaquoll), Clarion (Typic Hapludoll), and Storden (Typic Udorthent).
Adapted from Ref. 64.

subsurface (15 to 30 cm) samples of this soil.   Therefore, based on the data in Table 1, the amounts of phenolic acids detected in soils are too small to have a significant inhibitory effect on nitrification.

Rice and Pancholy [6] reported that the amounts of tannins in soil profiles under three climax stands ranged from 8 to 93 μg g^{-1} soil and that amounts as low as 2 μg ml^{-1} completely inhibited oxidation of NH_4^+ to NO_2^- in a nutrient solution suitable for growth of *Nitrosomonas* that had been inoculated with soil.   They concluded from these observations that tannins inhibited nitrification in soils under climax vegetation.   But McCarty and Bremner [67] found that tannins had very little, if any, inhibitory effect on nitrification in four soils treated with $(NH_4)_2SO_4$, even when they were applied at concentrations as high as 500 μg g^{-1} soil (Table 2). They concluded that the amounts of tannins present in the soils under the climax stands studied by Rice and Pancholy [6] were too small to have a significant inhibitory effect on nitrification in these soils.

Studies by Basaraba [81] of the effects of tannins on nitrification in soil are frequently cited in support of the hypothesis that climax vegetation produces phenolic compounds that inhibit nitrification in soil.   It is important to note, therefore, that these studies showed that tannins did not inhibit nitrification when added at a concentration of 2500 μg g^{-1} soil and that they retarded nitrification

Table 2   Effects of Tannins on Nitrification of NH_4^+ in Soils[b]

Source of tannin	Amount added ($\mu g\ g^{-1}$ soil)	% Inhibition of nitrification by tannin (4 soils)[b]	
		Range	Average
Mangrove	100	0	0
	250	0–2	1
	500	0–8	4
Quebracho	100	0	0
	250	0	0
	500	0–3	2
Mimosa	100	0	0
	250	0–1	0
	500	0–3	1
Chestnut	100	0	0
	250	0–2	1
	500	0–7	3
Sumac	100	0	0
	250	0–3	1
	500	0–7	3

[a]10-g samples of soil were incubated (30°C; 4 ml water) for 14 d after treatment with 2 mg N as $(NH_4)_2SO_4$ and 0, 1.0, 2.5, or 5.0 mg of the tannin specified.
[b]Soil used were Harps (Typic Calciaquoll), Canisteo (Typic Haplaquoll), Clarion (Typic Hapludoll), and Storden (Typic Udorthent).
Adapted from Ref. 64.

for only a few days when they were applied at concentrations as high as 5000 or 10,000 $\mu g\ g^{-1}$ soil. Moreover, Bohlool et al. [82] found that the nitrification potential of intertidal sediments was not significantly reduced when the sediments were highly enriched with tannins and other phenolic compounds released from a chipmill, and they demonstrated that nitrification of NH_4^+ by *Nitrosomonas* in a nutrient solution was not affected by the addition of 5 mg ml^{-1} of a pine bark tannin.

Further evidence that phenolic acids derived from plants or plant residues do not inhibit nitrification in soils has been provided by a recent study [83] of the effects of three phenolic acids (ferulic acid, caffeic acid, and p-coumaric acid) on NO_2^- production by three genera of terrestrial autotrophic nitrifying bacteria (*Nitrosospira, Nitrosomonas, Nitrosolobus*) grown on a defined medium containing NH_4^+. This study showed that production of NO_2^- by *Nitrosospira* was not inhibited by ferulic acid, caffeic acid, or p-coumaric acid at concentrations of 10^{-6} or 10^{-5} M and was only slightly inhibited when these acids were at a concentration of 10^{-4} M (Table 3). It also showed that ferulic acid did not markedly inhibit NO_2^- production by the three genera studied even when its concentrations were as high as 10^{-3} M (Table 4). These observations invalidate the hypothesis tested because the phenolic acids studied did not significantly retard oxidation of NH_4^+ by autotrophic bacteria even when their concentration in cultures of these bacteria greatly exceeded their concentration in soils.

The phenolic hypothesis of Rice and Pancholy was based to a significant extent on observations that populations of *Nitrosomonas* and *Nitrobacter* tend to be lower, and the ratio of NH_4^+ to NO_3^- tends to be higher, in soil under climax vegetation than in soil under nonclimax vegetation. Robertson and Vitousek [84] criticized the use of such observations to support the phenolic hypothesis. Their arguments were that the size of the nitrifier population is

Table 3 Effects of Different Concentrations of Phenolic Acids on Nitrification of NH_4^+ by *Nitrosospira* sp. Strain AV

Phenolic acid	Concentration (M)	Nitrite produced (nmol/ml/h)	Percent inhibition of nitrification
None (control)	—	26.9	—
Ferulic acid	10^{-6}	27.2	0
	10^{-5}	26.8	<1
	10^{-4}	26.3	2
Caffeic acid	10^{-6}	27.1	0
	10^{-5}	27.3	0
	10^{-4}	24.5	8
p-Coumaric acid	10^{-6}	27.3	0
	10^{-5}	27.9	0
	10^{-4}	22.6	15
	LSD(0.05)	2.4	

From Ref. 80.

Table 4   Effects of Different Concentrations of Ferulic Acid
on Nitrification of NH_4^+ by *Nitrosospira* sp. Strain
AV (A), *Nitrosomonas europeae* (B), and
Nitrosolobus sp. Strain 24C (C)

Concentration of ferulic acid (M)	Nitrite produced (nmol/ml/h)			Percent inhibition of nitrification		
	A	B	C	A	B	C
0 (control)	26.9	39.3	12.1	—	—	—
10^{-5}	26.8	39.6	12.6	<1	0	0
10^{-4}	26.3	38.8	11.4	2	1	6
10^{-3}	25.7	37.8	10.3	5	4	14
LSD(0.05)	1.1	2.1	0.7			

From Ref. 80.

not a good indicator of a soil's ability to nitrify NH_4^+ and that the
size of the inorganic N pool is not a good indicator of a soil's nitrifying potential.   Moreover, as noted by Schmidt [85], the methods used by Rice and Pancholy to count nitrifying bacteria in climax and nonclimax soils are open to criticism and could give misleading results (see Section VI,D).
    Several workers have questioned the theoretical basis used by
Rice and Pancholy [5] for their hypothesis concerning inhibition of
nitrification by late-succession vegetation (see Section III).   For
example, Reiners [86] questioned if there is sufficient ecological
pressure in successional systems for the development of late-succession vegetation that inhibits nitrification in soil, and he suggested that the pattern of nitrification during succession may be
more closely related to the pattern of NH_4^+ availability in soil than
to the ability of vegetation to inhibit nitrification by soil microorganisms.   Robertson and Vitousek [84] found no evidence for progressive allelochemical inhibition of nitrification in successional ecosystems, and they concluded that the pattern of nitrification in
these ecosystems is primarily controlled by the rate of N mineralization.
    Although some workers [87,88] have observed little, or no, ni-
trification in apparently mature, undisturbed ecosystems, as predicted by the Rice-Pancholy hypothesis, other workers [88–90]
have found that nitrification proceeds actively in some old-growth

forests. Moreover, there is good evidence from field and laboratory studies [91,92] and from studies of NO_3^- levels in stream-
waters [93,94] that nitrification increases throughout succession.
There is conflicting evidence, therefore, regarding the effects of
late-succession and climax vegetation on nitrification. As noted by
Runge [95], the generalization that inhibition of nitrification is
characteristic of climax ecosystems "does not seem permissible in
consideration of the almost exclusive NO_3^- production in the soils
of numerous forest ecosystems."

To summarize, although there has been wide interest in the
Rice-Pancholy hypothesis that climax vegetation inhibits nitrification in soil by producing phenolic compounds that inhibit oxidation
of NH_4^+ by nitrifying microorganisms, the evidence for this hypothesis has serious defects, and recent work leaves very little
doubt that the hypothesis is invalid.

V. EFFECTS OF TERPENOIDS ON NITRIFICATION

White [8,9] hypothesized that the vegetation in ponderosa pine
(*Pinus ponderosa* Dougl.) ecosystems inhibits nitrification in these
ecosystems by releasing volatile terpenoids that retard oxidation of
NH_4^+ by nitrifying microorganisms. This hypothesis was based
largely on two observations: that exposure of a ponderosa pine
ecosystem to fire decreased the amount of volatile organic compounds and stimulated production of NH_4^+ and NO_3^- in the forest
floor, and that exposure of soil samples to vapors from a terpenoid mixture decreased the production of NH_4^+ and NO_3^- by mineralization of organic N in these samples. Bremner and McCarty
[96] evaluated this hypothesis by studying the effects of different amounts of six terpenoids (α-terpinene, limonene, myrcene,
α-pinene, β-pinene, and α-phellandrene) on nitrification in soils
treated with $(NH_4)_2SO_4$. They found that the terpenoids studied
(Figure 2) did not significantly affect nitrification in soils, even
when the amounts applied greatly exceeded those reported to occur in soils (Table 5). White [9] detected α-pinene, β-pinene,
and limonene in soils under ponderosa pine in amounts ranging
from 0.0135 to 0.0454 $\mu g\ g^{-1}$ soil. Table 5 shows that much
larger amounts of these terpenoids had no inhibitory effect
on nitrification in soil.

Bremner and McCarty [96] studied the possibility that the apparent inhibition of nitrification observed by White when soils were
exposed to terpenoid vapors was the result of immobilization of
NH_4^+-N by microbial activity stimulated by the organic carbon (C)
in these vapors. They compared the effects of different amounts
of organic C as terpenoids and glucose on the amounts of NH_4^+
and NO_3^- detected after incubation of soils treated with $(NH_4)_2SO_4$.

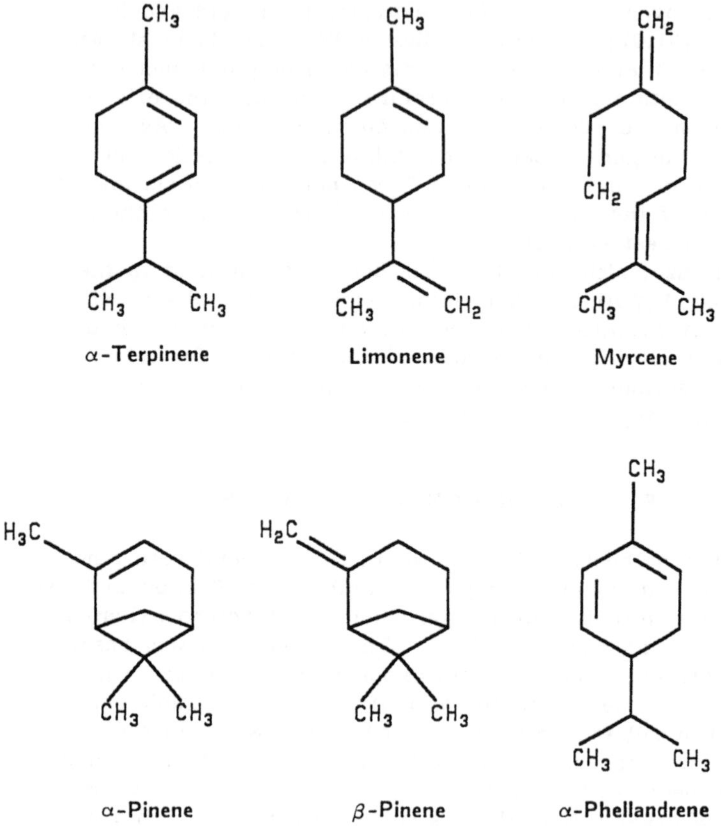

Figure 2   Structures of terpenoids studied.   Adapted from Ref. 93.

They found that terpenoid-C was as effective as glucose-C for inducing immobilization of NH_4^+-N and that the recovery of added
NH_4^+ as $(NH_4^+ + NO_3^-)$-N decreased with increase in the amount
of organic C added as terpenoids or glucose (Table 6).   They also
found that although addition of organic C in the form of terpenoids
decreased the recovery of NH_4^+-N as $(NH_4^+ + NO_3^-)$-N, most of
the NH_4^+-N recovered as $(NH_4^+ + NO_3^-)$-N was in the form of
NO_3^--N, which supported their conclusion that terpenoids have no
direct effect on nitrification.   Further evidence for this conclusion
was provided by experiments showing that almost all of the NH_4^+-N
added to soils as $(NH_4)_2SO_4$ was immobilized when the soils were
exposed to terpenoid vapors (Table 7).   Visual evidence that the

Table 5 Comparison of Effects of Terpenoids and Commercial Nitrification Inhibitors on Nitrification of NH_4^+ in Soils[a]

Compound	Amount added ($\mu g\ g^{-1}$ soil)	% Inhibition of nitrification by compound (3 soils)[b] Range	Average
Terpenoids			
α-Terpinene	10	0−3	1
	100	0−3	1
	250	0−10	4
Limonene	10	0−2	1
	100	0−3	1
	250	0−4	1
Myrcene	10	0−6	2
	100	0−8	3
	250	0−10	3
α-Pinene	10	0−3	1
	100	0−7	2
	250	0−11	5
β-Pinene	10	0−3	1
	100	0−5	2
	250	0−9	5
α-Phellandrene	10	0−2	1
	100	0−8	4
	250	2−13	7
Commercial nitrification inhibitors			
Nitrapyrin	1	45−59	52
	5	55−93	72
	10	92−99	97
Etridiazole	1	51−62	56
	5	71−96	86
	10	94−99	97

[a]10-g samples of soil were incubated (25°C; 4 ml water) for 14 d after treatment with 2 mg N as $(NH_4)_2SO_4$ and the amount of compound specified.
[b]Soils used were Harps (Typic Calciaquoll), Webster (Typic Haplaquoll), and Storden (Typic Udorthent).
Adapted from Ref. 93.

Table 6   Effects of Different Amounts of Organic C as Terpenoids and Glucose on the Amounts of NH_4^+-N and NO_3^--N Found after Incubation of a Harps Soil Treated with $(NH_4)_2SO_4$[a]

Compound added	Amount of organic C added ($\mu g\ g^{-1}$ soil)	N found after 14 d ($\mu g\ g^{-1}$ soil)		
		NH_4^+-N	NO_3^--N	Total[b]
None (control)	—	2	206	208
α-Terpinene	500	1	201	202
	1000	1	172	173
	2500	2	155	157
	5000	2	33	35
Limonene	500	1	196	197
	1000	1	177	178
	2500	2	159	161
	5000	2	110	112
Myrcene	500	1	202	203
	1000	1	184	185
	2500	2	159	161
	5000	2	120	122
α-Pinene	500	1	200	201
	1000	1	159	160
	2500	2	139	141
	5000	2	93	95
β-Pinene	500	1	201	202
	1000	2	188	190
	2500	4	169	173
	5000	5	85	90
α-Phellandrene	500	1	200	201
	1000	0	193	193
	2500	2	154	156
	5000	2	94	96
Glucose	500	0	199	199
	1000	0	170	170
	2500	0	148	148
	5000	0	102	102

[a]10-g samples of Harps soil (Typic Calciaquoll) were incubated (25°C; 4 ml water) for 14 d after treatment with 2 mg N as $(NH_4)_2SO_4$ and the amount of compound specified.
[b]$(NH_4^+ + NO_3^-)$-N. No NO_2^--N could be detected.
Adapted from Ref. 93.

Table 7 Effects of Terpenoid Vapors on the Amounts of NH_4^+-N and NO_3^--N Found after Incubation of a Harps Soil Treated with $(NH_4)_2SO_4$[a]

Terpenoid	N found after 14 d ($\mu g\ g^{-1}$ soil)		
	NH_4^+-N	NO_3^--N	Total[b]
None (control)	1	207	208
α-Terpinene	1	2	3
Limonene	1	0	1
Myrcene	2	0	2
α-Pinene	1	0	1
β-Pinene	1	0	1
α-Phellandrene	1	0	1
Mixture[c]	1	0	1

[a]Vials containing 0 or 1 ml of the terpenoid specified were placed in 250-ml bottles containing 10-g samples of Harps soil (Typic Calciaquoll) treated with 4 ml water containing 2 mg N as $(NH_4)_2SO_4$, and the bottles were stoppered and placed in an incubator at 25°C for 14 d.
[b]$(NO_3^- + NH_4^+)$-N. No NO_2^--N could be detected.
[c]Terpenoid mixture containing α-terpinene, limonene, myrcene, α-pinene, β-pinene, and α-phellandrene (0.17 ml of each compound).
Adapted from Ref. 93.

terpenoid vapors acted as sources of organic C readily utilized by microorganisms was provided by the observation of extensive fungal growth in soil samples exposed to these vapors.

To summarize, the studies reported by Bremner and McCarty [96] showed that terpenoids did not significantly affect nitrification in soil, even when the amounts applied greatly exceeded the amounts reported to occur in soils. These studies also left little doubt that the apparent inhibition of nitrification observed by White [9] when soils were exposed to terpenoid vapors was the result of immobilization of NH_4^+-N by microbial activity stimulated by the organic C in these vapors. In responding to the results of these studies, White [97] did not question the conclusion that the apparent inhibition of nitrification he observed with terpenoid vapors was caused by immobilization of NH_4^+-N, but he suggested that terpenoids may inhibit nitrification in some ecosystems that have low levels of NH_4^+-N and low populations of nitrifying bacteria.   In responding to White's comments, Bremner and McCarty [98] pointed out that the conditions in such ecosystems would promote immobilization of NH_4^+-N rather than inhibition of nitrification by terpenoids, even if terpenoids are suicide inhibitors of the ammonia monooxygenase (AMO) enzyme responsible for oxidation of NH_4^+ as postulated by White [97,99].   They also pointed out that any hypothesis concerning the mode of inhibition of nitrification by terpenoids is pure speculation in the absence of any evidence that terpenoids inhibit oxidation of NH_4^+ by nitrifying microorganisms. A recent study of the effects of three terpenoids on oxidation of NH_4^+ by three genera of nitrifying microorganisms (*Nitrosospira, Nitrosomonas, Nitrosolobus*) showed that the terpenoids tested (limonene, myrcene, α-phellandrene) did not significantly inhibit oxidation of NH_4^+ by pure cultures of these microorganisms, even when their concentrations in the cultures exceeded their reported concentrations in soils (G. W. McCarty, J. M. Bremner, and E. L. Schmidt, unpublished data).

VI.   ALTERNATIVE EXPLANATIONS FOR SLOW RATES OF NITRIFICATION IN SOME ECOSYSTEMS

A.   Soil pH and Fertility

Several workers [91,100–102] have suggested that the slow rate of nitrification in soils of low pH and/or low fertility is due not to allelopathic inhibition of nitrification but to the effects of low soil pH and/or low soil fertility on the activity of nitrifying microorganisms.

It has been commonly observed that the potential for nitrification in soils decreases with decrease in soil pH [85,103–106], and most observations indicate that the lower limit of pH for nitrification in soil is 4.0 [85].   Brar and Giddens [100] suggested that the effect of acidity on nitrification may be an expression of Al toxicity.   An alternative explanation of the effect of acidity was provided by work by Suzuki et al. [107] indicating that NH_3 rather than NH_4^+ is the substrate for the ammonia monooxygenase enzyme responsible for oxidation of NH_4^+ by chemoautotrophic microorganisms.   This suggests that the limited ability of autotrophic nitrifiers to oxidize NH_4^+ in acid soils may be largely due to substrate limitation because the ratio of NH_3 to NH_4^+ is very small in acid environments.

There is considerable evidence that nitrification in acid forest soils is often limited by low soil pH.   For example, several workers

have reported that nitrification in such soils can be greatly stimulated by amendment with $CaCO_3$ (or Na_2CO_3) [54,91,108–114], and
others have reported that the addition of urea to such soils increased both the pH and the rate of NH_4^+ oxidation [58,109,113,
115–119]. Studies by Martikainen [118] indicated that the increased rate of nitrification that resulted from addition of urea to
acidic soils was due to increased soil pH and not to increased NH_4^+
concentration via urea hydrolysis because addition of NH_4NO_3 did
not increase the rate of nitrification.

Although it has been demonstrated that low soil pH can limit the
rate of NH_4^+ oxidation in many soils, there are reports that nitrification can occur in some acid soils at pH levels far below the reported acid tolerance of chemoautotrophic nitrifiers [58,120,121].
Evidence that chemoautotrophic nitrifiers are responsible for NH_4^+
oxidation in these soils has remained inconclusive. Very few NH_4^+-
oxidizing autotrophic microorganisms have been isolated from acid
soils [122], but they have included representatives of the four major genera of terrestrial autotrophic NH_4^+-oxidizing microorganisms
(*Nitrosospira, Nitrosolobus, Nitrosomonas, Nitrosovibrio*) [49,106,
123–126]. Of these genera, *Nitrosospira* has been the most frequent isolate from acid soils [106,123,124,126]. Hankinson and
Schmidt [106] found that the strain of *Nitrosospira* they isolated
from an acid forest soil (pH 3.9 to 4.4) did not appear to be acid
tolerant because it could not be cultured on media with pH < 6.2,
and they concluded that autotrophic NH_4^+ oxidizers may be restricted to circumneutral microsites in acid soil. Similarly, Allison
and Prosser [125] reported that the strains of *Nitrosospira* and *Nitrosomonas* they isolated from acid soils were not acid tolerant.
They found, however, that most of the strains of NH_4^+-oxidizing
microorganisms isolated from acid soils were capable of hydrolyzing
urea and that strains of *Nitrosospira* could grow with urea as the
sole N source at pH 5.5. They concluded that although there was
no evidence for acidophilic strains of autotrophic NH_4^+ oxidizers,
urea hydrolysis may increase soil pH and thereby permit autotrophic nitrification in acid soils.

de Boer and co-workers [127,128] reported evidence that urea-
stimulated nitrification in acid soils may be due to increased activity of urease-positive autotrophic nitrifiers in these soils. They
also reported evidence that NO_3^- production in an acid heath soil
(pH 3.6) was due to autotrophic nitrification and was not limited by
NH_4^+ but was related to N mineralization [129]. They suggested
that the liberation of NH_3 by N mineralization in acid soils may increase soil pH to levels that are favorable for autotrophic nitrification.

Several workers have suggested that heterotrophic microorganisms contribute to NO_3^- production in soils, but the significance
of heterotrophic nitrification remains controversial. Focht and

Verstraete [130] suggested that heterotrophic nitrification may be responsible for formation of NO_3^- in strongly acidic soils, but Schmidt [85] reviewed the literature on this topic and concluded that although heterotrophic bacteria and fungi isolated from soils can be cultured under conditions that lead to production of NO_2^- and NO_3^-, there is no evidence that any of these heterotrophs can nitrify in soils under natural conditions. Recent work, however, has suggested that NO_3^- production in some acid soils may be largely due to heterotrophic nitrifiers. For example, indirect evidence for heterotrophic nitrification has been provided by studies showing that NO_3^- production in some acid soils was not inhibited by selective inhibitors of autotrophic nitrification such as acetylene and nitrapyrin [58,131–133]. Furthermore, Shimel et al. [131] obtained evidence that the apparent heterotrophic nitrification detected in a coniferous forest soil by such studies was predominantly by fungi because NO_3^- production was retarded by a fungicide (cycloheximide), but was not significantly reduced by a bacteriocide (streptomycin). Additional evidence for heterotrophic nitrification by fungi in acid environments has been provided by reports that fungal isolates from acid forest soils can produce NO_2^- or NO_3^- under acid conditions [134,135].

Although it has been suggested that nitrification in soils of low fertility may be restricted by nutrient limitations, there is little evidence to support this suggestion. Purchase [136] and Hue and Adams [137] reported evidence that nitrification in soils with a low content of phosphorus was limited by phosphorus deficiency, and Cole and Heil [138] hypothesized that both N mineralization and nitrification in many terrestrial ecosystems are controlled by the availability of phosphorus. However, Sahrawat et al. [102] found that the addition of phosphorus had little, if any, effect on nitrification in forest soils.

B. Soil Temperature and Water Potential

It is well established that soil temperature has an important effect on the rate of nitrification in soil. Mahendrappa et al. [139] found that soils from the northern regions of the United States had optimum temperatures for nitrification between 20 and 25°C, whereas soils from the southern regions nitrified most rapidly at temperatures around 35°C. Malhi and McGill [140] reported that the optimum temperature for nitrification in three Alberta soils was 20°C and that nitrifying activity in these soils almost ceased at 30°C. In contrast, Myers [141] found that the optimum temperature for nitrification in a tropical soil was about 35°C. Malhi and McGill [140] suggested that the relatively low temperature optimum for soils in the cool climate of Alberta was evidence for climatic selection of nitrifiers having a lower temperature optimum than the nitrifiers in soils in warmer climates. It is evident that significant nitrification by soil microorganisms can occur at extremes of temperature for biological activity because there are reports that some soils can nitrify at temperatures as low as 0°C [113,142] and as high as 45 to 60°C [141,143,144]. It seems likely, however, that the extremes in soil temperature in various climatic regions can at least periodically limit the rate of nitrification by soil microorganisms in these regions.

The water potential of a soil is an important factor with respect to nitrification by autotrophic microorganisms because it affects the metabolic activity of these microorganisms. As noted by Killham [58], the autotrophic nitrifiers in soil have relatively limited ability to adjust to conditions of low water potential and are among the soil microorganisms that are most sensitive to water stress. It is evident, however, that nitrification can occur in soils over a considerable range of water potentials because it has been observed that although the rate of nitrification in soil decreases with a decrease in the soil water potential from -33 to -1500 kPa, appreciable nitrification can occur in soils at the "permanent wilting point" (-1500 kPa) [140,145].

In addition to increasing the osmotic stress on nitrifying microorganisms, a decrease in soil water content may retard nitrification by decreasing the rate of diffusion of NH_4^+ toward the nitrifying microorganisms. Davidson et al. [52] calculated that a decrease in the volumetric water content of a grassland soil from 0.507 to 0.044 cm^3 cm^{-3} resulted in a decrease in the rate of diffusion of NH_4^+ from 14.56 to 0.12 mg m^{-2} s^{-1}. They concluded that by inducing such decreases in the rate of NH_4^+ diffusion, drying of a soil could result in a decrease in the population of nitrifying microorganisms.

C. Availability of Ammonium

Many workers have suggested that the rate of nitrification in soil under late-succession vegetation is limited not by the production by such vegetation of compounds that inhibit nitrification by microorganisms but by the rate of formation of NH_3 through mineralization of organic N by ammonifying microorganisms [41,42,84,90, 92,112,146–148]. For example, Robinson [41] and Purchase [42] could find no evidence that nitrification in grassland soils was inhibited by allelochemicals released from grass roots, and they suggested that nitrification in these soils was limited by the availability of NH_4^+. Support for this suggestion has been provided by studies showing that nitrification in forest soils was stimulated by the addition of NH_4^+ [149] and inhibited by the addition of organic amendments that promote immobilization of NH_4^+-N by microbial activity [92,112]. It has also been suggested that nitrification in some soils may be limited by the inability of nitrifying microorganisms to compete effectively with heterotrophic microorganisms for the NH_4^+ produced by mineralization of organic N [146,150].

D. Population of Nitrifying Microorganisms

O'Connor et al. [151] and Robinson [41] suggested that the slow rate of nitrification in grassland soils was linked to a low population of nitrifying bacteria caused by the scarcity of NH_4^+ ions in these soils and by the direct competition of grass roots for these ions. But Chase et al. [152] observed comparable populations of nitrifiers in urea-treated grass and fallow plots in Canada and concluded that grass roots do not compete for NH_4^+ to the detriment of nitrifiers and do not inhibit nitrifiers by other means.

Rice and Pancholy [5,6] enumerated nitrifiers by the most probable number (MPN) method in soil samples collected every 2 months from old-field and climax soils under three types of vegetation. They found that the counts of nitrifiers were about 10^2 to 10^3 g^{-1} of old-field soil and 10^1 to 10^2 g^{-1} of climax soil. Although these observations were consistent with the climax inhibition hypothesis, Schmidt [85] pointed out that estimates of nitrifier populations by the MPN method suffer from inherently high statistical uncertainty, that nitrifier populations of old-field soils may differ qualitatively from those of climax soils, and that the selectivity of a single MPN medium may result in failure to detect a predominant nitrifier in the climax soil [153].

Vitousek and Melillo [154] suggested that nitrification could be slow in an undisturbed forest as a result of competition between root-mycorrhizal complexes and nitrifiers for NH_4^+ and that after disturbance there could be a substantial lag in nitrification while the populations of nitrifying microorganisms become established. Matson and Vitousek [155] reported work that was consistent with this hypothesis because it indicated that the higher rates of nitrification in soils in clear-cut areas were the result of increased populations of nitrifying microorganisms in such soils. Other studies by Vitousek and Matson [114] indicated that competition for NH_4^+ resulted in low initial populations of nitrifying microorganisms and was the probable cause of delayed NO_3^- production in a soil collected from an Indiana pine forest.

VII. CONCLUSIONS

Although there is wide interest in hypotheses that plants and plant residues release allelochemicals that inhibit nitrification by soil microorganisms, there is little evidence to support these hypotheses, and alternative explanations are available to explain the slow rate of nitrification in some soils. The only hypotheses that attribute alleged allelopathic inhibition of nitrification in soil to specific chemicals are the phenolic and terpenoid hypotheses discussed in Sections IV and V, and it is clear from work cited in these sections that these hypotheses are invalid. It is also evident that the importance of phenolic acids as allelochemicals has been greatly overrated because, besides demonstrating that phenolic acids cannot be responsible for allelopathic inhibition of nitrification in soils, recent work has shown that there is no valid basis for the hypothesis that the adverse effect of plant residues on crop yield is due to phytotoxic compounds derived from these residues and that these phytotoxic compounds are phenolic acids such as p-coumaric acid, ferulic acid, and p-hydroxybenzoic acid [22,67]. This hypothesis has been supported by studies showing that phenolic acids known to occur in soil delay or inhibit seed germination and reduce seedling growth on germination paper or agar [156–163] or reduce growth of plants in nutrient solutions [74,80,160–162,164,165]. But recent work has shown that although phenolic acids postulated to be important allelochemicals affected germination and seedling growth on germination paper, they had no effect on seed germination, seedling growth, or early plant growth in soils, even when the amounts applied were much greater than the amounts detected in soils [166]. This is another illustration of the major defect of most studies of allelopathy, namely the lack of tests to demonstrate that the postulated allelochemicals have allelopathic effects when they are added to soils in amounts similar to those detected in soils under natural conditions. As noted in recent articles concerning allelopathy, "it is all too easy to invoke allelopathy as a theory for a lack of net nitrification without detailed consideration of the many other soil physicochemical and biological factors which must be taken into consideration" [58] and "rigorous proof of the validity of many widely recognized 'allelochemicals' is still lacking" [167]. We suggest that further reports of allelopathic inhibition of nitrification in soil be treated with considerable reserve if they are not accompanied by convincing evidence that they merit serious attention, the minimal evidence being that the postulated allelochemicals inhibit nitrification in soils at the concentrations at which they have been detected in soils.

REFERENCES

1. Putnam, A. R. 1983. Allelopathic chemicals. Chem. Eng. News 61:34–45.
2. Rice, E. L. 1984. Allelopathy, 2nd ed. Academic Press, New York.
3. Rice, E. L. 1986. Allelopathy effects on nitrogen cycling. Proc. Fourth Int. Symp. Microb. Ecol. 4:325–331.

4. Putnam, A. R., and L. A. Weston. 1986. Adverse impacts of allelopathy in agricultural systems, p. 43–56. *In* A. R. Putnam and C. S. Tang (eds.), The science of allelopathy. John Wiley & Sons, New York.

5. Rice, E. L., and S. K. Pancholy. 1972. Inhibition of nitrification by climax ecosystems. Am. J. Bot. 59:1033–1040.

6. Rice, E. L., and S. K. Pancholy. 1973. Inhibition of nitrification by climax ecosystems. II. Additional evidence and possible role of tannins. Am. J. Bot. 60:691–702.

7. Rice, E. L., and S. K. Pancholy. 1974. Inhibition of nitrification by climax ecosystems. III. Inhibitors other than tannins. Am. J. Bot. 61:1095–1103.

8. White, C. S. 1986. Effects of prescribed fire on rates of decomposition and nitrogen mineralization in a ponderosa pine ecosystem. Biol. Fertil. Soils 2:87–95.

9. White, C. S. 1986. Volatile and water-soluble inhibitors of nitrogen mineralization and nitrification in a ponderosa pine ecosystem. Biol. Fertil. Soils 2:97–104.

10. Molisch, H. 1937. Der Einfluss einer Pflanze auf die andere-Allelopathie. Fischer, Jena.

11. Rice, E. L. 1974. Allelopathy. Academic Press, New York.

12. Putnam, A. R., and C. S. Tang. 1986. Allelopathy: state of the science, p. 1–19. *In* A. R. Putnam and C. S. Tang (eds.), The science of allelopathy. John Wiley & Sons, New York.

13. Leather, G. R., and F. A. Einhellig. 1986. Bioassays in the study of allelopathy, p. 133–145. *In* A. R. Putnam and C. S. Tang (eds.), The science of allelopathy. John Wiley & Sons, New York.

14. Stowe, L. G. 1979. Allelopathy and its influence on the distribution of plants in an Illinois old-field. J. Ecol. 67:1065–1085.

15. Qasem, J. R., and T. A. Hill. 1989. On difficulties with allelopathy methodology. Weed Res. 29:345–347.

16. McCalla, T. M., and F. L. Duley. 1948. Stubble mulch studies: effect of sweetclover extract on corn germination. Science 108:163.

17. Nielsen, K. F., T. Cuddy, and W. Woods. 1960. The influence of the extract of some crops and soil residues on germination and growth. Can. J. Plant Sci. 40:188–197.

18. Grant, E. A., and W. G. Sallans. 1964. Influence of plant extracts on germination and growth of eight forage species. J. Br. Grassl. Soc. 19:191–197.

19. Lodhi, M. A. K., and G. L. Nickell. 1973. Effects of leaf extracts of *Celtis laevigata* on growth, water content and carbon dioxide exchange rates of three grass species. Bull. Torrey Bot. Club 100:159–165.

20. Colton, C. E., and F. A. Einhellig. 1980. Allelopathic mechanisms of velvetleaf (*Abutilon theophrasti* Medic. Malvaceae) on soybean. Am. J. Bot. 67:1407–1413.

21. Harper, S. H. T., and J. M. Lynch. 1982. The role of water-soluble components in phytotoxicity from decomposing straw. Plant Soil 65:11–17.

22. Krogmeier, M. J., and J. M. Bremner. 1989. Effects of water-soluble constituents of plant residues on water uptake by seeds. Commun. Soil Sci. Plant Anal. 20:1321–1333.

23. Hartley, R. D., and D. C. Whitehead. 1985. Phenolic acids in soils and their influence on plant growth and soil microbial processes, p. 109–149. *In* D. Vaughan and R. E. Malcolm (eds.), Soil organic matter and biological activity. Martinus Nijhoff/Dr. W. Junk, Dordrecht.

24. Kaminsky, R., and W. H. Muller. 1977. The extraction of soil phytotoxins using a neutral EDTA solution. Soil Sci. 124:205–210.

25. Whitehead, D. C., H. Dibb, and R. D. Hartley. 1981. Extractant pH and the release of phenolic compounds from soils, plant roots and leaf litter. Soil Biol. Biochem. 13:343–348.

26. Kaminsky, R., and W. H. Muller. 1978. A recommendation against the use of alkaline soil extractions in the study of allelopathy. Plant Soil 49:641–645.

27. Whitehead, D. C., H. Dibb, and R. D. Hartley. 1982. Phenolic compounds in soil as influenced by the growth of different plant species. J. Appl. Ecol. 19:579–588.

28. Stevenson, F. J. 1982. Humus chemistry. John Wiley & Sons, New York.

29. Richardson, H. L. 1938. The nitrogen cycle in grassland soils, with special reference to the Rothamsted park grass experiment. J. Agric. Sci. 28:73–121.

30. Theron, J. J. 1951. The influence of plants on the mineralization of nitrogen and the maintenance of organic matter in the soil. J. Agric. Sci. 41:289–296.

31. Soulides, D. A., and F. E. Clark. 1958. Nitrification in grassland soils. Soil Sci. Soc. Am. Proc. 22:308–311.

32. Clark, F. E., and E. A. Paul. 1970. The microflora of grassland. Adv. Agron. 22:375–435.

33. Harris, P. J. 1988. Microbial transformations of nitrogen, p. 609–651. *In* A. Wild (ed.), Russell's soil conditions and plant growth, 11th ed. Longman Scientific & Technical, London.

34. Greenland, D. J. 1958. Nitrate fluctuations in tropical soils. J. Agric. Sci. Camb. 50:82–92.

35. Meiklejohn, J. 1968. Numbers of nitrifying bacteria in some Rhodesian soils under natural grass and improved pastures. J. Appl. Ecol. 5:291–300.

36. Boughey, A. S., P. E. Munro, J. Meiklejohn, R. M. Strang, and M. J. Swift. 1964. Antibiotic reactions between African savanna species. Nature 203:1302–1303.

37. Munro, P. E. 1966. Inhibition of nitrite-oxidizers by roots of grass. J. Appl. Ecol. 3:227–229.

38. Munro, P. E. 1966. Inhibition of nitrifiers by grass root extracts. J. Appl. Ecol. 3:231–238.

39. Harmsen, G. W., and D. A. Van Schreven. 1955. Mineralisation of organic nitrogen in soil. Adv. Agron. 7:299–398.

40. Russell, E. W. 1961. Soil conditions and plant growth. John Wiley & Sons, New York.

41. Robinson, J. B. 1963. Nitrification in a New Zealand grassland soil. Plant Soil 2:173–183.

42. Purchase, B. S. 1974. Evaluation of the claim that grass root exudates inhibit nitrification. Plant Soil 41:527–539.

43. Stiven, G. 1952. Production of antibiotic substances by the roots of a grass (*Trachypogon plumosus* (H. B. K.) Nees) and of *Pentanisia variabilis* (E. Mey.) Harv. (Rubiaceae). Nature 170:712–713.

44. Rice, E. L. 1964. Inhibition of nitrogen-fixing and nitrifying bacteria by seed plants (I). Ecology 45:824–837.

45. Rice, E. L. 1965. Inhibition of nitrogen-fixing and nitrifying bacteria by seed plants. II. Characterization and identification of inhibitors. Physiol. Plant 18:255–268.

46. Rice, E. L. 1965. Inhibition of nitrogen-fixing and nitrifying bacteria by seed plants. III. Comparison of three species of *Euphorbia*. Proc. Okla. Acad. Sci. 45:43–44.

47. Neal, J. L. 1969. Inhibition of nitrifying bacteria by grass and forb root extracts. Can. J. Microbiol. 15:633–635.

48. Moore, D. R. E., and J. S. Waid. 1971. The influence of washings of living roots on nitrification. Soil Biol. Biochem. 3:69–83.

49. Vitousek, P. M., J. R. Gosz, C. C. Grier, J. M. Melillo, and W. A. Reiners. 1982. A comparative analysis of potential nitrification and nitrate mobility in forest ecosystems. Ecol. Monogr. 52:155–177.

50. Gosz, J. R., and C. S. White. 1986. Seasonal and annual variation in nitrogen mineralization and nitrification along an elevational gradient in New Mexico. Biogeochemistry 2:281–297.

51. Robertson, G. P. 1989. Nitrification and denitrification in humid tropical ecosystems: potential controls on nitrogen retention, p. 55–69. *In* J. Proctor (ed.), Mineral nutrients in tropical forest and savannah ecosystems. British Ecological Society Special Publication Number 9. Blackwell Scientific, Oxford.

52. Davidson, E. A., J. M. Stark, and M. K. Firestone. 1990. Microbial production and consumption of nitrate in an annual grassland. Ecology 71:1968–1975.

53. Schimel, J. P., L. E. Jackson, and M. K. Firestone. 1986. Control of nitrification and denitrification in a California annual grassland. Proc. Fourth Internat. Symp. Microb. Ecol. Proc., pp. 645–649.

54. Viro, P. J. 1962. Factorial experiments on forest humus decomposition. Soil Sci. 95:24–30.

55. Lodhi, M. A. K. 1978. Comparative inhibition of nitrifiers and nitrification in a forest community as a result of the allelopathic nature of various tree species. Am. J. Bot. 65:1135–1137.

56. Smith, W., F. H. Borman, and G. E. Likens. 1968. Response of chemoautotrophic nitrifiers to forest cutting. Soil Sci. 106:471–473.

57. Likens, G. E., F. H. Bormann, and N. M. Johnson. 1969. Nitrification: importance to nutrient losses from a cutover forested ecosystem. Science 163:1205–1206.

58. Killham, K. 1990. Nitrification in coniferous forest soils. Plant Soil 128:31–44.

59. Russell, E. J. 1914. The nature and amount of the fluctuations in nitrate contents of arable soils. J. Agric. Sci. 6:50–53.

60. Alsaadawi, I. S., J. K. Al-Uqaili, A. J. Al-Rubeaa, and S. M. Al-Hadithy. 1986. Allelopathic suppression of weeds and nitrification by selected cultivars of *Sorghum bicolor* (L.) Moench. J. Chem. Ecol. 12:1737–1745.

61. Alsaadawi, I. S. 1988. Biological suppression of nitrification by selected cultivars of *Helianthus annuus* L. J. Chem. Ecol. 14:733–741.

62. Lodhi, M. A. K. 1977. The influence and comparison of individual forest trees on soil properties and possible inhibition of nitrification due to intact vegetation. Am. J. Bot 64:260–264.

63. Lodhi, M. A. K., and K. T. Killingbeck. 1980. Allelopathic inhibition of nitrification and nitrifying bacteria in a ponderosa pine (*Pinus ponderosa* Dougl.) community. Am. J. Bot. 67:1423–1429.

64. Baldwin, I. T., R. K. Olson, and W. A. Reiners. 1983. Protein binding phenolics and the inhibition of nitrification in subalpine balsam fir soils. Soil Biol. Biochem. 15:419–423.

65. Olson, R. K., and W. A. Reiners. 1983. Nitrification in subalpine balsam fir soils: tests for inhibitory factors. Soil Biol. Biochem. 14:413–418.

66. Rice, E. L. 1979. Allelopathy–an update. Bot. Rev. 45:15–109.

67. McCarty, G. W., and J. M. Bremner. 1986. Effects of phenolic compounds on nitrification in soil. Soil Sci. Soc. Am. J. 50:920–923.

68. Winter, A. G. 1961. New physiological and biological aspects in the interrelationships between higher plants. Symp. Soc. Exp. Biol. 15:229–244.

69. del Moral, R., and C. H. Muller. 1970. The allelopathic effects of *Eucalyptus camaldulensis*. Am. Midl. Nat. 83:254–282.

70. Patrick, Z. A. 1971. Phytotoxic substances associated with the decomposition in soil of plant residues. Soil Sci. 111:13–18.

71. Chou, C. H., and Z. A. Patrick. 1976. Identification and phytotoxic activity of compounds produced during decomposition of corn and rye residues in soil. J. Chem. Ecol. 2:369–387.

72. del Moral, R., R. J. Willis, and D. H. Ashton. 1978. Suppression of coastal heath vegetation by *Eucalyptus baxteri*. Aust. J. Bot. 26:203–219.

73. Henderson, M. E. K., and V. C. Farmer. 1955. Utilization by soil fungi of *p*-hydroxybenzaldehyde, ferulic acid, syringaldehyde and vanillin. J. Gen. Microbiol. 12:37–46.

74. Sparling, G. P., and D. Vaughan. 1981. Soil phenolic acids and microbes in relation to plant growth. J. Sci. Food Agric. 32:625–626.

75. Sparling, G. P., B. G. Ord, and D. Vaughan. 1981. Changes in microbial biomass and activity in soils amended with phenolic acids. Soil Biol. Biochem. 13:455–460.

76. Vaughan, D., G. P. Sparling, and B. G. Ord. 1983. Amelioration of the phytotoxicity of phenolic acids by some soil microbes. Soil Biol. Biochem. 15:613–614.

77. Huang, P. M., T. S. C. Wang, M. K. Wang, M. H. Wu, and N. W. Hsu. 1977. Retention of phenolic acids by noncrystalline hydroxy-aluminum and -iron compounds and clay minerals of soils. Soil Sci. 123:213–219.

78. Whitehead, D. C. 1964. Identification of *p*-hydroxybenzoic, vanillic, *p*-coumaric, and ferulic acids in soils. Nature 202:417–418.

79. Guenzi, W. D., and T. M. McCalla. 1966. Phytotoxic substances extracted from soil. Soil Sci. Soc. Am. Proc. 30:214–216.

80. Wang, T. S. C., T. K. Yang, and T. Chuang. 1967. Soil phenolic acids as plant growth inhibitors. Soil Sci. 103:239–246.

81. Basaraba, J. 1964. Influence of vegetable tannins on nitrification in soil. Plant Soil 21:8–16.

82. Bohlool, B. B., E. L. Schmidt, and C. Beasley. 1977. Nitrification in the intertidal zone: influence of effluent type

and effect of tannin on nitrifiers. Appl. Environ. Microbiol. 34:523–528.

83. McCarty, G. W., J. M. Bremner, and E. L. Schmidt. 1991. Effects of phenolic acids on ammonia oxidation by terrestrial autotrophic nitrifying microorganisms. FEMS Microbiol. Ecol. 85:345–349.

84. Robertson, G. P., and P. M. Vitousek. 1981. Nitrification potentials in primary and secondary succession. Ecology 62: 376–386.

85. Schmidt, E. L. 1982. Nitrification in soil. *In* F. J. Stevenson (ed.), Nitrogen in agricultural soils. Agronomy 22:253–288.

86. Reiners, W. A. 1981. Nitrogen cycling in relation to ecosystem succession. F. E. Clark and T. Rosswall (eds.), Terrestrial nitrogen cycles. Ecol. Bull. (Stockh.) 33:507–528.

87. Jordan, C. F., R. L. Todd, and G. Escalante. 1979. Nitrogen conservation in a tropical rain forest. Oecologia 39:123–128.

88. Richards, B. N., J. E. N. Smith, G. J. White, and J. L. Charley. 1985. Mineralization of soil nitrogen in three forest communities from the New England region of New South Wales. Aust. J. Ecol. 10:429–441.

89. Runge, M. 1974. Die Stickstoff-Mineralization im Boden eines Sauerhumus-Buchenwaldes. II. die Nitratproduktion. Oecol. Plant 9:219–230.

90. Adams, M. A., and P. M. Attiwill. 1982. Nitrogen mineralization and nitrate reduction in forests. Soil Biol. Biochem. 14:197–202.

91. Montes, R. A., and N. L. Christensen. 1979. Nitrification and succession in the Piedmont of North Carolina. Forest Sci. 25:287–297.

92. Lamb, D. 1980. Soil nitrogen mineralisation in a secondary rainforest succession. Oecologia 47:257–263.

93. Vitousek, P. M., and W. A. Reiners. 1975. Ecosystem succession and nutrient retention: a hypothesis. BioScience 25: 376–381.

94. Vitousek, P. M. 1977. The regulation of element concentrations in mountain streams in the northeastern United States. Ecol. Monogr. 47:65–87.

95. Runge, M. 1983. Physiology and ecology of nitrogen nutrition, p. 163–200. *In* O. L. Lange, P. S. Nobel, C. B. Osmond, and H. Ziegler (eds.), Encyclopedia of plant physiology, Vol. 12C, Physiological plant ecology. III: Response to chemical and biological environment. Springer-Verlag, Berlin.

96. Bremner, J. M., and G. W. McCarty. 1988. Effects of terpenoids on nitrification in soil. Soil Sci. Soc. Am. J. 52:1630–1633.

97. White, C. S. 1990. Comments on "Effects of terpenoids on nitrification in soil." Soil Sci. Soc. Am. J. 54:296–297.

98. Bremner, J. M., and G. W. McCarty. 1990. Reply to "Comments on 'Effects of terpenoids on nitrification in soil'." Soil Sci. Soc. Am. J. 54:297–298.

99. White, C. S. 1988. Nitrification inhibition by monoterpenoids: theoretical mode of action based on molecular structures. Ecology 69:1631–1633.

100. Brar, S. S., and J. Giddens. 1968. Inhibition of nitrification in Bladen grassland soil. Soil Sci. Soc. Am. Proc. 32: 821–823.

101. Dierberg, F. E., and P. L. Brezonik, 1982. Nitrifying population densities and inhibition of ammonium oxidation in natural and sewage-enriched cypress swamps. Water Res. 16: 123–126.

102. Sahrawat, K. L., D. R. Keeney, and S. S. Adams. 1985. Rate of aerobic nitrogen transformations in six acid climax forest soils and the effect of phosphorus and $CaCO_3$. Forest Sci. 31:680–684.

103. Dancer, W. S., L. A. Peterson, and G. Chesters. 1973. Ammonification and nitrification of N as influenced by soil pH and previous N treatments. Soil Sci. Soc. Am. Proc. 37: 67–69.

104. Alexander, M. 1977. Introduction to soil microbiology, 2nd ed. John Wiley & Sons, New York.

105. Sarathchandra, S. U. 1978. Nitrification activities and the changes in the populations of nitrifying bacteria in soil perfused at two different H-ion concentrations. Plant Soil 50: 99–111.

106. Hankinson, T. R., and E. L. Schmidt. 1984. Examination of an acid forest soil for ammonia- and nitrite-oxidizing autotrophic bacteria. Can. J. Microbiol. 30:1125–1132.

107. Suzuki, I., U. Dular, and S. C. Kwok. 1974. Ammonia or ammonium ion as substrate for oxidation by *Nitrosomonas europaea* cells and extracts. J. Bacteriol. 120:556–558.

108. Corke, C. T. 1958. Nitrogen transformations in Ontario forest podzols, p. 116–121. *In* T. D. Stevens and R. L. Cook (eds.), Proceedings of the First North American Forest Soils Conference. Michigan State University Agricultural Experimental Station, East Lansing.

109. Williams, B. L. 1972. Nitrogen mineralisation and organic matter decomposition in Scots pine humus. Forestry 45:177–188.

110. van Praag, H. J., and F. Weissen. 1973. Elements of a functional definition of oligotroph humus based on the nitrogen nutrition of forest stands. J. Appl. Ecol. 10:569–583.

111. Keeney, D. R. 1980. Prediction of soil nitrogen availability in forest ecosystems: a literature review. Forest Sci. 26: 159–171.
112. Robertson, G. P. 1982. Factors regulating nitrification in primary and secondary succession. Ecology 63:1561–1573.
113. Martikainen, P. J. 1984. Nitrification in two coniferous forest soils after different fertilization treatments. Soil Biol. Biochem. 16:577–582.
114. Vitousek, P. M., and P. A. Matson. 1985. Causes of delayed nitrate production in two Indiana forests. Forest Sci. 31:122–131.
115. Overrein, L. N. 1971. Isotope studies on nitrogen in forest soil. I. Relative losses of nitrogen through leaching for forty months. Rep. Norw. For. Res. Inst. 114:261–280.
116. Heilman, P. 1974. Effect of urea fertilization on nitrification in forest soils of the Pacific North West. Soil Sci. Soc. Am. Proc. 38:664–667.
117. Popovic, B. 1977. Effect of ammonium nitrate and urea fertilizers on nitrogen mineralization, especially nitrification, in a forest soil. Royal College of Forestry, Stockholm. Departments of Forest Ecology and Forest Soils. Research Notes 30:1–33.
118. Martikainen, P. J. 1985. Numbers of autotrophic nitrifiers and nitrification in fertilized forest soil. Soil Biol. Biochem. 17:245–248.
119. White, C. S., and J. R. Gosz. 1987. Factors controlling nitrogen mineralization and nitrification in forest ecosystems in New Mexico. Biol. Fertil. Soils 5:195–202.
120. Weber, D. F., and P. J. Gainey. 1962. Relative sensitivity of nitrifying organisms to hydrogen ions in soils and in solutions. Soil Sci. 94:138–145.
121. Federer, C. A. 1983. Nitrogen mineralization and nitrification: depth variation in four New England forest soils. Soil Sci. Soc. Am. J. 47:1008–1014.
122. Hankinson, T. R., and E. L. Schmidt. 1988. An acidophilic and a neutrophilic *Nitrobacter* strain isolated from the numerically predominant nitrite-oxidizing population of an acid forest soil. Appl. Environ. Microbiol. 54:1536–1540.
123. Bhuiya, Z. H., and N. Walker. 1977. Autotrophic nitrifying bacteria in acid soils from Bangladesh and Sri Lanka. J. Appl. Bacteriol. 42:253–257.
124. Walker, N., and K. N. Wickramasinghe. 1979. Nitrification and autotrophic nitrifying bacteria in acid tea soils. Soil Biol. Biochem. 11:231–236.
125. Allison, S. M., and J. I. Prosser. 1991. Urease activity in neutrophilic autotrophic ammonia-oxidizing bacteria isolated from acid soils. Soil Biol. Biochem. 23:45–51.

126. Martikainen, P. J., and E. L. Nurmiaho-Lassila. 1985. *Nitrosospira*, an important ammonium-oxidizing bacterium in fertilized coniferous forest soil. Can. J. Microbiol. 31:190–197.

127. de Boer, W., H. Duyts, and H. J. Laanbroek. 1989. Urea stimulated autotrophic nitrification in suspensions of fertilized, acid heath soil. Soil Biol. Biochem. 21:349–354.

128. de Boer, W., and H. J. Laanbroek. 1989. Ureolytic nitrification at low pH by *Nitrosospira* spec. Arch. Microbiol. 152: 178–181.

129. de Boer, W., H. Duyts, and H. J. Laanbroek. 1988. Autotrophic nitrification in a fertilized acid heath soil. Soil Biol. Biochem. 20:845–850.

130. Focht, D. D., and W. Verstraete. 1977. Biochemical ecology of nitrification and denitrification. Adv. Microb. Ecol. 1: 135–214.

131. Schimel, J. P., M. K. Firestone, and K. S. Killham. 1984. Identification of heterotrophic nitrification in a Sierran forest soil. Appl. Environ. Microbiol. 48:802–806.

132. Adams, J. A. 1986. Identification of heterotrophic nitrification in strongly acid larch humus. Soil Biol. Biochem. 18: 339–341.

133. Killham, K. 1987. A new perfusion system for the measurement and characterization of potential rates of soil nitrification. Plant Soil 97:267–272.

134. Lang, E., and G. Jagnow. 1986. Fungi of a forest soil nitrifying at low pH values. FEMS Microbiol. Ecol. 38:257–265.

135. Stroo, H. F., T. M. Klein, and M. Alexander. 1986. Heterotrophic nitrification in an acid forest soil and by an acid-tolerant fungus. Appl. Environ. Microbiol. 52:1107–1111.

136. Purchase, B. S. 1974. The influence of phosphate deficiency on nitrification. Plant Soil 41:541–547.

137. Hue, N. V., and F. Adams. 1984. Effect of phosphorus level on nitrification rates in three low-phosphorus ultisols. Soil. Sci. 137:324–331.

138. Cole, C. V., and R. D. Heil. 1981. Phosphorus effects on terrestrial nitrogen cycling. *In* F. E. Clark and T. Rosswall (eds.), Terrestrial nitrogen cycles. Ecol. Bull. (Stockh.) 33:363–374.

139. Mahendrappa, M. K., R. L. Smith, and A. T. Christiansen. 1966. Nitrifying organisms affected by climatic region in Western United States. Soil. Sci. Soc. Am. Proc. 30:60–62.

140. Malhi, S. S., and W. B. McGill. 1982. Nitrification in three Alberta soils: effect of temperature, moisture and substrate concentration. Soil Biol. Biochem. 14:393–399.

141. Myers, R. J. K. 1975. Temperature effects on ammonification and nitrification in a tropical soil. Soil Biol. Biochem. 7:83–86.

142. Malhi, S. S., and M. Nyborg. 1979. Nitrate formation during winter from fall-applied urea. Soil Biol. Biochem. 11: 439—441.
143. Schloesing, M. T., and A. Muntz. 1879. Recherches sur la nitrification. Acad. Sci. Paris C.R. 89:1074—1077.
144. Russell, J. C., G. Jones, and T. M. Bahrt. 1925. The temperature and moisture factors in nitrate production. Soil Sci. 19:381—398.
145. Dubey, H. D. 1968. Effect of soil moisture levels on nitrification. Can. J. Microbiol. 14:1348—1350.
146. Johnson, D. W., and N. T. Edwards. 1979. The effects of stem girdling on biogeochemical cycles within a mixed deciduous forest in eastern Tennessee. II. Soil nitrogen mineralization and nitrification rates. Oecologia 40:259—271.
147. Donaldson, J. M., and G. S. Henderson. 1990. Nitrification potential of secondary-succession upland oak forests. I. Mineralization and nitrification during laboratory incubations. Soil Sci. Soc. Am. J. 54:892—897.
148. Donaldson, J. M., and G. S. Henderson. 1990. Nitrification potential of secondary-succession upland oak forests. II. Regulation of ammonium-oxidizing bacteria populations. Soil Sci. Soc. Am. J. 54:898—902.
149. Montagnini, F., B. Haines, and W. Swank. 1989. Factors controlling nitrification in soils of early successional and oak/hickory forests in the southern Appalchians. For. Ecol. Manag. 26:77—94.
150. Jones, J. M., and B. N. Richards. 1977. Effect of reforestation on turnover of ^{15}N-labelled nitrate and ammonium in relation to changes in soil microflora. Soil Biol. Biochem. 9: 383—392.
151. O'Connor, K. F., J. B. Robinson, and R. H. Jackman. 1962. Bacterial conditions and nutrient availability in a tussock grassland soil under different cultural treatments. Trans. Int. Soil Conf., pp. 177—182.
152. Chase, F. E., C. T. Corke, and J. B. Robinson. 1967. Nitrifying bacteria in soil, p. 593—611. *In* T. R. Gray and D. Parkinson (eds.), Ecology of soil bacteria, Liverpool University Press, Liverpool.
153. Belser, L. W., and E. L. Schmidt. 1978. Diversity in the ammonia oxidizing nitrifier population of a soil. Appl. Environ. Microbiol. 36:584—588.
154. Vitousek, P. M., and J. M. Melillo. 1979. Nitrate losses from disturbed forests: patterns and mechanisms. Forest Sci. 25:605—619.
155. Matson, P. A., and P. M. Vitousek. 1981. Nitrogen mineralization and nitrification potentials following clearcutting in the Hoosier national forest, Indiana. Forest Sci. 27:781—791.

156. Mayer, A. M., and M. Evenari. 1952. The relation between the structure of coumarin and its derivatives and their activity as germination inhibitors. J. Exp. Bot. 3:246–252.

157. Mayer, A. M., and M. Evenari. 1953. The activity of organic acids as germination inhibitors and its relation to pH. J. Exp. Bot. 4:256–263.

158. Varga, M., and E. Köves. 1959. Phenolic acids as growth and germination inhibitors in dry fruits. Nature 183:401.

159. Guenzi, W. D., and T. M. McCalla. 1966. Phenolic acids in oats, wheat, sorghum, and corn residues and their phytotoxicity. Agron. J. 58:303–304.

160. Hennequin, J. R., and C. Juste. 1967. Presence d'acide phénols libres dans le sol: Ètude de leur influence sur la germination et la croissance des végétaux. Ann. Agron. 18:545–569.

161. Rasmussen, J. A., and F. A. Einhellig. 1977. Synergistic inhibitory effects of p-coumaric and ferulic acids on germination and growth of grain sorghum. J. Chem. Ecol. 3:197–205.

162. Einhellig, F. A., and J. A. Rasmussen. 1978. Synergistic inhibitory effects of vanillic and p-hydroxybenzoic acids on radish and grain sorghum. J. Chem. Ecol. 4:425–436.

163. Williams, R. D., and R. E. Hoagland. 1982. The effects of naturally occurring phenolic compounds on seed germination. Weed Sci. 30:206–212.

164. Einhellig, F. A., and J. A. Rasmussen. 1979. Effects of three phenolic acids on chlorophyll content and growth of soybean and grain sorghum seedlings. J. Chem. Ecol. 5:815–824.

165. Patterson, D. T. 1981. Effects of allelopathic chemicals on growth and physiological responses of soybean (*Glycine max*). Weed Sci. 29:53–59.

166. Krogmeier, M. J., and J. M. Bremner. 1989. Effects of phenolic acids on seed germination and seedling growth in soil. Biol. Fertil. Soils 8:116–122.

167. Tang, C. S. 1986. Continuous trapping techniques for the study of allelochemicals from higher plants, p. 113–131. *In* A. R. Putnam and C. S. Tang (eds.), The science of allelopathy. John Wiley & Sons, New York.

6

Bacterial Mineralization of Organic Carbon Under Anaerobic Conditions

HENRY L. EHRLICH   *Rensselaer Polytechnic Institute,*
Troy, New York

I. INTRODUCTION

In microbiology, mineralization of organic matter generally refers to its complete conversion to inorganic matter through oxidation. Some, however, have used the term to refer to an entirely different process, namely mineral formation, such as calcium carbonate formation by calcareous algae and foraminifera (protozoa). The following discussion is about mineralization from the standpoint of degradation of organic matter. It is an essential process in the cycling of carbon wherever life occurs on Earth. Some organic matter is mineralized abiologically in nature, chiefly by combustion in wildfires in terrestrial environments and by photochemical reactions in terrestrial and aquatic environments [1]. Most organic matter is, however, mineralized biologically, chiefly through microbial activity. Although it was thought at one time that biological mineralization was possible only aerobically, it is now recognized that effective biological mineralization also occurs anaerobically.

II. THE PHYSIOLOGY OF MINERALIZATION

Both aerobic mineralization and anaerobic mineralization by microbes are, with one exception, forms of respiration. They require externally supplied electron acceptors for oxidative degradation. In aerobic respiration, the acceptor is usually oxygen, but other inorganic compounds such as NO_3^- and MnO_2 may also function as supplemental acceptors in special instances [2–6] In anaerobic respiration, the acceptors may be inorganic species such as Fe(III), manganese(IV) oxide, NO_3^-, NO_2^-, SO_4^{2-}, S, U(VI), or CO_2 [7–14], SeO_4^{2-}, SeO_3^{2-} [15,16], AsO_4^{2-} (reviewed in Ref. 8), CrO_4^{2-} [17–20], TeO_4^{2-}, TeO_3^{2-} (reviewed in Ref. 8), and VO_3^- [21]. They can also be organic compounds such as some chloroaromatics [22]. A rare instance of mineralization of organic compounds by anaerobic phototrophic metabolism has been reported by Kamal and Wyndhdam [23]. They observed mineralization of 3-chlorobenzoate by *Rhodopseudomonas palustris* strain WS17 in the light in the presence of benzoate as cosubstrate.

Not all anaerobically respiring microorganisms mineralize the organic compounds they attack. Some degrade them only partially; i.e., they produce organic products [24] (see also discussion in Section VI below).

Fermentation, while also degradative, never results in mineralization except for the conversion of acetate to methane and carbon dioxide by certain methanogens. This is because fermentation does not involve externally supplied electron acceptors but instead uses organic acceptors generated intracellularly from the compound being degraded. In a chemical sense, fermentation is a disproportionation or dismutation. One or more products in fermentations are always partially reduced organic compounds and are balanced by partially oxidized compounds. Although the homolactic fermentation of glucose, in which only a single product, lactate, is formed, may not appear to fit this concept, it does. The lactate is really the result of enzymatic disproportion of the intermediate triosephosphate into pyruvate and reduced pyridine nucleotide. The lactate results from the enzymatic reduction of pyruvate by the reduced pyridine nucleotide. Even though fermentation does not lead to complete degradation of organic compounds, it may have a very important role in anaerobic mineralization. In nature, many complex organics are mineralized by consortia in which fermenters convert the complex organics into simpler organic compounds that other microorganisms subsequently mineralize (see also discussion in Section VII below). Many anaerobic mineralizers are very restricted in the substrates that they can attack and thus depend on fermenters to produce these substrates from more complex compounds [25–27]. Some anaerobic bacteria can, however, mineralize complex organic substrates completely. For example, *Desulfobacterium anilini* has been shown to mineralize aniline [28].

In aerobic mineralization, organic carbon is always transformed into carbon dioxide, whereas in anaerobic mineralization, organic carbon is transformed into carbon dioxide or methane, depending on which external electron acceptors are available. When CH_4 is the product, it is frequently oxidized to CO_2 and H_2O by aerobic methanotrophs if it diffuses from its site of formation into an aerobic zone in soil or sediment [29]. However, strong indirect evidence has been obtained for anaerobic methane oxidation in marine sediments by microbes yet to be isolated and characterized [30–34]. Certain sulfate reducers may be involved in this anaerobic methane oxidation [34–36], although they may not oxidize the CH_4 directly [37].

In aerobic mineralization, organic nitrogen is more likely to accumulate as NO_3^- and organic sulfur as SO_4^{2-}, whereas in anaerobic mineralization, organic nitrogen accumulates as NH_4^+ and organic sulfur as HS^-. The higher oxidation state of the mineralized organic nitrogen and sulfur accumulating under aerobic conditions is the result of cooperation between one or more mineralizers and ammonia-nitrifying and sulfur-oxidizing bacteria, respectively. The latter two kinds of bacteria are strict aerobes, except for some sulfur oxidizers.

Aerobically, microbial mineralization of organic matter is dependent on the Krebs tricarboxylic acid cycle, whereby the acetate to which any organic matter is first converted is oxidized to CO_2 and H_2O,

$$CH_3COO^- + H^+ + 2O_2 \rightarrow 2CO_2 + 2H_2O$$

The aerobic pathways to acetate vary and depend on the nature of the organic substance being mineralized. Examples of aerobic metabolic pathways leading to acetate from specific organic compounds are listed in Table 1.

Anaerobically, microbial mineralization of organic matter depends on either the CO/acetyl-CoA pathway [38–42] or the Krebs tricarboxylic acid cycle [43–46] after its initial conversion to acetate. Which pathway is used depends on the organism involved. The CO/acetyl-CoA pathway, as proposed by Spormann and Thauer [41], is summarized in Figure 1. Anaerobic metabolic pathways leading to acetate from specific organic compounds are listed in Table 1 (see also Paul et al. [47] for pathways involved in denitrification).

III. BIOENERGETICS OF MINERALIZATION

Microbial mineralization is a form of catabolism. It releases free chemical energy, some of which is conserved by the organism through chemiosmosis in oxidative phosphorylation for use in anabolic processes. Using the mineralization of acetate as an example, free energy yields per mole of acetate consumed under standard conditions at pH 7 can be calculated from appropriate values of free energy of formation. These values show that the corresponding reactions are thermodynamically favorable under conditions of

Table 1   Some Degradative Metabolic Pathways in Microbes
Leading to Acetate

Pathways	Aerobes	Anaerobes
Sugars (via pyrovate)		
Glycolysis (EMP)	+	+
Entner-Doudoroff (ED)	+	+
Hexose monophosphate (HMP)	+	+
Fatty acids		
β oxidation	+	+
Amino acids		
Deamination, oxidation	+	−
Fermentation	−	+
Disproportionation	−	+
Alkanes		
Oxygenation, β oxidation	+	−
Aromatic hydrocarbons		
Aerobic ring cleavage (via catechol or protocatechuate)	+	−
Reductive ring cleavage	−	+

For details see Refs. 17, 127, 128, and 129.

aerobic respiration and several forms of anaerobic respiration. The
values in Table 2 show that anaerobic mineralization of 1 mole of
acetate at pH 7.0 by ferric ion [dissolved iron(III)] respiration
and denitrification yields almost as much free energy as aerobic
respiration and about 62% as much by MnO_2 respiration. On the
other hand, acetate mineralization by sulfate respiration yields only
5.6% and by iron respiration with crystalline $Fe(OH)_3$ only 2.7% as
much free energy as does aerobic respiration. Yet sulfate reduction has been found to be a very effective means of anaerobic mineralization of organic matter in sulfate-rich, anaerobic environments such as salt marshes or estuarine sediments. Furthermore,
iron respiration on some iron oxides has been found to be very
effective in some anaerobic sedimentary environments low in sulfate content (see Section VI below). Thus, it is more than just
the free energy yield that determines the dominant mineralization
process in a given environment.

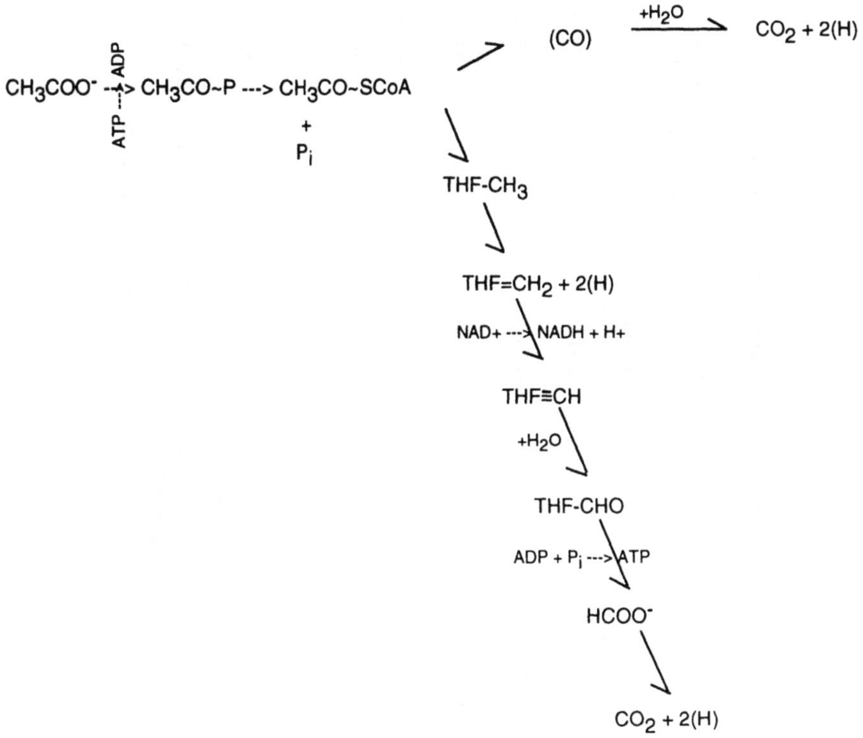

$$CH_3COO^- \xrightarrow[ATP]{ADP} CH_3CO\sim P \dashrightarrow CH_3CO\sim SCoA$$

$$+ \quad P_i$$

$$(CO) \xrightarrow{+H_2O} CO_2 + 2(H)$$

$$THF-CH_3$$

$$THF=CH_2 + 2(H)$$

$$NAD+ \dashrightarrow NADH + H+$$

$$THF\equiv CH$$

$$+H_2O$$

$$THF-CHO$$

$$ADP + P_i \dashrightarrow ATP$$

$$HCOO^-$$

$$CO_2 + 2(H)$$

Figure 1 The CO/acetyl-CoA pathway, as proposed by Spormann and Thauer [40]. TFH, tetrahydrofolate; (CO), bound carbon monoxide. Adapted from Fig. 6 in Spormann and Thauer, Ref. 40.

IV. ENVIRONMENTAL COMPATIBILITY OF DIFFERENT ANAEROBIC MINERALIZATION PROCESSES

An empirical correlation has frequently been noted between *standard oxidation potentials* of specific reductive half-reactions representing known respiratory processes at neutral pH (Table 3) and prevailing *environmental* redox potentials (e.g., Refs. 48, 49). Electron activities at standard conditions [pE(W)] are sometimes used in such correlations instead of oxidation or reduction potentials (e.g., Refs. 50–52). pE(W) is defined as $-\log \{e\} = E_H/2.3RTF^{-1}$, where E_H is the equilibrium redox potential of the reaction under consideration, referred to the hydrogen scale, R is

Table 2 Changes in Free Energies of Reaction for Different Microbial Mineralizations of Acetate

Reaction	Standard free energy changes of reaction at pH 7[a] ($\Delta G_r^{o'}$)	
	(kcal/mol)	(kJ/mol)
$CH_3COO^- + 2O_2 \rightarrow 2HCO_3^- + H^+$	-201.8	-843.5
$CH_3COO^- + 8Fe^{3+} + 4H_2O \rightarrow 2HCO_3^- + 8Fe^{2+} + 9H^+$	-193.4	-808.4
$CH_3COO^- + \frac{8}{5}NO_3^- + \frac{3}{5}H^+ \rightarrow 2HCO_3^- + \frac{4}{5}N_2 + \frac{4}{5}H_2O$	-189.3	-791.3
$CH_3COO^- + 4MnO_2 + 7H^+ \rightarrow 2HCO_3^- + 4Mn^{2+} + 4H_2O$	-125.3	-523.8
$CH_3COO^- + SO_4^{2-} \rightarrow 2HCO_3^- + HS^-$	-11.4	-47.7
$CH_3COO^- + H^+ \rightarrow CH_4 + CO_2^{b}$	-7.3	-30.5
$CH_3COO^- + 8Fe(OH)_3 + 15H^+ \rightarrow 2HCO_3^- + 8Fe^{2+} + 20H_2O$	-5.5	-23.0

[a]Calculated from free energies of formation given by Thauer et al. [130], except for values for crystalline Fe(OH)$_3$, and manganese, which were taken from Garrels and Christ [131].
[b]This is really a disproportionation reaction between two acetates, not a simple decarboxylation.

Table 3   Standard Oxidation Potentials for Some Overall
Reductive Half-Reactions in Anaerobic Respiration[a]

Reductive half-reactions	Standard oxidation potentials (in V at pH 7.0)
$O_2 + 4H^+ + 4e \rightarrow 2H_2O$	+0.82
$Fe^{3+} + e \rightarrow Fe^{2+}$	+0.77
$2NO_3^- + 12H^+ + 10e \rightarrow N_2 + 6H_2O$	+0.75
$MnO_2 + 4H^+ + 2e \rightarrow Mn^{2+} + 2H_2O$	+0.40
$SeO_4^{2-} + 8H^+ + 6e \rightarrow Se^o + 4H_2O$	+0.32
$SO_4^{2-} + 9H^+ + 8e \rightarrow HS^- + 4H_2O$	−0.22
$CO_2 + 8H^+ + 8e \rightarrow CH_4 + 2H_2O$	−0.23
$Fe(OH)_3 + 3H^+ + e \rightarrow Fe^{2+} + 3H_2O$	−0.25
$Fe_2O_3 + 6H^+ + 2e \rightarrow 2Fe^{2+} + 3H_2O$	−0.51

[a]Calculated from $\Delta G_0'$ values for respective half-reactions according to the relationship $E_0' = \Delta G_0'/23.06n$, where n is the number of electrons required for reduction.

the gas constant, T is the absolute temperature at which the reaction is run, and F is the Faraday [51]. Thus, in a sediment profile, aerobic respirers will be found in the oxidizing zone (uppermost portion of the profile), whereas nitrate respirers may be found in the uppermost part of the reducing zone, sulfate respirers below them, and methanogens at the bottom of the biologically active part of the sediment profile (e.g., Refs. 52, 53). The environmental redox potentials of these zones may range as follows: for denitrifiers, from +0.665 to −0.205 V; for sulfate reducers, from +0.115 to −0.450 V; and for methanogens, from +0.100 to −0.450 [53,54]. Optimal redox conditions are assumed to exist near the middle of each range. The implication of such correlations is that only respiratory reactions whose oxidation potentials match the redox potential of a given environmental niche closely can operate in it. Stated differently, environmental redox potentials determine which respiratory activities and, therefore, which organisms capable of them shall operate in the environment. However, such an inference is incorrect. Environmental redox potential is a symptom and not the cause of a given respiratory activity that occurs in a particular redox domain. Environmental redox potentials are only a measure of the dominant chemical oxidizing and reducing species present that can undergo reversible redox reactions. In biologically active environments, their redox state is affected by the biological activity.

Although the usual location of the sulfate reducing zone is under the zone of nitrate reduction or denitrification [55,56] and over the zone of methanogenesis [57–62] in marine sediment profiles, exceptions occur. Similarly, although the usual location of the zone of nitrate reduction/denitrification is below the zone of aerobic respiration in sediment and soil profiles [63,64], exceptions also occur. For example, in some sedimentary environments, extensive methanogenesis has been observed in the same niche in which sulfate reduction occurs [61,65,66]. Evidence has also been found that nitrate reduction can occur in the same niche in which aerobic respiration occurs [2,4,61,65,67–70]. Bacterial MnO_2 respiration has been noted in the presence of air [3,6,71–73]. It must be concluded that it is not environmental redox potential that is the determinant of whether one or more types of respiratory activity shall occur in a given environment but other chemical, physical, and biological factors.

As a generalization, respiratory activities with significantly different standard oxidation potentials can occur in the same redox niche when the two respiratory activities do not compete for the same electron donor (e.g., Ref. 65). Furthermore, respiratory activities with different standard oxidation potentials can occur in the same redox niche if the electron donor common to both respiratory processes, each in a different type of organism, is present in sufficient excess to eliminate competition for it. This can be inferred from the work of Kristjansson et al. [74], Lupton and Zeikus [75], and Schoenheit et al. [76]. It was experimentally demonstrated by Lovley et al. [77] in a case of concurrent methanogenesis and sulfate reduction with hydrogen in a eutrophic lake sediment and by Lovley and Phillips [78] in experiments with river sediment. In both studies, the sediment was supplied with excess hydrogen as electron donor for concurrent methanogenesis and Fe(III) and sulfate reduction. In the case of *Thiosphaera pantotropha*, which can perform nitrate respiration in air, a process that in *Pseudomonas stutzeri* is repressed by oxygen in excess of 5 mg L^{-1} [79], the compatibility with air seems to reside in the fact that nitrate reductase of this organism is not repressed or denatured by oxygen [4,5]. Oxygen is a repressor in most other nitrate reducers [80].

Recently, aerobic, i.e., oxygen-compatible, dissimilatory sulfate reduction has been detected during oxygenic photosynthesis in a hypersaline microbial mat dominated by the cyanobacterium

Microcoleus chthonoplastes [81]. Based on oxygen microsensor measurements in the mat, anaerobic microniches in the mat for this reduction were considered to be very unlikely sites for this activity
by the investigators. The organism(s) responsible for the sulfate
reduction has not yet been isolated and characterized. These observations are reminiscent of reports by Shturm [82] in 1948 and
by Bromfield [83] in 1953. Shturm noted aerobic production of
sulfide from sulfate by *Pseudomonas zelinskii* and Bromfield by
Bacillus megaterium. It is not known whether the sulfide in these
instances was produced by assimilatory sulfate reduction, which occurs in aerobes and anaerobes, or by dissimilatory sulfate reduction, which is thought to occur only anaerobically. If the sulfide
was produced by dissimilatory sulfate reduction, an enzyme system with unusual tolerance for oxygen or lack of competition with
aerobic respirers for electron donors, or both, could be the reason why the process can occur aerobically.

Spatial separation of different redox processes in a sediment or
soil column may result from spontaneous oxidation of a particular
electron donor at a more oxidized potential or from spontaneous
reduction of an electron acceptor at a more reduced potential.
Strict anaerobes do not function in aerobic environments because
oxygen may be toxic to them or may repress their respiratory metabolism. In the presence of an excess of an electron donor common to several organisms in a redox profile, spatial separation of
different respiratory process may be related to energy yield from
the different processes, with the process having the higher yield
outcompeting a process with a lower yield. On this basis, a sequence of anaerobic respiratory reactions on acetate in a reducing
sediment profile can be predicted from the free energy changes
listed in Table 2, assuming an abundance of acetate and appropriate electron acceptors and absence of chemical side reactions,
such as the reduction of nitrate to nitrite [84], NO_2^- to N_2O
[85] or MnO_2 to Mn^{2+} by Fe^{2+} [86,87]. Another interfering side
reaction can be the reduction of MnO_2 by H_2S [88]. Such a sequence of respiratory reactions, in which the electron acceptors
are consumed in descending order of the energy yield from the
oxidative reactions into which they enter, has been observed in
a natural environment containing a mixed population of anaerobic
respirers, an excess of organic matter, and a mixture of different
terminal electron acceptors [89].

Thus, the order in which different respiratory activities occur
in a soil or sediment profile may depend on (1) the types of organisms present; (2) available electron donors and the affinity of
the respiratory system of different organisms for them; (3) the
repressibility of an anaerobic respiratory pathway by oxygen; (4)
the stability of an anaerobic respiratory reductase in air; (5) the
relative abundance of different environmentally available electron

acceptors; (6) an extracellular, abiotic reaction between the product of respiration of one organism and the electron acceptor of
another, e.g., the reaction between Fe^{2+} formed in iron(III) respiration and MnO_2, the electron acceptor in Mn(IV) respiration
[86]; and (7) other environmental determinants such as the toxicity of a terminal electron acceptor or respiratory product to some
organisms.

The standard oxidation potentials at pH 7.0 for various respiratory half-reactions are listed in Table 3. Special notice should
be taken of the great difference in oxidation potentials of the
Fe^{3+}/Fe^{2+}, the $Fe(OH)_3/Fe^{2+}$, and the Fe_2O_3/Fe^{2+} couples. The
oxidation potential of crystalline ferric hydroxide is very close to
that of sulfate reduction, whereas the oxidation potential of Fe^{3+}
is very close to that of oxygen. Inspection of Table 2 shows that
when acetate oxidation is coupled to these reactions, three to four
times as much energy is released when ferric ion is the oxidant
than when sulfate or crystalline ferric hydroxide (or some other
solid ferric oxide) is the oxidant. On the basis of the oxidation
potentials of the reactions in Table 3, an anerobic sediment profile may exhibit iron reduction in a zone overlying a methanogenic zone if soluble ferric iron is an acceptor, but it may coincide with or underlie a methanogenic zone if a crystalline iron
oxide is the acceptor, as observed in nature [90]. Measured environmental potentials would be only a reflection of these respective activities and not their determinant (see discussion above).
Although the oxidation potentials in Table 3 also predict that iron
respiration of a crystalline oxide and sulfate reduction could occur in the same niche, this would not occur as the H_2S from sulfate respiration would tend to reduce chemically the ferric iron in
the oxide and thus interfere with its bacterial reduction. Hydrogen sulfide can also have this effect on the reduction of MnO_2
[88].

V. A BRIEF SURVEY OF BACTERIA RESPONSIBLE
FOR ANAEROBIC MINERALIZATION
OF ORGANIC MATTER

The bacteria that contribute to anaerobic mineralization are a very
diverse group. Those currently recognized are mostly gram-positive and gram-negative eubacteria, but some archaebacteria are
also able to mineralize organic matter anaerobically. Table 4 lists
some examples of bacteria capable of anaerobic mineralization of
organic matter.

As noted in Section II, the capacity to respire anaerobically
should not be equated with a capacity to mineralize organic matter
anaerobically, as a number of bacteria respire anaerobically but

are unable to oxidize the organic carbon completely to CO_2. A classic example of such an organism is *Desulfovibrio desulfuricans*, which oxidizes lactate to acetate and CO_2 with $SO_4{}^{2-}$ as the terminal electron acceptor [91]. Another example is *Shewanella* (formerly *Alteromonas putrefaciens*), which converts lactate and pyruvate to acetate and CO_2 with ferric iron as the terminal electron acceptor [92]. Both organisms appear to lack a mechanism for oxidizing acetate to CO_2, such as an anaerobically operating tricarboxylic acid (TCA) cycle or an acetyl-CoA/CO dehydrogenase pathway. By contrast, strain GS-15 oxidizes acetate by the TCA cycle with Fe(III) or Mn(IV) as the terminal electron acceptor [93], and *Desulfotomaculum acetoxidans* oxidizes acetate by the acetyl-CoA pathway with $SO_4{}^{2-}$ as the terminal electron acceptor [38]. Table 5 lists some representative types of organisms that are able to respire anaerobically but that cannot oxidize the organic matter completely.

Some extremely thermophilic archaebacteria seem to respire organic matter anaerobically by using elemental sulfur as terminal electron acceptor, but they do not do so. Examples of such organisms are *Pyrococcus furiosus*, *Thermotoga thermarum*, *Thermotoga neopolitana*, and *Thermococcus litoralis* [94–96]. They convert organic carbon that they use as an energy source to CO_2 and H_2 and detoxify the H_2, whose accumulation is inhibitory to them, by oxidizing it with the reduction of S^o to HS^-. No energy seems to be gained in the sulfur-reducing step [94]. However, it is not clear how these organisms can completely mineralize organic carbon, since the primary process by which they catabolize organic matter appears to be fermentation and not respiration.

VI. IN SITU STUDIES

A global assessment of the contribution that anaerobic mineralization of organic matter makes to the carbon economy in marine sediments was presented by Henrichs and Reeburgh [30]. They concluded that anaerobic mineralization rates are intrinsically similar to aerobic ones. However, because much of the organic carbon is consumed in the aerobic zone before it reaches the anaerobic zones, aerobic rates are greater near the sediment surface than anaerobic rates at depth. They estimated that global anaerobic carbon mineralization in marine sediments represents only 9% of the global aerobic mineralization, i.e., an average of 150 Tg C yr^{-1}. Exceptions do occur on a local scale. Jørgensen [97] found that in coastal sediments of the Kattegat and Skagerrak at the entrance to the Baltic from the North Sea and in some Danish fjords, sulfate reducers mineralized as much organic matter as did all the aerobic respirers, but the contribution of the $SO_4{}^{2-}$ reducers to total

Table 4   Some Bacteria Capable of Anaerobic Mineralization of Organic Matter

Name of organism	Electron acceptor/ electron donor	References
Facultative eubacteria		
Paracoccus denitrificans	Nitrate/glucose	132
Pseudomonas denitrificans	Nitrate/glucose	133
Pseudomonas stutzeri	Nitrate/glucose	134
Shewanella putrefaciens	Iron(III)/formate	92
	Manganese(IV) oxides/ formate	92
Obligately anaerobic eubacteria		
Desulfobacter postgatei	Sulfate/acetate	135
Desulfobacter hydrogenophilus	Sulfate/acetate	38
Desulfobacterium anilini	Sulfate/aniline, catechol, resorcinol, hydroquinone	28
Desulfobacterium autotrophicum	Sulfate lactate	38
Desulfobacterium catecholicum	Sulfate/catechol	108
Desulfobacterium indolicum	Sulfate/indol	136
Desulfobacterium phenolicum	Sulfate/phenol	137
Desulfobacterium vacuolatum	Sulfate/lactate	38
Desulfococcus biacatus	Sulfate/acetone	138
Desulfococcus niacini	Sulfate/pyruvate	139
Desulfococcus multivorans	Sulfate/pyruvate	38
Desulfonema limicola	Sulfate/fatty acids	140
Desulfonema magnum	Sulfate/fatty acids	140
Desulfosarcina variabilis	Sulfate/pyruvate	38
Desulfovibrio MB6	Sulfate/methanol	141
Desulfovibrio baarsii	Sulfate/butyrate	38
Desulfotomaculum acetoxidans	Sulfate/acetate	142
Desulfotomaculum thermo-acetoxidans	Sulfate/lactate, pyruvate	42
Strain EDK82	Sulfate/methanol	143
Desulfurella acetivorans	Sulfur/acetate	9,46
Desulfuromonas acetoxidans	Sulfur/acetate, ethanol, propanol	144
Strain GS-15	Iron(III)/acetate, butyrate, propionate, ethanol, toluene, phenol, paracresol	145

(continued)

Table 4 (Continued)

Name of organism	Electron acceptor electron donor	References
Obligately anaerobic eubacteria		
[Strain GS-15]	Manganese(IV) oxide/ acetate, ethanol, bu- tyrate, propionate	146
	Nitrate/[as for mangan- ese(IV) oxide]	146
Strain SeS[a]	Selenate/acetate	15
Obligately anaerobic archaebacteria		
Archeoglobus fulgidus	Sulfate/lactate	147
Methanosarcina barkeri	Acetate/acetate	148
Methanosarcina strain 227	Acetate/acetate	149,150
Methanothrix soehnsgenii	acetate/acetate	151,152

[a]Obligately anaerobic trait assumed.

mineralization of organic matter decreased to one-third toward the edge of the shelf area.

In contrast to marine sediments, which contain some persistent anaerobic regions, soil is generally a predominantly aerobic environment where mineralization of organic matter is the result of aerobically respiring bacteria and fungi [49,98]. Nevertheless, anaerobic microhabitats exist even in these soils. Water-saturated soil crumbs larger than 6 mm in diameter are anaerobic at their center (Ref. 99 as cited in Ref. 100). Denitrification, nitrate ammonification (reduction of nitrate to ammonia), and methanogenesis occur in these anaerobic microhabitats. Evidence for denitrification in anaerobic zones adjacent to aerobic zones in the rhizosphere of rice growing in wetland soil has been described by Reddy et al. [101]. The NO_3^- reduced in this process originated from bacterial NH_3 nitrification in the aerobic zones, and some of it diffused into the anaerobic zones. Oxygen transport through the stems and roots of the rice into the rhizosphere was responsible for the aerobic zones around the roots.

Although aerobic soils predominate globally, anaerobic soil environments such as those of rice paddies, swamps, and salt marshes that are extensive and predominate in some areas should not be ignored. In Asia, rice paddies constitute a significant portion of agricultural soils. Anaerobic mineralization in these soils, when

Table 5 Some Anaerobes That Can Respire But Cannot Mineralize the Substrate on Which They Respire

Organism	Electron acceptor	References
Acetobacterium woodii[a]	CO_2	153
Acetogenium kivui[a]	CO_2	154
Clostridium aceticum[a]	CO_2	155
Clostridium butyricum	NO_3^-	111
Clostridium perfringens	NO_3^-	156
Desulfobulbus propionicus	SO_4^{2-}	157,158
Desulfomonas pigra	SO_4^{2-}	159
Desulfovibrio desulfuricans	SO_4^{2-}	91
Desulfovibrio fructosovorans	SO_4^{2-}	160
Desulfovibrio sulfodismutants	SO_4^{2-}	161
Desulfotomaculum nigrificans	SO_4^{2-}	91
Desulfotomaculum orientis	SO_4^{2-}	162
Shewanella putrefaciens	Fe(III), MnO_2	92,163
Thermodesulfobacterium commune	SO_4^{2-}, $S_2O_3^{2-}$	164

[a]Homoacetic fermenter; facultative chemoautotroph on H_2 + CO_2.

flooded, appears to be dominated by nitrate respiration and methanogenesis, although iron respiration may be equally or more important [90].

Sulfate respiration is rare in most soils because sulfate is generally not abundant in them. Salt marshes are an exception, as they contain abundant sulfate that had its origin in seawater [102–104]. This does not mean, however, that SO_4^{2-} reducers cannot exist in soils with a low sulfate concentration. A few species of *Desulfovibrio* and *Desulfotomaculum* and *Desulfobacterium catecholicum* are able to use nitrate in place of sulfate as terminal electron acceptor. They reduce the nitrate to ammonia by nitrate ammonification [64, 105–107]. Such activity has been observed in organic-rich, reducing sediments [109,110]. *Clostridium butyricum* is another organism that reduces NO_3^- to NH_4^+ when growing under conditions of NO_3^- [111]. The extent to which nitrate ammonification occurs in more typical anerobic soils has not yet been determined. Alternatively, some SO_4^{2-} reducers have the capacity to live in association with methanogens in the absence of sulfate. The methanogens serve as scavengers of the H_2 produced by the SO_4^{2-} reducers in interspecies H_2 transfer [112–114]. Accumulation of the H_2 produced by the sulfate reducers in the absence of sulfate is inhibitory to them. Such consortia could exist in some anaerobic soils. However, they do not necessarily have to cause complete degradation of the organic matter attacked by the sulfate reducers.

Anaerobic iron and manganese respiration must now be added to denitrification, nitrate ammonification, and methanogenesis as processes that contribute significantly to anaerobic mineralization of organic matter in soils and sediments [88,115,116]. Indeed, because of the general abundance of Fe(III) and Mn(IV) in many soils and sediments, anaerobic mineralization based on Fe(III) and/or Mn(IV) respiration may predominate in them [90]. However, if iron and manganese oxides occur together, each in significant quantities, iron respiration may be the dominant process. This is because the ferrous iron produced in the respiratory process is likely to reduce the manganese oxide by a nonbiological chemical reaction [86]. Anaerobic mineralization in a soil by SeO_4^{2-} is globally unimportant, but it may be significant in some special localities from a bioremedial standpoint [15,16,56].

Microbial consortia can make important contributions to anaerobic mineralization of simple and complex organic matter and xenobiotics in soils or sediments. Wolin and Miller [117] have described the activity of some consortia of fermenting and methanogenic bacteria in converting organic matter to methane in sulfate-deficient anaerobic environments. The success of such consortia may be based on interspecies H_2 transfer, in which fermenters transform their substrate to acetate, H_2, and CO_2. Methanogens may then generate methane from the acetate and from the H_2 and CO_2 [e.g., 118]. Such consortia may also be based on acetate formation from substrates such as glucose or fructose by acetogens, which methanogens then convert into methane by hydrogen reduction [119] or fermentation [120]. Terrestrially, rice paddies and swamps are important sites of this methanogenic activity.

Mineralization by sulfate respiration often involves consortia, some of which may be consortia of different SO_4^{2-} reducers. For instance, *Desulfovibrio desulfuricans* and *Desulfotomaculum nigrificans* oxidize lactate only as far as acetate with SO_4^{2-} as the terminal electron acceptor. This is, as previously explained, because they have neither the enzymes for an anaerobic tricarboxylic acid cycle nor the enzymes for the acetyl ~CoA/CO dehydrogenase pathway. For complete mineralization of lactate, these organisms need to associate with acetate-oxidizing sulfate reducers such as *Desulfotomaculum acetoxidans*, *Desulfobacter postgatei*, *Desulfobacter hydrogenophilus*, or other acetate-mineralizing bacteria (see Tables 4 and 5). For other organic compounds, different consortia of $SO_4{}^{2-}$ reducers or of fermenters plus $SO_4{}^{2-}$ reducers may be required [98].

Young and Fraser [121] have discussed the evidence for the coupling of anaerobic degradation of lignin and lignin-derived compounds to denitrification and methanogenesis, an activity that is very important in the soil environment, where lignin from woody plants is a major source of organic matter. This anaerobic degradation of lignin may involve microbial consortia.

Lovley and Phillips [122] have described microbial associations in which iron respiration formed the terminal step in organic matter mineralization. They found that in sediments that support iron respiration, glucose was first fermented to fatty acids. These were then oxidized to CO_2 by way of iron respiration. This phenomenon is likely to apply to some iron oxide—rich soils as well, as suggested by early observations by Bromfield [123]. Iron minerals such as amorphous ferric oxides, lepidocrocite (gamma-$FeOOH$), hematite (Fe_2O_3) and goethite (alpha-$FeOOH$), that may occur in some soils can serve as terminal electron acceptors in anaerobic iron respiration processes [124—126].

VII. CONCLUSIONS

Anaerobic mineralization of organic matter is important in the carbon cycle because it prevents relatively permanent anoxic environments in the biosphere from becoming permanent sinks for organic matter. Instances in which organic matter is not completely degraded in anoxic environments and thus accumulates, as in organic soils, may be due to limited availability of external electron acceptors, the relative refractoriness of some of the organic matter, an accumulation rate of organic matter that is significantly greater than the rate of mineralization, or a combination of these.

ACKNOWLEDGMENT

The author is greatly indebted to Derek R. Lovley for a critical reading of a draft of this chapter and for several valuable suggestions for improvement.

REFERENCES

1. Mopper, K., and X. Zhou. 1990. Hydroxyl radical photoproduction in the sea and its potential impact on marine processes. Science 250:661—664.

2. Davies, K. J. P., D. Lloyd, and L. Boddy. 1989. The effect of oxygen on denitrification in *Paracoccus denitrificans* and *Pseudomonas aeruginosa*. J. Gen. Microbiol. 135:2445–2451.

3. Ehrlich, H. L. 1987. Manganese oxide reduction as a form of anaerobic respiration. Geomicrobiol. J. 5:423–431.

4. Robertson, L. A., and J. G. Kuenen. 1984. Aerobic denitrification–old wine in new bottles? Antonie van Leeuwenhoek 50:525–544.

5. Robertson, L. A., and J. G. Kuenen. 1984. Aerobic denitrification: a controversy revived. Arch. Microbiol. 139: 351–354.

6. Di-Ruggerio, J., and A. M. Gounot. 1990. Microbial manganese reduction mediated by bacterial strains isolated from aquifer sediments. Microb. Ecol. 20:53–63.

7. Tomlinson, G. A., L. L. Jahnke, and L. I. Hochstein. 1986. *Halobacterium denitrificans* sp. nov., an extremely halophilic denitrifying bacterium. Int. J. Syst. Bacteriol. 36:66–70.

8. Ehrlich, H. L. 1990. Geomicrobiology, 2nd ed. Marcel Dekker, New York.

9. Bonch-Osmolovskaya, E. A., T. G. Sokolova, N. A. Kostrikina, and G. A. Zavarzin. 1990. *Desulfurella acetivorans* gen. nov. and sp. nov.–a new thermophilic sulfur-reducing eubacterium. Arch. Microbiol. 153:151–155.

10. Criddle, C. S., J. T. DeWitt, D. Grbić-Galić, and P. L. McCarty. 1990. Transformation of carbon tetrachloride by *Pseudomonas* sp. strain KC under denitrifying conditions. Appl. Environ. Microbiol. 56:3240–3246.

11. Haeggblom, M. M., and L. Y. Young. 1990. Chlorophenol degradation coupled to sulfate reduction. Appl. Environ. Microbiol. 56:3255–3260.

12. Jones, A. M., and R. Knowles. 1990. Denitrification in *Flexibacter canadensis*. Can. J. Microbiol. 36:430–434.

13. Frunzke, K., and O. Meyer. 1990. Nitrate respiration, denitrification, and utilization of nitrogen sources by aerobic carbon-monoxide-oxidizing bacteria. Arch. Microbiol. 154: 168–174.

14. Lovley, D. R., E. J. P. Phillips, Y. A. Gorby, and E. R. Landa. 1991. Microbial reduction of uranium. Nature (London) 350:413–416.

15. Oremland, R. S., J. T. Hollibaugh, A. S. Maest, T. S. Presser, L. G. Miller, and C. W. Culbertson. 1989. Selenate reduction to elemental selenium by anaerobic bacteria in sediments and culture: biogeochemical significance of a novel, sulfate-independent respiration. Appl. Environ. Microbiol. 55:2333–2343.

16. Steinberg, N. A., and R. S. Oremland. 1990. Dissimilatory selenate reduction potentials in a diversity of sediment types. Appl. Environ. Microbiol. 56:3550–3557.

17. Bopp, L. W., and H. L. Ehrlich. 1988. Chromate resistance
     and reduction in *Pseudomonas fluorescens* strain LB 300. Arch.
     Microbiol. 150:426–431.
18. Lebedeva, E. V., and N. N. Lyalikova. 1979. Reduction of
     crocoite by *Pseudomonas chromatophila* sp. nov. Mikrobio-
     logiya 48:517–522 (Engl. transl. pp. 405–410).
19. Romanenko, V. I., and V. N. Koren'kov. 1977. A pure cul-
     ture of bacteria utilizing chromates and bichromates as hydro-
     gen acceptors in growth under anaerobic conditions. Mikro-
     biologia 46:414–417 (Engl. transl. pp. 329–332).
20. Wang, P.-C., T. Mori, K. Komori, M. Sasatsu, K. Toda, and
     H. Ohtake. 1989. Isolation and characterization of an *En-
     terobacter cloacae* strain that reduces hexavalent chromium
     under anaerobic conditions. Appl. Environ. Microbiol. 55:
     1665–1669.
21. Yurkova, N. A., and N. N. Lyalikova. 1990. New vanadate-
     reducing facultative chemolithotrophic bacteria. Mikrobiologiya
     59:968–975 (Engl. transl. pp. 672–677).
22. Dolfing, J. 1990. Reductive dechlorination of 3-chlorobenzo-
     ate is coupled to ATP production and growth in an anerobic
     bacterium, Strain DCB-1. Arch. Microbiol. 153:264–266.
23. Kamal, V. S., and R. C. Wyndham. 1990. Anaerobic photo-
     trophic metabolism of 3-chlorobenzoate by *Rhodopseudomonas
     palustris* WS17. Appl. Environ. Microbiol. 56:3871–3873.
24. Heijthuisen, J. H. F. G., and T. A. Hansen. 1989. Anaero-
     bic degradation of betaine by marine *Desulfobacterium* strains.
     Arch. Microbiol. 152:393–396.
25. Pfennig, N., F. Widdel, and H. G. Trueper. 1981. The dis-
     similatory sulfate reducing bacteria, p. 926–940. *In* M. P.
     Starr, H. Stolp, H. G. Trueper, A. Balows, and H. G.
     Schlegel (eds.), The prokaryotes. Handbook on habitats,
     isolation, and identification of bacteria, Vol. 1. Springer-
     Verlag, Belrin.
26. Colberg, P. J. S. 1990. Role of sulfate in microbial trans-
     formations of environmental contaminants: chlorinated aromatic
     compounds. Geomicrobiol. J. 8:147–165.
27. Grbić-Galić, D. 1990. Methanogenic transformation of aro-
     matic hydrocarbons and phenols in groundwater aquifers.
     Geomicrobiol. J. 8:167–200.
28. Schnell, S., F. Bak, and N. Pfennig. 1989. Anaerobic deg-
     radation of aniline and dihydroxybenzenes by newly isolated
     sulfate-reducing bacteria and description of *Desulfobacterium
     anilini*. Arch. Microbiol. 152:556–563.
29. Whittenbury, R., and H. Dalton. 1981. Methylotrophic bac-
     teria, p. 894–912. *In* M. P. Starr, H. Stolp, H. G. Trueper,
     A. Balows, and H. G. Schlegel (eds.), The prokaryotes. A

handbook on habitats, isolation and identification of bacteria, Vol. 1. Springer-Verlag, Berlin.

30. Henrichs, S. M., and W. S. Reeburgh. 1987. Anaerobic mineralization of marine sediment organic matter: rates and the role of anaerobic processes in the oceanic carbon economy. Geomicrobiol. J. 5:191–237.
31. Iversen, N., and B. B. Jørgensen. 1985. Anaerobic methane oxidation rates at the sulfate-methane transition in marine sediments from Kattegat and Skagerrak (Denmark). Limnol. Oceanogr. 30:944–955.
32. Reeburgh, W. S. 1976. Methane consumption in Cariaco Trench waters and sediments. Earth Planet. Sci. Lett. 28: 337–344.
33. Reeburgh, W. S. 1980. Anaerobic methane oxidation: rate depth distribution in Skan Bay sediments. Earth Planet. Sci. Lett. 47:345–352.
34. Reeburgh, W. S., and D. T. Heggie. 1977. Microbial methane consumption reactions and their effect on methane distribution in freshwater and marine environments. Limnol. Oceanogr. 22:1–9.
35. Devol, A. H., and S. I. Ahmed. 1981. Are the high rates of sulfate reduction associated with methane oxidation? Nature (London) 291:407–408.
36. Devol, A. H., J. J. Andersen, K. Kuivila, and J. W. Murray. 1984. A model for coupled sulfate reduction and methane oxidation in the sediments of Saanich Inlet. Geochim. Cosmochim. Acta 48:993–1004.
37. Alperin, M. J., and W. S. Reeburgh. 1984. Geochemical observations supporting anaerobic methane oxidation, p. 282–298. *In* R. L. Crawford and R. S. Hanson (eds.), Microbial growth on C-1 compounds. American Society for Microbiology, Washington, DC.
38. Schauder, R., B. Eikmanns, R. K. Thauer, F. Widdel, and G. Fuchs. 1986. Acetate oxidation to CO_2 in anaerobic bacteria via a novel pathway not involving reactions of the citric acid cycle. Arch. Microbiol. 145:162–172.
39. Schauder, R., A. Preuss, M. Jetten, and G. Fuchs. 1989. Oxidative and reductive acetyl CoA/carbon monoxide dehydrogenase pathway in *Desulfobacterium autotrophicum*. 2. Demonstration of the enzymes of the pathway and comparison of CO dehydrogenase. Arch. Microbiol. 151:84–89.
40. Spormann, A. M., and R. K. Thauer. 1988. Anaerobic acetate oxidation to CO_2 by *Desulfotomaculum acetoxidans*. Demonstration of enzyme required for the operation of an oxidative acetyl-CoA/carbon monoxide dehydrogenase pathway. Arch. Microbiol. 150:374–380.

41. Spormann, A. M., and R. K. Thauer. 1989. Anaerobic acetate oxidation to CO_2 by *Desulfotomaculum acetoxidans*. Isotopic exchange between CO_2 and the carbonyl group of acetyl-CoA and topology of enzymes involved. Arch. Microbiol. 152: 189—195.

42. Min, H., and S. H. Zinder. 1990. Isolation and characterization of a thermophilic sulfate-reducing bacterium *Desulfotomaculum thermoacetoxidans* sp. nov. Arch. Microbiol. 153: 399—404.

43. Brandis-Heep, A., N. A. Gebhardt, R. K. Thauer, F. Widdel, and N. Pfennig. 1983. Anaerobic acetate oxidation to CO_2 by *Desulfobacter postgatei*. 1. Demonstration of all enzymes required for the operation of the citric acid cycle. Arch. Microbiol. 136:222—229.

44. Beghardt, N. A., D. Lindner, and R. K. Thauer. 1983. Anaerobic acetate oxidation to CO_2 by *Desulfobacter postgatei*. 2. Evidence from ^{14}C-labelling studies for the operation of the citric acid cycle. Arch. Microbiol. 136:230—233.

45. Moeller, D., R. Schauder, G. Fuchs, and R. K. Thauer. 1987. Acetate oxidation to CO_2 via a citric acid cycle involving an ATP-citrate lyase: a mechanism for the synthesis of ATP via substrate-level phosphorylation in *Desulfobacter postgatei* growing on acetate and sulfate. Arch. Microbiol. 148:202—207.

46. Schmitz, R. A., E. A. Bonch-Osmolovskaya, and R. K. Thauer. 1990. Different mechanisms of acetate activation in *Desulfurella activorans* and *Desulfuromonas acetoxidans*. Arch. Microbiol. 154:274—279.

47. Paul, J. W., E. G. Beauchamp, and J. T. Trevors. 1989. Acetate, propionate, butyrate, glucose, and sucrose as carbon sources for denitrifying bacteria in soil. Can. J. Microbiol. 35:754—759.

48. Alexander, M. 1971. Microbiol Ecology. Wiley, New York.

49. Fenchel, T., and T. H. Blackburn. 1979. Bacteria and mineral cycling. Academic Press, London.

50. Nealson, K. H. 1983. Microbial oxidation and reduction of manganese and iron, p. 459—479. *In* P. Westbroek and E. W. De Jong (eds.), Biomineralization and biological metal accumulation. Biological and geological perspectives. D. Reidel, Dodrecht.

51. Stumm, W., and J. J. Morgan. 1981. Aquatic chemistry, 2nd ed. Wiley, New York.

52. Zehnder, A. J. B. 1978. Ecology of methane formation, p. 349—376. *In* R. Mitchell (ed.), Water pollution microbiology, Vol. 2. Wiley-Interscience, New York.

53. Baas Becking, L. G. M., I. R. Kaplan, and D. Moore. 1960. Limits of the natural environment in terms of pH and oxidation-reduction potentials. J. Geol. 68:243–284.

54. Atlas, R. M. 1988. Microbiology. Fundamentals and applications, 2nd ed. Macmillan, New York.

55. Sørensen, J., B. B. Jørgensen, and N. P. Revsbech. 1979. A comparison of oxygen, nitrate, and sulfate respiration in coastal marine sediments. Microb. Ecol. 5:105–115.

56. Oremland, R. S., N. A. Steinberg, A. S. Maestl, L. G. Miller, and J. T. Hollibaugh. 1990. Measurement of in situ rates of selenate removal by dissimilatory bacterial reduction in sediments. Environ. Sci. Technol. 24:1157–1164.

57. Abram, J. W., and D. B. Nedwell. 1978a. Hydrogen as substrate for methanogenesis and sulfate reduction in anaerobic saltmarsh sediment. Arch. Microbiol. 117:93–97.

58. Abram, J. W., and D. B. Nedwell. 1978b. Inhibition of methanogenesis by sulfate reducing bacteria competing for transferred hydrogen. Arch. Microbiol. 117:89–92.

59. Claypool, G. E., and I. R. Kaplan. 1974. The origin and distribution of methane in marine sediments. *In* I. R. Kaplan (ed.), Natural gases in marine sediments, p. 99–139. Plenum, New York (cited by Kuivila et al. [62]).

60. Martens, C. S., and R. A. Berner. 1974. Methane production in the interstitial waters of sulfate-depleted sediments. Science 185:1167–1158.

61. Oremland, R. S., and B. F. Taylor. 1978. Sulfate reduction and methanogenesis in marine sediments. Geochim. Cosmochim. Acta 42:209–214.

62. Kuivila, K. M., J. W. Murray, and A. H. Devol. 1989. Methane production, sulfate reduction and competition for substrates in the sediments of Lake Washington. Geochim. Cosmochim. Acta 53:409–416.

63. Alexander, M. 1977. Soil microbiology, p. 279–280. Wiley, New York.

64. Sørensen, J. 1987. Nitrate reduction in marine sediment: pathways and interactions with iron sulfur cycling. Geomicrobiol. J. 5:401–421.

65. Oremland, R. S., and S. Polcin. 1982. Methanogenesis and sulfate reduction; competitive and noncompetitive substrates in estuarine sediments. Appl. Environ. Microbiol. 44:1270–1276.

66. Oremland, R. S., L. M. Marsh, and S. Polcin. 1982. Meth- and production and simultaneous sulfate reduction in anoxic saltmarsh sediments. Nature (London) 296:143–145.

67. Bonin, P., M. Gilewicz, and J. C. Bertrand. 1989. Effects of oxygen on each step of denitrification on *Pseudomonas naustica*. Can. J. Microbiol. 35:1061–1064.

68. Brons, H. J., and A. J. B. Zehnder. 1990. Aerobic nitrate and nitrite reduction in continuous cultures of *Escherichia coli* E4. Arch. Microbiol. 153:531–536.

69. Lloyd, D., L. Boddy, and K. J. P. Davies. 1987. Persistence of bacterial denitrification capacity under aerobic conditions: the rule rather than the exception. FEMS Microbiol. Ecol. 45:185–190.

70. Robertson, L. A., and J. G. Kuenen. 1990. Combined heterotrophic nitrification and aerobic denitrification in *Thiosphaera pantotropha* and other bacteria. Antonie van Leeuwenhoek 57:139–152.

71. Ehrlich, H. L. 1980. Bacterial leaching of manganese ores, p. 609–614. *In* P. Trudinger, M. R. Walter, and B. J. Ralph (eds.), Biogeochemistry of ancient and modern environments. Australian Academy of Science, Canberra, and Springer-Verlag, Berlin.

72. Trimble, R. B., and H. L. Ehrlich. 1968. Bacteriology of manganese nodules. III. Reduction of MnO_2 by two strains of nodule bacteria. Appl. Microbiol. 16:695–702.

73. Troshanov, E. P. 1968. Iron- and manganese-reducing microorganisms in ore-containing lakes of the Karelian Isthmus. Mikrobiologiya 37:934–940 (Engl. transl. 786–791).

74. Kristjansson, J. K., P. Schoenheit, and R. K. Thauer. 1982. Different K_m values for hydrogen and methanogenic bacteria and sulfate reducing bacteria: an explanation for the apparent inhibition of methanogenesis by sulfate. Arch. Microbiol. 131:278–282.

75. Lupton, F. S., and J. G. Zeikus. 1984. Physiological basis for sulfate-dependent hydrogen competition between sulfidogens and methanogens. Curr. Microbiol. 11:7–12.

76. Schoenheit, P., J. K. Kristjansson, and R. K. Thauer. 1982. Kinetic mechanism for the ability of sulfate reducers to outcompete methanogens for acetate. Arch. Microbiol. 132:285–288.

77. Lovley, D. R., D. F. Dwyer, and M. J. Klug. 1982. Kinetic analysis of competition between sulfate reducers and methanogens for hydrogen in sediments. Appl. Environ. Microbiol. 43:1373–1379.

78. Lovley, D. R., and E. J. P. Phillips. 1987. Competitive mechanisms for inhibition of sulfate reduction and methane production in the zone of ferric iron reduction in sediments. Appl. Environ. Microbiol. 53:2636–2641.

79. Koerner, H., and W. G. Zumft. 1989. Expression of denitrification enzymes in response to the dissolved oxygen level

and respiratory substrate in continuous culture of *Pseudomonas stutzeri*. Appl. Environ. Microbiol. 55:1670–1676.

80. Knowles, R. 1982. Denitrification. Microbiol. Rev. 46:43–70.
81. Canfield, D. E., and D. J. Des Marais. 1991. Aerobic sulfate reduction in microbial mats. Science 251:1471–1473.
82. Shturm, L. D. 1948. Sulfate reduction by facultative aerobic bacteria. Mikrobiologiya 17:415–418.
83. Bromfield, S. M. 1953. Sulfate reduction in partially sterilized soil exposed to air. J. Gen. Microbiol. 8:378–390.
84. Obuekwe, C. D., D. W. S. Westlake, and F. D. Cook. 1981. Effect of nitrate on reduction of ferric iron by a bacterium isolated from crude oil. Can. J. Microbiol. 27:692–697.
85. Sorensen, J., and L. Thorling. 1991. Stimulation by lepidocrocite (gamma-FeOOH) of Fe(II)pdependent nitrite reduction. Geochim. Cosmochim. Acta 55:1289–1294.
86. Lovley, D. R., and E. J. P. Phillips. 1988. Manganese inhibition of microbial iron reduction in anaerobic sediments. Geomicrobiol. J. 6:145–155.
87. Myers, C. R., and K. H. Nealson. 1988. Microbial reduction of manganese oxides: interactions with iron and sulfur. Geochim. Cosmochim. Acta 52:2727–2732.
88. Burdige, D. J., and K. H. Nealson. 1986. Chemical and microbiological studies of sulfide-mediated manganese reduction. Geomicrobiol. J. 4:361–387.
89. Jacobsen, P., W. H. Patrick, Jr., and B. G. Williams. 1981. Sulfide and methane formation in soils and sediments. Soil Sci. 132:279–287.
90. Lovley, D. R. 1991. Dissimilatory Fe(III) and Mn(IV) reduction. Microbiol. Rev. 55:259–287.
91. Postgate, J. R. 1984. The sulfate-reducing bacteria, 2nd ed. Cambridge University Press, Cambridge, UK.
92. Lovley, D. R., E. J. P. Phillips, and D. J. Lonergan. 1989. Hydrogen and formate oxidation coupled to dissimilatory reduction of iron and manganese by *Alteromonas putrefaciens*. Appl. Environ. Microbiol. 55:700–706.
93. Champine, J. E., and S. Goodwin. 1991. Acetate catabolism in the dissimilatory iron-reducing isolate GS-15. J. Bacteriol. 173:2704–2706.
94. Fiala, G., and K. O. Stetter. 1986. *Pyrococcus furiosus* sp. nov. represents a novel genus of marine heterotrophic archaebacteria growing optimally at 100°C. Arch. Microbiol. 145:56–61.
95. Windberger, E., R. Huber, A. Trincone, H. Fricke, and K. O. Stetter. 1989. *Thermotoga thermarum* sp. nov. and *Thermotoga neopolitana* occurring in African continental solfataric springs. Arch. Microbiol. 151:506–512.

96. Neuner, A., H. W. Jannasch, S. Belkin, and K. O. Stetter.
     1990. *Thermococcus litoralis* sp. nov.: a new species of ex-
     tremely thermophilic marine archaebacteria. Arch. Microbiol.
     153:205—207.
97. Jorgensen, B. B.   1982.   Mineralization of organic matter in
     the sea bed—the role of sulfate reduction.   Nature (London)
     296:643—645.
98. Hungate, R. E.   1985.   Anaerobic biotransformations of or-
     ganic matter, p. 39–95.   *In* E. R. Leadbetter and J. S.
     Poindexter (eds.), Bacteria in nature, Vol. 1.   Bacterial
     activities in perspective.   Plenum, New York.
99. Harris, R. F.   1981.   Effect of water potential on microbial
     growth and activity, p. 23–95.   *In* J. Parr et al. (eds.),
     Water potential relation in soil microbiology.   Soil Science
     Society of America, Madison, WI.
100. Paul, E. A., and F. E. Clark.   1989.   Soil microbiology and
     biochemistry.   Academic Press, San Diego.
101. Reddy, K. R., V. H. Patrick, Jr., and C. W. Lindau.   1989.
     Nitrification-denitrification at the plant root—sediment inter-
     face in wetlands.   Limnol. Oceanogr. 34:1004—1013.
102. Howarth, R. W.   1979.   Pyrite: its rapid formation in a salt
     marsh and its importance in ecosystems metabolism.   Science
     203:49—51.
103. Howarth, R. W., and S. Merkel.   1984.   Pyrite formation and
     the measurement of sulfate reduction in salt marsh sediments.
     Limnol. Oceanogr. 29:598—608.
104. Howes, B. L., J. W. H. Dacey, and G. M. King.   1984.
     Carbon flow through oxygen and sulfate reduction pathways
     in salt marsh sediments.   Limnol. Oceanogr. 29:1037—1051.
105. Keith, S. M., G. T. McFarlane, and R. A. Herbert.   1982.
     Dissimilatory nitrate reduction by a strain of *Clostridium
     butyricum* isolated from estuarine sediments.   Arch. Micro-
     biol. 132:62—66.
106. McCready, R. G. L., W. D. Gould, and F. D. Cook.   1983.
     Respiratory nitrate reduction by *Desulfovibrio* sp.   Arch.
     Microbiol. 135:182—185.
107. Mitchell, G. J., J. G. Jones, and J. A. Cole.   1986.   Dis-
     tribution and regulation of nitrate and nitrite reduction by
     Desulfovibrio and *Desulfotomaculum* species.   Arch. Microbiol.
     144:35—40.
108. Szewzyk, R., and N. Pfennig.   1987.   Complete oxidation of
     catechol by strictly anaerobic sulfate-reducing *Desulfobac-
     terium catecholicum* sp. nov.   Arch. Microbiol. 147:163—168.
109. Koike, I., and A. Hattori.   1978.   Denitrification and am-
     monium formation in anaerobic coastal sediments.   Appl. En-
     viron. Microbiol. 35:853—857.

110. Sørensen, J. 1978. Capacity for denitrification and reduction of nitrate to ammonium in coastal marine sediment. Appl. Environ. Microbiol. 35:301–305.

111. Keith, S. M., G. T. MacFarlane, and R. A. Herbert. 1982. Dissimilatory nitrate reduction by a strain of *Clostridium butyricum* isolated from estuarine sediments. Arch. Microbiol. 132:62–66.

112. Bryant, M. P., L. L. Campbell, C. A. Reddy, and M. R. Crabill. 1977. Growth of *Desulfovibrio* in lactate or ethanol media low in sulfate in association with H_2-utilizing methanogenic bacteria. Appl. Environ. Microbiol. 33:1162–1169.

113. Cord-Ruwisch, R., B. Ollivier, and J.-L. Garcia. 1986. Fructose degradation by *Desulfovibrio* sp. in pure culture and in coculture with *Methanospirillum hungatei*. Curr. Microbiol. 13:285–289.

114. Cord-Ruwisch, R., H.-J. Seitz, and R. Conrad. 1988. The capacity of hydrogenotrophic anaerobic bacteria to compete for traces of hydrogen depends on the redox potential and the terminal electron acceptor. Arch. Microbiol. 149:350–357.

115. Lovley, D. R. 1987. Organic matter mineralization with the reduction of ferric iron: a review. Geomicrobial. J. 5:375–399.

116. Lovley, D. R., M. J. Baedecker, D. J. Lonergan, I. M. Cozzarelli, E. J. P. Phillips, and D. I. Siegel. 1989. Oxidation of aromatic contaminants coupled to iron reduction. Nature (London) 339:297–300.

117. Wolin, M. J., and T. L. Miller. 1987. Bioconversion of organic carbon to CH_4 and CO_2. Geomicrobial. J. 5:239–259.

118. McInerney, M. J., and M. P. Bryant. 1981. Anaerobic degradation of lactate by syntrophic association of *Methanosarcina barkeri* and *Desulfovibrio* species and effect of H_2 on acetate degradation. Appl. Environ. Microbiol. 41:346–354.

119. Zeikus, J. G., P. J. Weimer, D. R. Nelson, and L. Daniels. 1975. Bacterial methanogenesis: acetate as a methane precursor in pure culture. Arch. Microbiol. 104:129–134.

120. Winter, J. U., and R. S. Wolfe. 1979. Complete degradation of carbohydrate to carbon dioxide and methane by syntrophic cultures of *Acetobacterium woodii* and *Methanosarcina barkeri*. Arch. Microbiol. 121:97–102.

121. Young, L. Y., and A. C. Fraser. 1987. The fate of lignin and lignin-derived compounds in anaerobic environments. Geomicrobiol. J. 5:261–293.

122. Lovley, D. R., and E. J. P. Phillips. 1989. Requirement for a microbial consortium to completely oxidize glucose in

Fe(III)-reducing sediments. Appl. Environ. Microbiol. 55: 3234–3236.

123. Bromfield, S. M. 1954. Reduction of ferric compounds by soil bacteria. J. Gen. Microbiol. 11:1–6.

124. Munch, J. C., and J. C. G. Ottow. 1980. Preferential reduction of amorphous to crystalline iron oxides by bacterial activity. Soil Sci. 129:15–21.

125. Munch, J. C., and J. C. G. Ottow. 1982. Einfluss von Zellkontakt und Eisen(III)-Oxidform auf die bakterielle Eisenreduktion. Z. Pflanzern. Bodenkd. 145:66–77.

126. Munch, J. C., Th. Hillebrand, and J. C. G. Ottow. 1978. Transformations in the Fe_o/Fe_d ratio of pedogenic iron oxides affected by iron reducing bacteria. Can. J. Soil Sci. 58:475–486.

127. Doelle, H. W. 1975. Bacterial metabolism, 2nd ed. Academic Press, New York.

128. Berry, D. F., A. J. Francis, and J.-M. Bollag. 1987. Microbial metabolism of homocyclic and heterocyclic aromatic compounds under anaerobic conditions. Microbiol. Rev. 51:43–59.

129. Evans, W. C., and G. Fuchs. 1988. Anaerobic degradation of aromatic compounds. Annu. Rev. Microbiol. 42:289–317.

130. Thauer, R. K., K. Jungermann, and K. Decker. 1977. Energy conservation in chemotrophic anaerobic bacteria. Bacteriol. Rev. 41:100–180.

131. Garrels, R. M., and C. L. Christ. 1965. Solutions, minerals and equilibria. Harper & Row, New York.

132. Forget, P., and F. Pichinoty. 1965. Le cycle tricarboxylique chez une bactérie denitrificante obligatoire. Ann. Inst. Pasteur 108:364–377.

133. Delwiche, C. C. 1967. Energy relationships in soil biochemistry, p. 173–193. *In* A. D. McLaren and G. H. Peterson (eds.), Soil biochemistry, Vol. 1. Marcel Dekker, New York.

134. Spangler, W. J., and C. M. Gilmour. 1966. Biochemistry of nitrate respiration in *Pseudomonas stutzeri*. I. Aerobic and nitrate respiration routes of carbohydrate catabolism. J. Bacteriol. 91:245–254.

135. Widdel, F., and N. Pfennig. 1981. Studies on dissimilatory sulfate-reducing bacteria that decompose fatty acids. I. Isolation of new sulfate-reducing bacteria enriched with acetate from saline environments. Description of *Desulfobacter postgatei* gen. nov., sp. nov. Arch. Microbiol. 129:395–400.

136. Bak, F., and F. Widdel. 1986. Anaerobic degradation of indolic compounds by sulfate-reducing enrichment cultures, and description of *Desulfobacterium indolicum* gen. nov., sp. nov. Arch. Microbiol. 146:170–176.

137. Bak, F., and F. Widdel. 1986. Anaerobic degradation of phenol and phenol derivatives by *Desulfobacterium phenolicum* sp. nov. Arch. Microbiol. 146:177–180.
138. Platen, H., A. Temmes, and F. Schink. 1990. Anaerobic degradation of acetone by *Desulfococcus biacatus* spec. nov. Arch. Microbiol. 154:355–361.
139. Imhoff-Stuckle, D., and N. Pfennig. 1983. Isolation and characterization of a nicotinic acid–degrading sulfate-reducing bacterium, *Desulfococcus niacini* sp. nov. Arch. Microbiol. 136:194–198.
140. Widdel, F., G.-W. Kohring, and F. Mayer. 1983. Studies on dissimilatory sulfate-reducing bacteria that decompose fatty acids. III. Characterization of the filamentous gliding *Desulfonema limicola* gen. nov. sp. nov., and *Desulfonema magnum* sp. nov. Arch. Microbiol. 134:286–294.
141. Braun, M., and H. Stolp. 1985. Degradation of methanol by a sulfate reducing bacterium. Arch. Microbiol. 142:77–80.
142. Widdel, F., and N. Pfennig. 1977. A new anaerobic, sporing, acetate-oxidizing, sulfate-reducing bacterium, *Desulfotomaculum* (emend.) *acetoxidans*. Arch. Microbiol. 112:119–122.
143. Nanninga, H. J., and J. C. Gottschal. 1986. Isolation of a sulfate-reducing bacterium growing with methanol. FEMS Microbiol. Ecol. 38:125–130.
144. Pfennig, N., and H. Biebl. 1976. *Desulfuromonas acetoxidans* gen. nov. and sp. nov., a new anaerobic, sulfur-reducing, acetate-oxidizing bacterium. Arch. Microbiol. 110:3–12.
145. Lovley, D. R., and D. J. Longergan. 1990. Anaerobic oxidation of toluene, phenol, and p-cresol by the dissimilatory iron-reducing organism GS-15. Appl. Environ. Microbiol. 56:1858–1864.
146. Lovley, D. R., and E. J. P. Phillips. 1988. Novel mode of microbial energy metabolism: organic carbon oxidation coupled to dissimilatory reduction of iron or manganese. Appl. Environ. Microbiol. 54:1472–1480.
147. Moeller-Zinkhan, D., and R. K. Thauer. 1990. Anaerobic lactate oxidation to 3 CO_2 by *Archeoglobus fulgidus* via the carbon monoxide dehydrogenase pathway: demonstration of the acetyl-CoA carbon-carbon cleavage reaction in cell extracts. Arch. Microbiol. 153:215–218.
148. Blaut, M., and G. Gottschalk. 1982. Effect of trimethylamine on acetate utilization by *Methanobacterium barkeri*. Arch. Microbiol. 133:230–235.
149. Mah, R. A., M. R. Smith, and L. Baresi. 1978. Studies on an acetate-fermenting strain of *Methanosarcina*. Appl. Environ. Microbiol. 35:1174–1184.

150. Smith, M. R., and R. A. Mah. 1978. Growth and methano-
     genesis by *Methanosarcina* strain 227 on acetate and meth-
     anol. Appl. Environ. Microbiol. 36:870—879.
151. Zehnder, A. J. B., B. A. Huser, T. D. Brock, and K.
     Wuhrmann. 1980. Characterization of an acetate-decar-
     boxylating, non-hydrogen-oxidizing methane bacterium.
     Arch. Microbiol. 124:1—11.
152. Huser, B. A., K. Wuhrmann, and A. J. B. Zehnder. 1982.
     Methanothrix soehngenii gen. nov. sp. nov., a new aceto-
     trophic non-hydrogen-oxidizing methane bacterium. Arch.
     Microbiol. 132:1—9.
153. Balch, W. E., S. Schoberth, R. S. Tanner, and R. S.
     Wolfe. 1977. *Acetobacterium*, a new genus of hydrogen-
     oxidizing, carbon dioxide—reducing, anaerobic bacteria.
     Int. J. Syst. Bacteriol. 27:355—361.
154. Leigh, J. A., F. Mayer, and R. S. Wolfe. 1981. *Aceto-
     genium kivui*, a new thermophilic hydrogen-oxidizing, aceto-
     genic bacterium. Arch. Microbiol. 129:275—280.
155. Braun, M., F. Mayer, and G. Gottschalk. 1981. *Clostrid-
     ium aceticum* (Wieringa), a microorganism producing acetic
     acid from molecular hydrogen and carbon dioxide. Arch.
     Microbiol.1 28:288—293.
156. Hasan, S. M., and J. B. Hall. 1975. The physiological
     function of nitrate reduction in *Clostridium perfringens*. J.
     Gen. Microbiol. 87:120—128.
157. Laanbroek, H. J., T. Abee, and I. J. Voogd. 1982. Al-
     cohol conversions by *Desulfobulbus propionicus* Lindhorst in
     the presence and absence of sulfate and hydrogen. Arch.
     Microbiol. 133:178—184.
158. Widdel, F., and N. Pfennig. 1982. Studies on dissimilatory
     sulfate reducing bacteria that decompose fatty acids. II. In-
     complete oxidation of propionate by *Desulfobulbus propionicus*
     gen. nov. spec. nov. Arch. Microbiol. 131:360—365.
159. Moore, W. E. C., J. L. Johnson, and L. V. Holdeman. 1976.
     Emendation of *Bacteoidaceae* and *Butyrivibrio* and descrip-
     tions of *Desulfomonas* gen. nov. and ten new species in the
     genera *Desulfomonas*, *Butyrivibrio*, *Eubacterium*, *Clostridium*,
     and *Ruminococcus*. Int. J. Syst. Bacteriol. 26:238—252.
160. Ollivier, B., R. Cord-Ruwisch, E. C. Hatchikian, and J. L.
     Garcia. 1988. Characterization of *Desulfovibrio fructosovor-
     ans* sp. nov. Arch. Microbiol. 149:447—450.
161. Bak, F., and N. Pfennig. 1987. Chemolithotrophic growth
     of *Desulfovibrio sulfodismutans* sp. nov. by disproportiona-
     tion of inorganic sulfur compounds. Arch. Microbiol. 147:
     184—189.

162. Klemps, R., H. Cypionka, F. Widdel, and N. Pfennig. 1985. Growth with hydrogen, and further physiological characteristics of *Desulfotomaculum* species. Arch. Microbiol. 143: 203–208.

163. Myers, C. R., and K. H. Nealson. 1988. Bacterial manganese reduction and growth with manganese oxide as the sole electron acceptor. Science 240:1319–1321.

164. Zeikus, J. G., M. A. Dawson, T. E. Thompson, K. Ingvosen, and E. C. Hatchikian. 1983. Microbial ecology of volcanic sulfidogenesis: isolation and characterization of *Thermosulfidobacterium commune* gen. nov. and sp. nov. J. Gen. Microbiol. 129:1159–1169.

7

Immunological and Molecular Techniques for Studying the Dynamics of Microbial Populations and Communities in Soil

JONATHAN P. CARTER *University of East Anglia, Norwich, Norfolk, United Kingdom*

JAMES M. LYNCH *Horticulture Research International, Littlehampton, West Sussex, United Kingdom*

I. INTRODUCTION

One of the major difficulties encountered by microbial ecologists has been the inability to identify and study specific components of the microbial community in the presence of other members of the community. Various techniques have been developed to overcome this problem, including microscopic observations [1,2], separation of microbial fractions from soil [3,4], use of marker compounds [5,6], and culturing of cells from the environment [7,8]. Although these approaches have proved useful, they are limited when the soil microbial community is being studied at the level of its component species or even strains. The ability to study the complex soil community at this level becomes important where fluctuations resulting from deliberate attempts at manipulation, such as the use of microbial inoculants, or indirect changes resulting from variations in land use and agricultural practices are to be assessed. Microbial inoculants have been used as both agents for biological control of soilborne diseases and biofertilizers. The potential release of genetically engineered organisms (GEMs) for these purposes has added new challenges for soil microbial ecologists. It is no longer sufficient to be able to follow the fate of an introduced strain, as the recombinant DNA which it contains must also be tracked and any transfer to and replication of this DNA by the indigenous microbial community must be detectable. Changes in agricultural practice can change the relative abundance of microbial species [9], which may favor organisms detrimental to plant growth [10].

To obtain highly specific measurements in population biology and
to study the population genetics of soil microorganisms, new techniques are being developed. These new techniques include the use
of antibodies that recognize specific microbial antigens, nucleic acid
probes to the DNA and RNA of the community fraction of interest,
and genetic engineering of novel, easily identifiable phenotypes.

Antibodies have been used to identify specific fungal species [11]
and bacterial strains [12] against the background of other soil microorganisms. If antibodies are to be used, there is a choice between using polyclonal or monoclonal antibodies [13]. Polyclonal
antibody solutions contain a mixture of different antibodies with
different binding sites on antigenic molecules. In contrast, monoclonal antibody solutions contain only one type of antibody, capable
of binding to only one site.

Although polyclonal antibodies are easier to produce, there are
problems in obtaining the desired specificity when they are used to
study soil fungi. Many attempts, with varying degrees of success,
have been made to produce specific antibodies to detect fungal
pathogens of plants. In general, extracts from fungal mycelium
seem to be a poor source of specific antigens. To optimize the
chance of producing suitable antibodies, soluble antigens can be
made particulate [14,15], specific molecules may be purified [16],
or spores may be used as the immunogen if they will contain antigens also common to the mycelium [17]. The cross-reactivity of
antiserum to fungal antibodies remains a problem [18]. If cross-
reactive polyclonal antisera are obtained, cross-absorption with
antigens from the cross-reacting fungus may be used to increase
specificity, but this often results in a decreased antibody titer
[19]. In studies with plant roots, cross-reactivity to fungi incapable of colonizing the roots may not interfere with the assay.
Fallon and Newell [14] developed antibodies to *Phaeosphaeria typharum,* the dominant saprophytic fungus associated with *Spartina
alterniflora,* in a salt marsh ecosystem and successfully measured
the fungal biomass of this species in the system.

Although monoclonal antibodies can be used to develop more
specific assays, they may also pose problems when used with environmental samples. These problems are particularly acute when
the microorganism to be detected is a mycelial fungus. For example, monoclonal antibodies against fungal antigens have usually
been raised against spores [20], soluble low-molecular-weight compounds produced by the fungus [21], or fungal toxins [13].
These antigens would be of limited use in investigating the ecology of fungal mycelium. In one case, when spores from the vesicular-arbuscular mycorrhizal (VAM) fungus *Glomus occultum* were

used as the immunogen in monoclonal antibody production, an antibody that recognized both spores and mycelium was obtained [17].
This indicates that at least one antigen occurs on both the mycelium and the spores of this fungus under the growth conditions
used. Another problem is that the antigenicity of the mycelium
may vary with the age of the mycelium [22] and with the substrate on which the fungus is growing [15]. As each monoclonal
antibody will recognize only one binding site, a monoclonal-based
assay would be particularly susceptible to error. Loss of a single
binding site could result in loss of the ability of the antibody to
bind. In contrast, a polyclonal antibody would require the loss of
several binding sites before no reaction would occur. To avoid
this problem, very careful screening of the monoclonal antibody is
required. This screening will be dependent on the degree of specificity required in the assay. If a species-specific assay is required, the antibody should be tested not only for cross-reactivity to other species but also for its ability to detect the species
of interest under the variety of growth conditions and physiological states likely to be encountered in the sample to be tested.

DNA probes do not depend on expression of the DNA that they
recognize, so they do not depend on the phenotype of the microorganism to be detected. Using radiolabeled probes, 0.02 pg of
DNA has been detected [23], which was obtained from a population of 4.3×10^4 cells g^{-1} oven-dry soil. To increase the sensitivity of the assays, a probe that recognizes sequences that are
present in more than one copy, either plasmids with a multiple
copy number or chromosomal genes that are duplicated, can be
used. Because DNA probes can detect nonexpressed DNA, they
can detect the transfer of DNA from one strain to another if the
transferred DNA is not expressed. Care must be taken that the
specificity of the probe used is known. Longer probes may be
labeled with greater numbers of reported molecules, thereby increasing sensitivity, but shorter probes offer greater specificity
with less chance of binding to similar but not identical DNA sequences. Care must also be taken because some compounds, often
coextracted with DNA from environmental samples, also bind DNA
probes.

Another group of nucleic acid probes are those to RNA. Ribosomal RNA (rRNA) sequences have become an important aid in the
study of the evolution and taxonomy of microorganisms since the
demonstration of their use by Woesse [24]. The same properties
that made them useful for taxonomy also make rRNAs very useful
for the production of specific probes. rRNA contains sequences
of bases that are very highly conserved and other areas that are
more variable. A probe that binds to the conserved areas would
detect all microorganisms, but there is sufficient variation in the
variable regions for strain-specific probes to be produced. The

number of ribosomes contained in each cell depends on the physiological status of the cell. In rapidly growing cells of *Escherichia coli*, the RNA can account for up to 20% of the dry weight of the cell [25], but the percentage is much lower in a cell growing under oligotrophic conditions. This variation could be exploited by comparing the signal obtained from DNA probes that hybridize to the genes coding for the rRNA to the signal of the probe that hybridizes with the rRNA itself. This would provide some indication of how active specific components of the microbial community are at any time. It has also been suggested that probes against messenger RNA (mRNA) could be used to determine when genes are being expressed [26]. However, mRNA is likely to be present at very low levels. All RNA molecules require very careful extraction to avoid degradation.

Nucleic acid probes have the potential to distinguish between and provide information on the activity of strains, but their use is relatively time-consuming. Less time-consuming marker systems have been developed by genetically engineering novel phenotypes, allowing very rapid identification of strains of interest [27].

These new methods have been used in soil microbiology in a number of different ways. It is often necessary to couple the new methods with established ones. This chapter is divided into three sections on the basis of this coupling of methods. The first section describes approaches that depend on culture of microorganisms; the second, those dependent on enrichment of the microbial fraction from the bulk soil; and the final section, methods that may be used directly on the bulk soil.

II. METHODS REQUIRING THE CULTURE OF ORGANISMS

Microbial populations have been analyzed both quantitatively and qualitatively by use of techniques based on the ability to culture the microorganisms (Figure 1). These techniques make two key assumptions: all the microorganisms of interest are assumed to be able to grow under the test conditions, and the growth in culture must be related to some unit of population. These two assumptions have long been questioned, especially as the number of colonies obtained on dilution plates does not agree well with direct microscopic examination of the same sample, often being less by a factor of 10 or more. The source of this discrepancy is important if a new technique is designed to improve established culture-based assays. This type of assay may be inherently unsuitable for the quantitative detection of mycelial organisms, especially those which sporulate profusely, as each colony-forming unit (CFU) is more likely to derive from a spore than a fragment of active mycelium

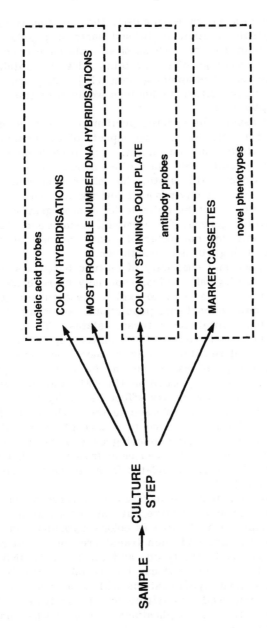

Figure 1 Culture-based assays.

[28]. For a single-celled microorganism, the situation is more complex. A considerable fraction of the soil bacterial population has a diameter of less than 0.4 μm. These small cells have been shown to have a very low percentage of culturability, despite containing sufficient DNA to indicate a complete genome [29]. Any study of natural populations using culture techniques would miss this component of the bacterial population.

Culture-based assays for specific portions of the bacterial population can be useful. However, some aspects of the behavior of the population under study must be known. *Vibrio cholerae* has been shown to enter a nonculturable state when introduced into the environment [30]. Once in this state, the cells can no longer be cultured directly from the environment but remain viable. Culture-based techniques would not be suitable for microorganisms exhibiting this trait. Other genera, for example, *Pseudomonas* [27], do not appear to exhibit this nonculturable state and may be more suitable for study using culture techniques.

The relative speed and simplicity of culture-based assays are very attractive in ecological studies, especially as replicate samples may be examined. One way to exploit these advantages is in monitoring of introduced bacterial populations, assuming that the isolates used do not exhibit nonculturable states. To follow a specific strain against a background of the native population, antibiotic and heavy metal resistance markers have been used. These approaches share two problems; there may be a high background of resistance in the natural population [31–33], and the introduction of resistance into a strain may affect its ability to compete with native strains when it is released to the environment [34]. Selective culture media may be used to reduce problems with native resistance, but media that support growth of healthy cells may not be suitable for isolating stressed cells from the environment, necessitating the subculture of isolates from recovery media onto more selective media [35].

Techniques are now available to circumvent these problems with background resistance and selective but not optimal media. One approach is to adopt the technique of colony hybridization. Bacteria are cultured on agar and then transferred onto a nitrocellulose membrane. The cells are lysed, and the released DNA is bound to the membrane, where it can be probed for specific sequences using DNA–DNA hybridizations with a labeled DNA probe. This method has been used extensively for the analysis of DNA in molecular biology, although complications may arise when using cells from the environment and a more rigorous lysis step may have to be included [36,37]. There may also be some interference from the extracellular polysaccharides often associated with environmental isolates. This method has been used to confirm the presence of phage in investigations of transduction in soil [38] and to trace the movement of a genetically engineered *Pseudomonas* in a soil microcosm [39].

Despite problems with culturing representative samples of microorganisms from the soil environment, colony hybridizations have been used to study natural populations. Colony hybridizations using DNA probes specific for catabolic genes have been performed to study the natural occurrence of these genotypes in the environment [40]. An advantage of this technique is that the colony carrying the gene detected by the probe may be identified, enabling a comparison between the genotype and the actual phenotype of the colony. In a study of bacteria capable of degrading polychlorinated biphenyls (PCBs) from polluted sites, more isolates were shown to carry DNA homologous to genes for the degradation of PCBs than were actually able to degrade PCBs as tested in a clearing zone assay [41]. Recently, the highly conserved *nif-DK* and *nif-H* sequences of the genes encoding the nitrogenase enzyme system have been used successfully to probe rhizosphere populations by colony hybridization (R. Kimura and J. M. Lynch, unpublished). This showed that far more colonies carry the genes than appear to express them.

The initial culture step may also be carried out in liquid culture, as in the most probable number—DNA hybridization technique [42]. In this method, the initial culture step occurs in liquid medium, where a selection pressure for the organisms of interest may be used. The soil sample is suspended in culture medium containing antibiotics, and serial dilutions are carried out in the same medium in microtiter plates that are then incubated for 3 to 4 days. After this incubation, the contents of each well of the microtiter plate are filtered onto a membrane, the cells are lysed, and the DNA is fixed onto the membrane. After DNA-DNA hybridization using a ^{32}P-labeled probe, a most probable number (MPN) value for the original population size can be calculated, and extremely small populations, as low as 10 cells g^{-1} soil can be detected. When this method was compared with fluorescent antibody epifluorescence-microscopic counting of cells, the fluorescent antibody technique showed a greater population than was detected by MPN-DNA hybridization in a study of the survival of *Rhizobium leguminosarum* in soil [42]. This could have resulted from the presence of a viable but nonculturable subpopulation of the test bacteria or from labeling with the fluorescent antibody of nonviable cells whose antibody binding sites remained intact.

A combination of culture and antibody-based detection has been used in a colony-staining pour-plate method. By incorporating the sample to be tested in the agar rather than spreading a small volume over the agar surface, assay sensitivity can be improved [43]. The agar plate, after being dried, may be incubated with fluorescently labeled antibodies and colonies of interest identified. In addition to providing an increase in sensitivity by a factor of 10 in the detection of a test bacterium against the background of a native population of 10^8 cells ml^{-1} in cattle slurry (8.3% solids) [44], this technique avoids staining of nonviable cells, which could interfere with fluorescent antibody epifluorescence-microscopic cell counting.

Another technique involves the introduction into strains of "marker cassettes," which preferably are not based on resistance genes. This approach involves genetic manipulation of the strain, which would currently limit its use as the result of restrictions on the release to the environment of genetically engineered microorganisms. In contained facilities and microcosm studies, this form of marking has the advantage of speed, the potential for low backgrounds from native populations, and slight or no effect on the competitive ability of the strain [45]. A good example of an introduced marker gene is the *xylE* gene-based system. The *xylE* gene codes for the enzyme catechol 2,3-dioxygenase, which catalyzes the conversion of catechol to 2-hydroxymuconic semialdehyde. Any colony expressing the gene, without the rest of the degradative pathway, will accumulate the yellow 2-hydroxymuconic semialdehyde after spraying with catechol [27]. Continuous expression of a marker gene may place the strain to be detected under a much higher metabolic burden than just carrying the gene itself [46]. To reduce this burden, Winstanley et al. [47] constructed a cassette in which the *xylE* gene is expressed from the bacteriophage lambda P$_L$ promoter under control of the temperature-sensitive lambda repressor c1867. The marker gene in this construct will be expressed only after the temperature has been raised to 42°C. This provides a rapid detection system with little or no background interference from the native soil population (personal observation).

Other marker systems are based on the *lac* operon genes from *E. coli* and bacterial bioluminescence. The two genes involved in the *lac* operon are *lacZ* and *lacY*, encoding β-galactosidase and lactose permease, respectively. These two genes share the advantage of the *xylE* gene of having been sequenced and having well-characterized products, so sequences on the marker cassette may be used as areas for DNA probes and antibody and enzyme assays may be used to detect the gene products. The *lacZY* marker system has been used with fluorescent *Pseudomonas* sp., as this important group of bacteria is unable to utilize lactose as the sole source of carbon [48]. If only the *lacZ* gene is used, the marked strain is able to cleave the substrate, X-Gal. Both genes are required for the marked cells to grow on lactose as the sole carbon source. Although this system did suffer from background interference from the natural soil population, this was overcome by the addition of a rifampicin resistance marker.

The use of bacterial bioluminescence as a reporter gene–encoded phenotype, coupled with broth enrichment culturing or low-light camera imaging, has negligible background problems and is also relatively quick to set up and measure. The incorporation of the *luxCDABE* operon into *Xanothomas campestris* pv. *campestris* allowed it to be monitored on cabbage plants and in soil [49]. The same technique has been used to track *E. coli* in soil [50].

Once a marker cassette has been designed, it can be incorporated into the chromosome of a microorganism or introduced on a plasmid. Plasmid-carried markers can be easier to introduce into a range of strains [47], but plasmid stability and the possibility of plasmid transfer could affect the interpretation of results. A plasmid-based marker system carrying genes from the *E. coli lac* operon was shown to be stable for 3 to 4 weeks when the marked *Pseudomonas* strain was growing on soybean roots but to be less stable over longer periods [48]. One way to introduce a marker directly onto the chromosome is to use a transposon as the marker. Frederickson et al. [42] introduced the transposon Tn5 into the chromosome of *Pseudomonas putida* and *Rhizobium* spp. Tn5 confers resistance to the antibiotic kanamycin and also provides a DNA sequence that may be probed for. One limitation of this approach is the difficulty in demonstrating that the transposon is maintained not only in liquid culture but also after the marked strain has been introduced into soil.

Some predictions may be made about the stability of introduced DNA and its ability to be transferred to another host. Important factors are the length of any introduced sequence, whether the sequence is plasmid or chromosomally encoded, and where on a plasmid or the chromosome the sequence is located [35]. The stability of any introduced DNA in a host strain may vary between growth in pure culture and after the strain has been introduced into soil. A deletion in a chromosomal marker has been shown to occur at a much higher frequency in soil than in culture under optimal conditions [51]. When the pure culture conditions were changed, by using either suboptimal media or growth temperatures, the frequency of the deletion was increased. The growth of cells in pure culture under conditions of stress might be a better indicator of marker stability than growth under optimal conditions before systems are tested under environmental conditions.

III. METHODS REQUIRING ENRICHMENT OF THE MICROBE/SOIL RATIO

Many of the problems associated with the use of molecular techniques in soil microbial ecology stem from the soil matrix itself.

Disassociation of this matrix would require disruption of the interactions between microorganisms and soil particles and separation of
the two fractions or, at least, enrichment of the microbial fraction
(Figure 2). This has to be accomplished without selectively enriching elements of the microbial community and, ideally, be suited
for use with a wide range of soil types. Faegri et al. [3] described a method which employs repeated homogenizations to disperse the soil and release the associated bacterial cells, followed
by differential centrifugation to separate the two fractions. Although this proved effective with organic soils, being able to
achieve recovery rates of 55 to 88% of bacteria, it proved less effective with mineral soil, with recovery rates dropping to around
30% [23,52]. An alternative method was proposed by MacDonald
[4] in which the soil is dispersed in the presence of an ion exchange resin and the microbial fraction is enriched by elutriation.
The use of elutriation as a separation technique has been criticized on both theoretical and practical grounds [53]. However,
the use of ion exchange resins coupled to differential centrifugation was found to produce a fraction highly enriched for microbial cells, although complete separation from soil particles could
not be achieved [54]. These techniques, although enriching for
single-celled microorganisms, are not effective in enriching for
those which have mycelial growth [3,54] and can be used only to
study the spores of mycelial organisms [55]. A method for enriching for fungal mycelium from soil has been described by
Bingle and Paul [56], but the hyphae may be fragmented in this
method and the degree of leakage of the cell contents has not
been investigated.

A number of techniques may be used to examine a dispersed
and enriched microbial fraction from soil. The use of DNA probes
with extracts obtained directly from the soil is described in the
next section, but soil fractions enriched for bacterial cells have
also been used as a source of DNA to be probed. Despite the
drop in sensitivity and possible selective enrichment for certain
fractions of the bacterial community, this approach can be adopted
to reduce the difficulty of obtaining DNA of sufficient purity for
subsequent analysis.

Holben et al. [23] produced an enriched bacterial fraction using
a fractionation-centrifugation technique based on that of Faegri et
al. [3]. The cells in this fraction were lysed, and the DNA was
purified by two cesium chloride equilibrium density centrifugations.
The DNA obtained was concentrated by ethanol precipitation and
shown to be of sufficient purity to be digested by restriction enzymes. The products of the digests could be separated by agarose gel electrophoresis and, after transfer onto nitrocellulose by
Southern blotting, probed with a specific DNA gene probe. This
allowed the detection of more than one organism in each sample

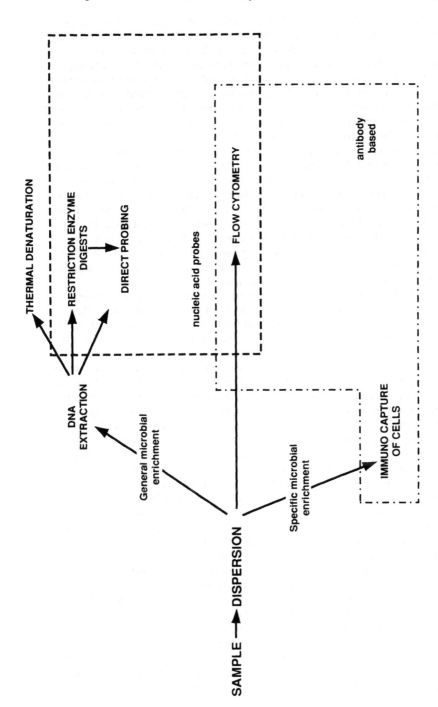

Figure 2 Enrichment-based assays.

and also provided a method for detecting some genetic rearrangements [51]. Alternatively, DNA obtained by this method could be bound directly to nitrocellulose in a slot-blot before probing. This method is faster and allows more samples to be handled, although only one probe may be used at a time and genetic rearrangements will not be detected. A cell separation technique involving sucrose gradients has also been used to purify bacterial cells from soil, and DNA of sufficient purity for use in the amplifying polymerase chain reaction (PCR) could be obtained [57].

In addition to enabling the study of specific genes or isolates, bacterial DNA obtained by the cell extraction-lysis method has been used to study the bacterial community as a whole. A total bacterial DNA extract was used in a thermal denaturation and reassociation experiment to determine the genetic diversity of the community [58]. The length of time required for the denatured DNA to reassociate indicated a very high degree of diversity in the DNA extracted. This diversity was so great that it could not be accurately determined, and it was much greater than would be expected from the dilution plating of soil samples.

A technique that requires efficient dispersion but only partial separation of microorganisms and soil particles is flow cytometry. Flow cytometers enable the rapid identification, sorting, and enumeration of different cell types. A description of the machinery involved may be found in Shapiro [59]. To use this technique with soil microorganisms, a range of fluorescent markers and dyes must be employed. The fraction of the microbial community identified is selected by using a fluorescent marker with the correct specificity. Fluorescent dyes, such as acridine orange, and fluorescently labeled antibodies have been used in conjunction with ultraviolet (UV) microscopy to identify various fractions of the microbial community. The use of these methods of identification with a flow cytometer would enable measurements to be taken much more rapidly. When flow cytometry was used to enumerate a strain of *Flavobacterium* added to soil, using a fluorescent monoclonal antibody to label the introduced cells, this technique was shown to be less accurate but much faster than direct counts [60]. One other group of fluorescent markers, fluorescently labeled oligonucleotides, will enable studies to be extended still further. Whole cells are able to take up oligonucleotides coupled to a fluorescent label, and these probes will allow strain identification in mixed populations [61]. As techniques for sample preparation improve, the speed of a flow cytometer and its ability to separate labeled from unlabeled cells should make it a powerful tool in microbial ecology.

Specific subpopulations from the microbial fraction may be further enriched by the use of immunocapture strategies. These strategies depend on the availability of a specific antibody which will bind to a site on the cell surface. These antibodies are bound to a solid support. Both enzyme-linked immunosorbent assay (ELISA) plates [44] and magnetic beads [62] have been used with environmental samples. The magnetic beads approach has the advantage that a larger sample volume may be used. The magnetic bead is first coated with antibody and then introduced into the suspension containing the target microorganism. The antibody then interacts with the cell surface antigen, and the bead-microorganism complex is separated from the suspension by using a magnet. This method was able to enrich for a strain of *P. putida* after it had been introduced into a lakewater microcosm [62] and also for spores of *Streptomyces* sp. from soil [63]. This type of enrichment avoids the need for culture techniques, which were adopted to enable the detection of Tn5 mutants [64,42].

IV. DIRECT ANALYSIS FROM THE ENVIRONMENT

The final group of methods allows direct analysis of environmental samples, either by direct observation or by the extraction of marker molecules directly from soil without any enrichment (Figure 3). As well as being useful for studies of bacterial populations, these methods are suitable for the study of mycelial organisms.

Direct microscopic observations will remain important, as they provide spatial as well as quantitative data on the microbial community. The greatest problems with microscopy are differentiating the fraction of the community under study both from other microorganisms and from soil particles and the time-consuming nature of the work. Improvements in image analysis may decrease the time required to process samples. For light and UV microscopy, the different ways to detect fractions of the microbial community are analogous to the methods used in flow cytometry and depend on the use of fluorescent dyes, conjugated antibodies, or oligonucleotide-DNA probes,

As an alternative to fluorescent markers, colloidal gold particles have been bound to antibodies. The advantages of this are that normal light microscopes may be used together with a wide range of light microscopic techniques and that the label will not fade. These advantages have to be offset against the disadvantages of lower sensitivity and less intense staining compared to fluorescent markers [65]. The electron density of colloidal gold means that these markers are also suitable for use with electron microscopy. Using colloidal gold–labeled antibodies to molecules of fungal origin, enzymes have been located with respect to both fungal hyphae and the cell walls of the beech heartwood substrate [66], as well as a phytotoxic glycopeptide produced by

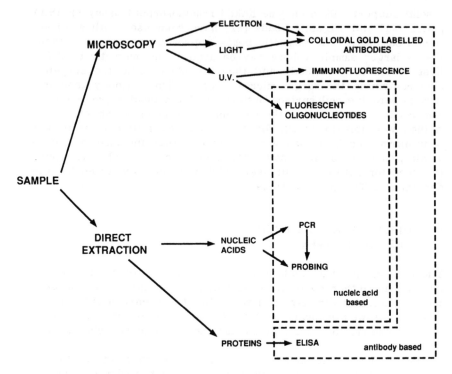

Figure 3 Direct analysis.

Ophiostoma ulmi in both host elm tissue and the cells of the pathogen [13].   Provided that specific antisera can be obtained against
enzymes and toxins of interest, this method could provide very
precise details of the site of processes occurring in soil.

Direct extraction of nucleic acids from soil samples poses a number of problems.   The microbial cells must be lysed, the released
DNA must be prevented from binding to soil particles and be protected from degradation, and as little contaminating material as possible should be extracted.

Ogram et al. [67] described a method for extracting DNA from
sediment that used 2.5% sodium dodecyl sulfate (SDS) at 70°C together with a bead mill to ensure efficient cell lysis.   The extracts
obtained were concentrated by precipitation with ethanol or polyethylene glycol, contaminating organic matter was removed by precipitation with potassium acetate, and the DNA was further purified either by cesium chloride gradient centrifugation or hydroxy-
apatite chromatography.   By omitting the lysis step, this method

was also able to measure extracellular DNA in the sample. Although the method was effective in extracting DNA (e.g., 26 μg of intracellular DNA was obtained from 1 g of sediment), the DNA obtained was difficult to purify. Including polyvinyl polypyrolidine (PVPP) at the lysis stage decreased the amount of humic matter contaminants [68]. In a comparison of this direct extraction technique with the cell extraction followed by lysis method of Holben et al. [23], the direct lysis method yielded more than 10 times as much DNA. Comparison of the bases in the DNA extracted showed that the DNA extracts obtained were also qualitatively different [68]. There is increased interest in DNA extractions from smaller samples of soil, down to 1 g in size [69], although the problem of purifying direct extracts remains.

RNA has also been extracted directly from soil. Ribosomal RNA was extracted by sonicating soil in a guanidine hydrochloride–Tris buffer at room temperature, and the extract was purified by ethanol precipitation and phenol-chloroform extraction. This method also yielded more of the target nucleic acid than the cell extraction followed by lysis method [70]. Using a ribosomal RNA probe, the soil sample size could be as small as 1 g fresh weight and a detection limit of 10^4 cells g^{-1} was estimated.

One problem encountered in any method using direct nucleic acid extraction is the difficulty in detecting small quantities of target nucleic acid against a background of nontarget nucleic acid. This problem is apparent when nucleic acid samples are bound on solid supports before probing and when sequences have to be cloned to produce a library of cDNA. Large numbers of clones have to be screened to detect those carrying the nucleic acid sequences of interest [71]. To get around this problem without having to adopt enrichment culture techniques, a selective recovery technique or the PCR can be employed. Weller and Ward [72] used a selective recovery technique based on the cloning of 16S rRNA sequences obtained from a cyanobacterial mat. This selective cloning has the advantage over the original shotgun cloning proposed by Pace et al. [71] that it enables a smaller library to be screened, and the likelihood of detecting organisms present only in low numbers is increased. The selectivity of the cloning was achieved by using a primer at the start of the cloning procedure that bound to one of the highly conserved regions of the 16S rRNA. The alternative method, which employs no cloning, is to use the PCR, which allows a sequence of DNA to be amplified many times. The DNA is first heated until it becomes single stranded. It is then allowed to cool and annealed to oligonucleotide primers that bind to either side of the area to be amplified. A temperature-stable DNA polymerase is then used, together with free nucleotides, and the primers are extended across the target region. The reaction mixture is then heated again and the cycle repeated, resulting in exponential amplification of the target sequence. This method has been used with strain-specific oligonucleotide probes to identify strains of *Frankia* sp. in root nodules [73] and to detect a genetically engineered strain of *Pseudomonas cepacia* that had been introduced into sediment microcosms [52]. In the latter case, it was possible to detect 0.3 pg of target DNA against a background of 20 µg of nontarget DNA. In the study with *Frankia* sp., competition studies could be carried out that previously were not possible. Some interference with the activity of the enzyme, *TaqI* DNA polymerase, by phenolic compounds was noted.

Antigens may also be directly extracted and quantified in ELISA. The use of an ELISA to distinguish between different vesicular-arbuscular mycorrhizal (VAM) fungi was described by Aldwell et al. [74]. This technique has also been used to detect infection of wheat roots by the causative agent of take-all, the fungus *Gaeumannomyces graminis* var. *tritici* [75]. Both of these studdies used polyclonal antisera. The problem of trying to detect an antigen that is present at a low concentration against a background of similar compounds was solved by El-Nashaar et al. [75] by using a double-antibody sandwich ELISA. In this technique, antiserum is bound to a solid support and the sample to be tested is allowed to react with it. Any bound antigen is detected by a second incubation with antiserum bound to a reporter molecule. Only antigens reacting with the antibody attached to the solid support are bound, eliminating the competition for binding sites between different antigens that would occur if the sample was directly immobilized to the solid support as in indirect ELISA. This problem of competition for binding sites also occurs with most DNA hybridization procedures.

Extracted antigens may also be assayed after immobilization on membranes. A polyvinylidene difluoride (PVDF) membrane has been used to enable a simple, rapid "dipstick" assay for detecting fungal antigens without the requirement for complex apparatus [21]. PVDF proved to be a better membrane material than nitrocellulose in this case. Nitrocellulose membranes have been used in an assay for antigens from *Trichoderma harzianum*. After extraction from a peat-sand plant growth medium, the antigens were slot blotted onto the membrane and assayed using a polyclonal antiserum raised agaisnt *T. harzianum* mycelium grown on straw (J. P. Carter and J. M. Lynch, unpublished). This test system showed much lower levels of cross-reaction with other fungi than an ELISA using polystyrene plates.

V. CONCLUSIONS

A wide variety of new methods are being developed that will help answer questions about the activity, composition, and size of the soil microbial community. Many of these methods have initially been applied to the study of the ecology of soil bacteria, which probably reflects the influence on this area of the assessment of the risk of introducing genetically engineered bacteria to the environment. Extension of the new methods to the study of soil fungi will be of great importance because of the greater fungal than bacterial biomass in most soils and the crucial roles of this biomass in many geochemical cycles, soil fertility, and crop production. DNA probes have now been developed for at least one soilborne fungal plant pathogen [76].

The ability to analyze a sample from the environment with accuracy means that sampling precision will have to be improved [77]. Many of the soil extraction techniques require large samples, which prevents investigation of the importance of microsites in microbial interactions. More work is needed to improve the methods for extracting microorganisms and marker compounds from the soil matrix.

It is clear that soil microorganisms can have much greater longevities than would be predicted from the energy available to them [78]. The capacity to monitor populations with greater precision will provide a sounder basis for understanding soil processes, such as nutrient cycling.

The nature of the study being undertaken will determine which of the new techniques are suitable. The inability to culture the majority of the microorganisms in soil is a great limitation to the study of natural populations. However, the speed of culture techniques may leave them with a role in analyzing the survival of bacteria added to the soil if the strains used do not enter a nonculturable stage.

It is important to note that many of the more interesting studies do not depend on the use of only one method. DNA analysis may provide information on the genetic potential of the soil microbial community, but other methods are required to determine whether this potential is being expressed. In this chapter we have tried to outline some of the principles involved in the selection of some of the newer methods to study soil populations. For more detailed accounts of the molecular techniques, the reader is referred to a recent chapter in this series [79]. Fine-tuning of the methodologies involves, for example, considerations of the labels used on probes [80]. It seems likely that the combination of DNA-, RNA-, and antibody-based assays with more established biochemistry- and physiology-based assays will enable a more complete analysis of any situation. Such analyses should herald an exciting new era in the study of microbial populations and biology in natural ecosystems and should enable new theoretical concepts that have been emerging to be tested by experiment. This should also be beneficial for the development of soil biotechnology [81], where attempts to manipulate the soil population balance can be monitored.

ACKNOWLEDGMENT

This work was partly supported by a grant to J. P. C. from the Department of the Environment and funds to J. M. L. from the Agriculture and Food Research Council.

REFERENCES

1. Jones, P. C. T., and J. E. Mollison. 1948. A technique for the quantitative estimation of soil microorganisms. J. Gen. Microbiol. 2:54–69.
2. Hanssen, J. F., T. F. Thingstad, and J. Goksøyr. 1974. Evaluation of hyphal lengths and fungal biomass in soil by a membrane filter technique. Oikos 25:102–107.
3. Faegri, A., V. L. Torsvik, and J. Goksøyr. 1977. Bacterial and fungal activities in soil: separation of bacteria by a rapid fractionated centrifugation technique. Soil Biol. Biochem. 9: 105–112.
4. MacDonald, R. M. 1986. Sampling soil microfloras: optimization of density gradient centrifugation in Percoll to separate microorganisms from soil suspensions. Soil Biol. Biochem. 18: 407–410.
5. West, A. W., W. D. Grant, and G. P. Sparling. 1987. Use of ergosterol, diaminopimelic acid and glucosamine contents of soils to monitor changes in microbial populations. Soil Biol. Biochem. 19:607–612.
6. Lumsden, R. D., J. P. Carter, J. M. Whipps, and J. M. Lynch. 1990. Comparison of biomass and viable propagule measurements in the antagonism of *Trichoderma harzianum* against *Pythium ultimum*. Soil Biol. Biochem. 22:187–194.
7. Waksman, S. A. 1944. Three decades with soil fungi. Soil Sci. 58:89–115.
8. Olsen, B. H. 1978. Enhanced accuracy of coliform testing in seawater by a modification of the most probable number method. Appl. Environ. Microbiol. 36:438–444.
9. Sumner, D. R., B. Doupnik, and M. G. Boosalis. 1981. Effects of reduced tillage and multiple cropping on plant diseases. Annu. Rev. Phytopathol. 19:167–187.
10. Elliott, L. F., and J. M. Lynch. 1985. Plant growth-inhibitory pseudomonads colonizing winter wheat (*Triticum aestivum* L.) roots. Plant Soil 84:57–65.
11. Frankland, J. C., and A. D. Bailey. 1981. Development of an immunological technique for estimating mycelial biomass of *Mycena galopus* in leaf litter. Soil Boil. Biochem. 13:87–92.

12. Postma, J., J. D. van Elsas, J. M. Govaert, and J. A. van Veen. 1988. The dynamics of *Rhizobium leguminosarum* biovar *trifoli* introduced into soil as determined by immunofluorescence and selective plating techniques. FEMS Microb. Ecol. 53:251–260.
13. Ouellette, G. B., and N. Benhamou. 1987. Use of monoclonal antibodies to detect molecules of fungal plant pathogens. Can. J. Plant Pathol. 9:167–176.
14. Fallon, R. D., and S. Y. Newell. 1989. Use of ELISA for fungal biomass measurement in standing-dead *Spartina alterniflora* Loisel. J. Microbiol. Methods 9:239–252.
15. Carter, J. P., and J. M. Lynch. 1991. Substrate-dependent variation in the protein profile and antigens of *Trichoderma harzianum*. Enz. Microb. Technol. 13:537–543.
16. Cleyet-Marel, J. C., N. Bousquet, and D. Mousain. 1989. The immunochemical approach for the characterization of ectomycorrhizal fungi. Agric. Ecosyst. Environ. 28:79–83.
17. Wright, S. F., J. B. Morton, and J. E. Sworobuk. 1987. Identification of a vesicular-arbuscular mycorrhizal fungus by using monoclonal antibodies in an enzyme-linked immunosorbent assay. Appl. Environ. Microbiol. 53:2222–2225.
18. Dewey, F. M. 1988. Development of immunological diagnostic assays for plant pathogens. Brighton Crop Protection Conference — Pests and Diseases 2:777–786.
19. Gerik, J. S., S. A. Lommel, and O. C. Huisman. 1987. A specific serological staining procedure for *Verticillium dahliae* in cotton root tissue. Phytopathology 77:261–266.
20. Wong, W. C., M. White, and I. G. Wright. 1988. Production of monoclonal antibodies to *Fusarium oxysporum* f. sp. *cubense* race 4. Lett. Appl. Microbiol. 6:39–42.
21. Dewey, F. M., M. M. MacDonald, and S. I. Pillips. 1989. Development of monoclonal-antibody-ELISA, -dot-blot and -dip-stick immunoassays for *Humicola lanuginosa* in rice. J. Gen. Microbiol. 135:361–374.
22. Burrell, R. G., C. W. Clayton, M. E. Gallegly, and V. G. Lilly. 1966. Factors affecting the antigenicity of the mycelium of three species of *Phytophthora*. Phytopathology 56:422–426.
23. Holben, W. E., J. K. Jansson, B. K. Chelm, and J. M. Tiedje. 1988. DNA probe method for the detection of specific microorganisms in the soil bacterial community. Appl. Environ. Microbiol. 54:703–711.
24. Woesse, C. R. 1977. The concept of cellular evolution. J. Mol. Evol. 10:1–6.
25. Ingraham, J. L., O. Maaloe, and F. C. Neidhart. 1983. Growth of the bacterial cell. Sinauer, Sunderland, MA.

26. Tsai, Y.-L., M. J. Park, and B. H. Olson. 1991. Rapid method for direct extraction of mRNA from seeded soils. Appl. Environ. Microbiol. 57:765–768.
27. Morgan, J. A. W., C. Winstanley, R. W. Pickup, J. G. Jones, and J. R. Saunders. 1989. Direct phenotypic and genotypic detection of a recombinant pseudomonad population released into lake water. Appl. Environ. Microbiol. 55:2537–2544.
28. Wainwright, M. 1982. Origin of fungal colonies on dilution and soil plates determined using nonanoic acid. Trans. Br. Mycol. Soc. 79:178–179.
29. Bakken, L. R., and R. A. Olsen. 1989. DNA-content of soil bacteria of different cell size. Soil Biol. Biochem. 6:789–793.
30. Colwell, R. R., P. R. Brayton, D. J. Grimes, D. B. Roszak, S. A. Huq, and L. M. Palmer. 1985. Viable but non-culturable *Vibrio cholerae* and related pathogens in the environment: implications for release of genetically engineered microorganisms. Biotechnology 3:817–820.
31. Kelch, W. J., and J. S. Lee. 1978. Antibiotic resistance patterns of gram-negative bacteria isolated from environmental sources. Appl. Environ. Microbiol. 36:450–456.
32. Jones, J. G. S., S. Gardener, B. M. Simon, and R. W. Pickup. 1986. Antibiotic resistant bacteria in Windermere and two remote upland tarns in the English Lake District. J. Appl. Bacteriol. 60:443–453.
33. Jones, J. G. S., S. Gardener, B. M. Simon, and R. W. Pickup. 1986. Factors affecting the measurement of antibiotic resistance in bacteria isolated from lake water. J. Appl. Bacteriol. 60:455–462.
34. Turco, R. F., T. B. Moorman, and D. F. Bezdicek. 1986. Effectiveness and competitiveness of spontaneous antibiotic resistant mutants of *Rhizobium leguminosarum* and *Rhizobium japonicum*. Soil Biol. Biochem. 18:259–262.
35. Stotzky, G., M. A. Devanas, and L. R. Zeph. 1990. Methods for studying bacterial gene transfer in soil by conjugation and transduction. Adv. Appl. Microbiol. 35:57–169.
36. Datta, A. R., B. A. Wentz, and W. E. Hill. 1987. Detection of hemolytic *Listeria monocytogenes* by using DNA colony hybridisation. Appl. Environ. Microbiol. 53:2256–2259.
37. Trevors, J. T., and J. D. van Elsas. 1989. A review of selected methods in environmental microbial genetics. Can. J. Microbiol. 35:895–902.
38. Zeph, L. R., and G. Stotzky. 1989. Use of a biotinylated DNA probe to detect bacteria transduced by bacteriophage P1 in soil. Appl. Environ. Microbiol. 55:661–665.

39. Trevors, J. T., J. D. van Elsas, L. S. van Overbeek, and M.-E. Starodub. 1990. Transport of a genetically engineered *Pseudomonas fluorescens* strain through a soil microcosm. Appl. Environ. Microbiol. 56:401–408.
40. Sayler, G. S., M. S. Shields, E. T. Tedford, A. Breen, S. W. Hooper, K. M. Sirotkin, and J. W. Davis. 1985. Application of DNA-DNA colony hybridizations to the detection of catabolic genotypes in environmental samples. Appl. Environ. Microbiol. 49:1295–1303.
41. Walia, S., A. Khan, and N. Rosenthal. 1990. Construction and applications of DNA probes for detection of polychlorinated biphenyl–degrading genotypes in toxic organic–contaminated soil environments. Appl. Environ. Microbiol. 56: 254–259.
42. Frederickson, J. K., D. F. Bezdicek, F. J. Brockman, and S. W. Li. 1988. Enumeration of Tn5 mutant bacteria in soil by using a most-probable-number-DNA hybridization procedure and antibiotic-resistance. Appl. Environ. Microbiol. 54:446–453.
43. van Vuurde, J. W. L. 1989. Immunostaining of colonies for sensitive detection of viable bacteria in sample extracts and on plant parts. Proc. 7th Int. Conf. Plant Path. Bact., p. 907–912, Budapest, Hungary.
44. van Vuurde, J. W. L, and N. J. M. Roozen. 1990. Comparison of immunofluorescence colony-staining in media, selective isolation on pectate medium, ELISA and immunofluorescence cell staining for detection of *Erwinia carotovora* subsp. *atroseptica* and *E. chrysanthemi* in cattle manure slurry. Neth. J. Pl. Path. 96:75–89.
45. Van Elsas, J. D., L. S. van Overbeek, A. M. Feldman, A. M. Dullemans, and O. de Leeuw. 1991. Survival of genetically engineered *Pseudomonas fluorescens* in soil in competition with the parent strain. FEMS Microb. Ecol. 85:73–80.
46. Sayler, G. S., C. Harris, C. Pettigrew, D. Pacia, A. Breen, and K. M. Sirotkin. 1987. Evaluating the maintenance and effects of genetically engineered microorganisms. Dev. Ind. Microbiol. 27(J. Ind. Microbiol. Suppl. 1):135–149.
47. Winstanley, C., J. A. W. Morgan, R. W. Pickup, J. G. Jones, and J. R. Saunders. 1989. Differential expression of lambda P_L and P_R promoters by a cl repressor in a broad-host-range thermoregulated plasmid marker system. Appl. Environ. Microbiol. 55:771–777.
48. Drahos, D. J., B. C. Hemming, and S. McPherson. 1986. Tracking recombinant organisms in the environment: β-galactosidase as a selectable non-antibiotic marker for fluorescent pseudomonads. Biotechnology 4:439–444.

49. Shaw, J. J., F. Dane, D. Geiger, and J. W. Kloepper. 1992.
     Use of bioluminescence for detection of genetically engineered
     microorganisms released into the environment. Appl. Environ.
     Microbiol. 58:267–273.

50. Rattray, E. A. S., J. I. Prosser, K. Killham, and L. A.
     Glover. 1990. Luminescence-based nonextractive technique
     for *in situ* detection of *Escherichia coli* in soil. Appl. En-
     viron. Microbiol. 56:3368–3374.

51. Jansson, J. K., W. E. Holben, and T. J. Tiedje. 1989. De-
     tection in soil of a deletion in an engineered DNA sequence
     by using DNA probes. Appl. Environ. Microbiol. 55:3022–
     3025.

52. Steffan, R. J., and R. M. Atlas. 1988. DNA amplification
     to enhance detection of genetically engineered bacteria in en-
     vironmental samples. Appl. Environ. Microbiol. 54:2185–2191.

53. Hopkins, D. W., A. G. O'Donnell, and S. J. MacNaughton.
     1991. Evaluation of a dispersion and elutriation technique
     for sampling microorganisms from soil. Soil Biol. Biochem.
     23:217–226.

54. Hopkins, D. W., A. G. O'Donnell, and S. J. MacNaughton.
     1991. A dispersion and differential centrifugation technique
     for representatively sampling microorganisms from soil. Soil
     Biol. Biochem. 23:227–232.

55. Herron, P. R., and E. M. H. Wellington. 1990. New method
     for extraction of streptomyces spores from soil and application
     to the study of lysogeny in sterile-amended and nonsterile
     soil. Appl. Environ. Microbiol. 56:1406–1412.

56. Bingle, W. H., and E. A. Paul. 1986. A method for sep-
     arating fungal hyphae from soil. Can. J. Microbiol. 32:62–
     66.

57. Pillai, S. D., K. L. Josephson, R. L. Bailey, C. P. Gerba,
     and I. L. Pepper. 1991. Rapid method for processing soil
     samples for polymerase chain reaction amplification of specific
     gene sequences. Appl. Environ. Microbiol. 57:2283–2286.

58. Torsvik, V., J. Goksøyr, and F. L. Daae. 1990. High di-
     versity in DNA of soil bacteria. Appl. Environ. Microbiol.
     56:782–787.

59. Shapiro, H. M. 1990. Flow cytometry in laboratory micro-
     biology: new directions. ASM News 56:584–588.

60. Page, S., and R. G. Burns. 1991. Flow cytometry as a
     means of enumerating bacteria introduced into soil. Soil
     Biol. Biochem. 23:1025–1028.

61. Amann, R. I., B. J. Binder, R. J. Olson, S. W. Chisholm,
     R. Devereaux, and D. A. Stahl. 1990. Combination of 16S
     rRNA–targeted oligonucleotide probes with flow cytometry
     for analyzing mixed microbial populations. Appl. Environ.
     Microbiol. 56:1919–1925.

62. Morgan, J. A. W., C. Winstanley, R. W. Pickup, and J. R. Saunders. 1991. Rapid immunocapture of *Pseudomonas putida* cells from lake water by using bacterial flagella. Appl. Environ. Microbiol. 57:503–509.

63. Wellington, E. M. H., N. Cresswell, P. R. Herron, L. J. Clewlow, V. A. Saunders, and A. Wipat. 1990. Gene transfer between streptomycetes in soil, p. 216–231. *In* J. C. Fry and M. J. Day (eds.), Bacterial genetics in natural environments. Chapman & Hall, London.

64. Bentjen, S. A., J. K. Fredrickson, P. van Voris, and S. W. Li. 1988. Intact soil-core microcosms for evaluating the fate and ecological impact of the release of genetically engineered microorganisms. Appl. Environ. Microbiol. 55:198–202.

65. Underberg, H., and J. W. L. van Vuurde. 1989. *In situ* detection of *Erwinia chrysanthemi* on potato roots using immunofluorescence and immunogold staining. Proc. 7th Int. Conf. Plant. Path. Bact., p. 937–942. Budapest, Hungary.

66. Gallagher, I. M., M. A. Fraser, C. S. Evans, P. T. Atkey, and D. A. Wood. 1989. Ultrastructural localization of lignocellulose degrading enzymes, p. 426–442. *In* N. G. Lewis and M. G. Paice (eds.), Plant cell wall polymers, biogenesis and biodegradation. ACS Symposium Series No. 339. American Chemical Society, Washington, DC.

67. Ogram, A., G. S. Sayler, and T. Barkay. 1987. The extraction and purification of microbial DNA from sediments. J. Microbiol. Methods 7:57–66.

68. Steffan, R. J., J. Goksøyr, A. K. Bej, and R. M. Atlas. 1988. Recovery of DNA from soils and sediments. Appl. Environ. Microbiol. 54:2908–2915.

69. Porteous, L. A., and J. L. Armstrong. 1991. Recovery of bulk DNA from soil by a rapid, small-scale extraction method. Curr. Microbiol. 22:345–348.

70. Hahn, D., R. Kester, M. J. C. Starrenburg, and A. D. L. Akkermans. 1990. Extraction of ribosomal RNA from soil for detection of *Frankia* with oligonucleotide probes. Arch. Microbiol. 154:329–335.

71. Pace, N. R., D. A. Stahl, D. J. Lane, and G. J. Olsen. 1986. The analysis of natural microbial populations by ribosomal RNA sequences. Adv. Microb. Ecol. 9:1–55.

72. Weller, R., and D. M. Ward. 1989. Selective recovery of 16S rRNA sequences from natural microbial communities in the form of cDNA. Appl. Environ. Microbiol. 55:1818–1822.

73. Simonet, P., P. Normand, A. Moiroud, and R. Bardin. 1990. Identification of *Frankia* strains in nodules by hybridization of polymerase chain reaction products with strain-specific oligonucleotide probes. Arch. Microbiol. 153:235–240.

74. Aldwell, F. E. B., I. R. Hall, and J. M. B. Smith. 1983. Enzyme linked immunosorbent assay (ELISA) to identify endomycorrhizal fungi. Soil Biol. Biochem. 15:377–378.

75. El-Nashaar, H. M., L. W. Moore, and R. A. George. 1986. Enzyme linked immunosorbent assay quantification of initial infection of wheat by *Gaeumannomyces graminis* var. *tritici* as moderated by biocontrol agents. Phytopathology 76:1319–1322.

76. Henson, J. M. 1989. DNA probe for identification of the take-all fungus, *Gaeumannomyces graminis*. Appl. Environ. Microbiol. 55:284–288.

77. Hirano, S. S. 1990. Appropriate monitoring strategies for released microorganisms. Risk Assessment in Agricultural Biotechnology: Proceedings of the International Conference, p. 146–152. University of California.

78. Lynch, J. M. 1990. Longevity of bacteria: considerations in environmental release. Curr. Microbiol. 20:387–390.

79. Sayler, G. S., K. Nikbakht, J. T. Fleming, and J. Packhard. 1992. Applications of molecular techniques to soil biochemistry, p. 131–172. *In* G. Stotzky and J.-M. Bollag (eds.), Soil biochemistry, Vol. 7. Marcel Dekker, New York.

80. Zeph, L. R., and G. Stotzky. 1991. Comparison of three nonradioactive and a radioactive DNA probe for the detection of target DNA by DNA hybridization. Curr. Microbiol. 22:79–84.

81. Lynch, J. M. 1983. Soil biotechnology. Microbiological factors in crop productivity. Blackwell Scientific Publications, Oxford.

8

Pesticide Effects on Enzyme
Activities in the Soil Ecosystem

ANDREAS SCHÄFFER *CIBA-GEIGY Ltd., Basel, Switzerland*

Based on the multidisciplinary studies on the biology, chemistry, and physics of soil in the last 100 years, which in themselves are a reflection of the complexity and heterogeneity of the system, soil can be regarded as living tissue [1] on an inorganic and organic support. Enzyme activities in this environment have been the topic of intensive research since the middle of this century, and excellent reviews are available as comprehensive introductions and surveys of significant publications [1–5]. In this chapter, an attempt is made to present a selective overview of recent literature on the effects of agrochemicals on enzyme activities in soil. Earlier studies of this aspect have been summarized elsewhere [6–11].

Section I is an introduction to the three subjects soil, enzyme activities in soil, and pesticides in soil. In Section II experimental data selected from about the last decade on interactions between pesticides and soil enzyme activities are presented in chronological order for each enzyme class, i.e., for various oxidoreductases, ester hydrolases, glycoside hydrolases, and amidohydrolases. In several of the works cited, the authors studied the effects of agrochemicals by comparing various enzymatic activities. Such studies are presented as they are discussed in the original papers to prevent repetition of experimental conditions in the individual sections devoted to a specific enzyme. Cross-references in the respective individual enzyme sections are provided to compile all essential information. By this means it is possible to correlate the effects of a certain pesticide on various microbiological activities. All experimental data on soil enzymes influenced by pesticides are additionally summarized in tabular form.

Many of the available reports on interactions between pesticides and soil enzymes are restricted to observations of stimulative or repressive effects of the chemicals on biological activities. Thus, many results are of an empirical character with rare interpretations of the basic mechanisms responsible for the observed effects. Since most of the described interactions can be reduced to a small number of reaction types, they will be comprehensively discussed in the conclusion, Section III.

I. THE SOIL ECOSYSTEM

A. The Soil Matrix

The primary function of soil is related directly to its biomass, which, as a continually cycled pool of nutrients, is the prerequisite for the growth of plants. For the nutrients to be taken up, organic matter entering soil has to be degraded to inorganic components. This degradation is brought about by the indigenous microbiological activity of soil. These manifold and multistep processes represent the link between soil and the global carbon and nutrient cycles [12]. The biological activity comprises the contributions of soil animals, plant roots, and microorganisms, with bacteria and fungi being the most numerous soil organisms overall. Estimates range from 10^9 to 10^{12} bacteria per kg of soil, which is equivalent to approximately 300 kg bacteria per ha [13]. Almost all soil microorganisms live under conditions of starvation and are largely dormant: even under favorable conditions, the amount of cells that are active is estimated to be only 15 to 30% [14]. The localization of microbial activity and the characterization of the microenvironment in which it occurs need further research, but it is assumed that microbes reside primarily in close proximity to humic substances and organic substrates, as well as to clay minerals [15,16].

The inorganic part of soil is a composite of clays, silt, sand, and stones interspersed with air- or water-filled micro- and macropores. The overall negative charge of clays is responsible for the high activity of H_3O^+ ions in their immediate surface vicinity, which can reduce the local pH up to three units [2]. Clay has a high capacity for binding all types of natural and anthropogenic chemicals that enter soil. With respect to agricultural chemicals, the continuous partitioning between the adsorption sites on clay minerals and the liquid soil phases will affect their availability for transformation processes. Active structural domains of the inorganic soil components, on the other hand, may lower the activation energies for certain chemical reactions, a well-known but scarcely defined phenomenon known as inorganic catalysis [1,17].

Soil organic matter is composed of plant and animal debris, high- and low-molecular-weight organic compounds, and humic material. The latter can be partitioned between water and organic solvents as a function of the pH of the aqueous phase and, thus, separated into fulvic acids, humic acids, and humin. This insoluble residue, which remains undissolved in the alkali extract of humic matter, contains lipids, bound humic acids, and inorganic residues [18]. The soluble humic acids are predominantly of aromatic character with a distinct oxygen-containing functionality where the aromatic groups are linked by long-chain alkyl structures to form a flexible network. Covalent binding, intercalation, or entrapment in the voids of the sponge-like structure are possible modes for the binding of a wide variety of chemicals, including carbohydrates, amino acids, peptides, nonenzymatic and enzymatic proteins, nucleic acids, and xenobiotics. In addition, inorganic soil material can be associated with humic material to form organomineral complexes [19–21].

It is generally accepted that the sorption potential of soil is related to its organic matter content (soil organic matter, SOM) based on the enormous apparent SOM surface area of 560 to 800 m^2/g, as determined by the ethylene glycol retention method that assumes monolayer surface adsorption [22]. In a study applying the standard Brunauer-Emmett-Teller (BET) method, which is based on liquid nitrogen adsorption, SOM of soil with a high organic matter content was determined to be less than 1 m^2/g [23]. This discrepancy has been discussed in terms of inherent differences of the two methods applied [24]. Thus, it is known that the internal surface area of clays will increase considerably in presence of polar liquids, such as ethylene glycol, as interlamellar surfaces are created by lattice swelling and solvation of exchangeable cations. On the other hand, weakly adsorbed molecules, such as N_2, do not penetrate the interlayer surfaces, which would explain the observed differences in the total, i.e., external and internal, surface areas. Restricting the term "surface areas" to those that exist before the measurement and do not change during the course of the experiment, the authors concluded that sorption of nonionic organic pollutants, including pesticides, by SOM may have been overestimated in previous studies [23,25].

B. Enzyme Activities in Soil

Microorganisms are the dominant producers of enzyme activities in soil. The overall enzymatic activity of soil can be divided into 10, reasonably distinct, categories [26]:

1. Intracellular, cytoplasmic enzymes
2. Periplasmic enzymes
3. Cell membrane–associated enzymes
4. Secreted, extracellular enzymes present in the soil solution
5. Enzymes within nonproliferating cells

6. Enzymes attached to dead cells or cell debris
7. Intracellular enzymes leaked from cells
8. Enzymes complexed to substrates
9. Enzymes adsorbed on soil inorganic matter
10. Enzymes associated with humic material

Originally, only cell-free "abiotic" enzymes were considered to be soil enzymes [27], whereas in the more comprehensive "accumulated enzyme" concept both microbial enzymes (i.e., those in dead and viable but nonproliferating cells) and abiontic enzymes are defined as soil enzymes [3].

To test whether a certain activity is of cellular origin or the result of accumulated enzymes, bacteriostatic reagents are often included in the assay buffer to stop proliferation of cells and, thus, neosynthesis of extracellular activity [28]. However, the mode and extent of enzyme inhibition of the added chemicals have to be investigated as well. It is known that enzymes are inhibited by organic solvents—e.g., aryl sulfatase, glucosidase, and catalase by dimethyl sulfoxide (DMSO) [28] and laccase and peroxidase by toluene [29]. Toluene limits the growth of microorganisms, but it may artificially increase exoenzyme activities as the result of microbial lysis. Bacterial and fungal activities can be selectively inhibited by the addition of streptomycin and cycloheximide, respectively [30]. Nonchemical alternatives for testing predominantly exoenzyme activity are either short incubation times, to keep proliferation to a minimum, or physical sterilization of the soil, e.g., by gamma irradiation. For the latter, balancing the degree of damage of microbial activity and the retention of enzyme activities is a function of the dose applied. To evaluate adequate amounts of irradiation, several parameters, such as the type of soil, its volume and moisture content, and the genera of microorganisms present, have to be considered.

Cell-free enzymes remaining in the soil solution will sooner or later become degraded by proteases or denatured in the inhospitable soil environment. However, exoenzymes may become stabilized by association with clay minerals or humus fractions [5]. The negatively charged clays will interact with the ionizable functional groups of proteins and especially tight complexes may be formed below their isoelectric point. As a result of binding on clays, the retention of catalytic activity and the denaturation and concomitant deactivation of enzymes have both been reported [31], but usually their stability is increased [5]. Due to the Brönsted acidity of clay surfaces, the apparent pH optimum of enzymes can be shifted to more alkaline values by up to 2 units [32,33]. Jackbean urease immobilized on kaolinite has a pH optimum similar to that in the free form (7.0), but on montmorillonite the optimum pH is shifted to 8.5 [34].

Usually, the recovery of exoenzymes from soil even after vigorous grinding or sonication is low [35]. Interactions of proteins with humic substances include entrapment, hydrophobic interactions, or intercalation, as well as covalent linkage. The latter may involve hydroxyl, thiol, or amino functional groups of enzymes, leading to active immobilized derivatives if the binding groups are not part of the active sites. Generally, the immobilized derivatives are considerably more resistant to proteolysis and environmental stresses [36]: under favorable conditions extremely long persistences of enzymatic activities in soil have been reported [1]. As the result of stable organomineral enzyme complexes, some of which have been isolated [37], soil possesses a persistent enzymatic capacity that is, at least temporarily, independent of cell growth. Accordingly, the biochemical activity of accumulated enzymes for certain reactions has been estimated to be more important than that of the microbial cells [38].

One general, but not easy to meet, criterion for assaying soil enzyme activities is the correlation of activities under field and laboratory conditions, especially with regard to the substrate concentration and the temperature, moisture content, and aeration of the soil. Inasmuch as soil is a heterogeneous medium, wherein all reactions occur at interfaces between solid, liquid, and gaseous phases, the methodological problems are more difficult than in classical enzyme tests performed in homogeneous aqueous solutions [32]. Thus, the adsorption of substrates to the soil matrix will affect the turnover rate and binding constants of the enzymes involved. As an example, Cervelli et al. investigated the hydrolysis of urea by urease in presence of herbicide inhibitors. By combining the adsorption behavior of the herbicides, characterized by the Freundlich isotherm, and classical Michaelis-Menten kinetics, they determined the inhibition constants of urease by the respective pesticide inhibitors [39].

To maintain the pH during assays, buffers should be used, especially if products with acidic or alkaline properties are formed. The optimal buffer concentration and pH can be tested first by varying the conditions, but the latter has to be close to the original soil pH. General and specific instructions for performing enzyme assays are available [1—4,35]. Some details will also be provided for specific enzymes as they are discussed below.

C. Pesticides in Soil

Despite the beneficial impact of pesticides in improving and stabilizing agricultural productivity by the control of obnoxious weeds, fungi and insects, these allochthonous organic chemicals have been in the center of debates about the contamination of environmental ecosystems for a long time.

With an annual production of ca. 10^6 tons of active ingredients worldwide [40] and assuming a total arable land area of 13.8×10^6 km² on earth [41], the, admittedly unrealistic, even distribution of pesticides would amount to about 0.7 kg/ha/year. Usual application rates recommended by the producers are within 0.1–10 kg/ha; even lower dosages of up to 0.01 kg/ha are sufficient for some specific herbicides, such as sulfonylureas. Although these figures appear small compared to the quantity of organic carbon in topsoil of up to 10^5 kg/ha, the inherent bioactivity of agrochemicals demands conscientious and careful use. It is, however, fair to state that pesticides represent a class of industrial chemicals with extensively studied and controlled ecological and ecotoxicological profiles. During the last decade, the number of new commercial agrochemicals decreased as a result of the enormous difficulties in meeting both biologically and environmentally favorable behavior. Strategies such as integrated pest management (IPM) represent significant improvements in well-balanced and sound usage of pesticides, as they are based on chemicals with optimized ratios of biospecificity and dosage, favor the use of as little as possible and only as much as necessary, and recommend the supplementary utilization of nonchemical alternatives for pest control.

The behavior of pesticides in soil is governed by various parameters. Of primary importance are the chemical, structural, and physical properties of the molecules themselves, such as their water solubility, dissociation constants, and vapor pressure. Several nondegradative mass transport phenomena are related to the octanol-water partition coefficient; examples are adsorption to soil matter and desorption into the soil solution, partitioning between soil and air and between water and air, and uptake by plants, i.e., their systemic properties [42–44]. Evaporation of agrochemicals is restricted to those located at the soil surface, but pesticides from deeper soil layers become volatile as they return to the surface soil layer and as a result of capillary flow under drying conditions [45]. Mass transport by leaching depends on the equilibria between the dissolved and sorbed state and is driven by gravitational movement. A potentially significant contribution to the latter is ascribed to so-called preferential flow, a phenomenon currently intensely studied and defined as flow through matrix regions with higher than average permeability and through macropores created by plant roots, soil animals, or successive drastic changes in soil temperature and moisture [46].

Pesticide degradation in soil is the result of a combination of physical, chemical, and, predominantly, biological events [47]. Although the influences of the chemical structure of the pesticides, the physical properties of both pesticides and soil, and even the climate on degradation rates can be accounted for in modeling approaches [44], prediction of the resistance of a chemical to abiotic and biotic transformations is far too complex to be of any practical use. Abolishing microbial activity by soil sterilization gives some mechanistic indication of the extent to which the pesticide is transformed abiologically. However, depending on the methodology applied, the residual activity of accumulated enzymes may still contribute to the observed degradation. Usually, abiotic transformations lead to products that have higher polarity than the parent compound and are thus more susceptible to microbial attack. The usual abiotic degradation pathways in soil, with or without inorganic catalysis, include photolysis at the soil surface, hydrolysis, reduction and oxidation, and, to a lesser extent, isomerizations [43,48].

As for biological ways to degrade xenobiotics, there is a plethora of microbial soil metabolism studies, including those aimed at specific degradation of pesticides by genetically engineered microbes [49–51]. However, only a limited number of papers describe detailed biochemical interactions between pesticides and soil enzymes. Thus, the extent to which accumulated enzymes contribute to degradation pathways has not been conclusively defined, although it is estimated to be significant [38]. In general, microbial activities require the bioavailability of substrates, i.e., material dissolved in the soil solution. Sorbed or immobilized pesticides seem to become metabolized only after desorption has occurred [52,53]. The principal mechanisms of microbial activity are the catabolic and cometabolic degradation of substrates, their cross-coupling to organic matter, and their incorporation and accumulation [47]. Catabolism is referred to as utilization of substrates as a source of energy and growth, and it will lead to mineralization of either all or some portion of the molecules, with the rest accumulating in soil. Cometabolism, on the other hand, refers to degradation without energy being derived from substrate oxidation to support growth [54]. The latter is the prevalent form of microbial metabolism of pesticides [55].

The structural relation of many agrochemicals, with their aromatic moieties, to soil humic matter and the chemical reactivity of their functional groups readily explain their incorporation by oxidative coupling or other mechanisms to form nonextractable, bound residues [56]. Often, bound residues are formed to a certain extent in native soils, whereas in sterilized soils under otherwise identical conditions, the amount of bound residues is significantly smaller (A. Schäffer, unpublished), indicating that biological mechanisms are involved in the process of immobilization. The immobilized pesticides, to a certain extent still as the parent compounds, remain, in part, available, as they may be released by microbial attack or taken up by plant roots [56–58]. In this respect, they must be looked upon as sources of slowly released substances but with only a minor deleterious impact on ecosystems, as they are continuously deactivated by the mechanisms discussed above.

Repeated applications of the same pesticide or of derivatives closely related structurally often lead to accelerated degradation rates, as a result of adapted and optimized microbial catabolism. This "enhanced biodegradation" is an unwelcome agricultural phenomenon that leads to less efficient pest control and reductions in crop yield [59]. However, the adapted microorganisms can be utilized advantageously for the decontamination of soil sites polluted with pesticides [60].

II. INTERACTIONS BETWEEN PESTICIDES AND SOIL ENZYME ACTIVITIES

The data presented refer to laboratory experiments, unless otherwise indicated. Common names of the pesticides are used according to customary nomenclature, e.g., in *The pesticide manual* [61]. Concentrations are given in mg/kg soil or in kg/ha and are based on the active ingredients (a.i.)—that is, on the pure chemicals, irrespective of their application as a.i. or as commercial formulations that contain various solvents and additives for optimized agricultural performance. Available data about the effects of the formulation additives themselves on enzyme activities in soil are included.

For each enzyme discussed, data are summarized in Table 1, pp. 316-326.

A. Oxidoreductases

Dehydrogenases

Dehydrogenases conduct a broad range of oxidative activities that are responsible for the degradation, i.e., the dehydrogenation, of organic matter. In a cascade of events involving specific carriers, electrons are transferred from substrates to oxygen as the final acceptor. Many specific dehydrogenases transfer the liberated hydrogen atoms to either NAD or NADP, thus taking part in the oxidoreductive processes of biosynthesis. Unlike the other enzymes discussed in this chapter, they do not accumulate extracellularly in soil and are invariably linked to the viability of intact cells. Therefore, based on the abiontic enzyme concept, dehydrogenases are not soil enzymes. However, as they represent a class of enzymes that has been most intensely studied, to gain information about the influence of natural environmental conditions and xenobiotics on activities, they are discussed here. More than 200 papers on the interactions of pesticides with dehydrogenases have been published in the last decade, and as a result of the sound basis of available experimental data, this class of microbial activities has been recommended as a useful indicator for testing the side effects of agrochemicals [62]. Clearly, it is neither feasible nor meaningful to present a complete survey of the relevant literature; rather, selected examples are presented to demonstrate the potential and shortcomings of such investigations. Dehydrogenases can be regarded as indicative of the overall microbial activities of soil, and numerous studies reveal correlations of the activity with several biomass parameters, such as microbial numbers, soil respiratory activity, ATP concentration, other enzyme activities, carbon and nitrogen cycling, and organic matter content [15,63]. To what extent the often contradictory results reduce the significance of such relations must be carefully evaluated by comparison of the respective experimental conditions under which the results were obtained.

Assays of dehydrogenases are based on the spectrophotometric quantification of a formazan dye formed by reduction of tetrazolium salts concurrent with the oxidation of a usually unknown substrate. The most common tetrazolium derivatives used are TTC, triphenyltetrazolium chloride [63]; INT, 2-(p-iodophenyl)-3-(p-nitrophenyl)-5-phenyl-tetrazolium chloride [64]; and MTT, 3-(4,5-dimethylthiazol-2-yl)-2,5-diphenyl-tetrazolium bromide [65]. Advantages of using INT as an assay reagent are the stability of the formazan complex and its high redox potential, rendering it independent of the presence or absence of oxygen and a better competitor for electrons than TTC. When assaying dehydrogenases, several potential interferences have to be taken into account:

Adsorption of tetrazolium salts to soil inorganic and organic matter

pH dependence of formazan reduction, requiring the use of buffers

Interference of certain anions (NO_3^-, SO_4^{2-}, PO_4^{3-}, Cl^- and chemicals (Fe_2O_3 or glucose) with the reduction [63]

Reaction of formazans specifically with Cu ions, e.g., in Cu-contaminated soils [66]

Inhibition of the activity by bacteriostatic agents

In a sandy soil, with 3.0% organic matter and a pH of 7.5, 5 mg/kg atrazine reduced the dehydrogenase activity by about 40%, but after 30 days the activity was similar to that of the control. However, no complete recovery of activity within this period was observed at concentrations above 50 mg/kg. A less pronounced effect was observed with fluometuron, i.e., a reversible 30% reduction at 50 mg/kg, and trifluralin, which inhibited the activity by less than 30% at 500 mg/kg. All pesticides tested were formulated products. Different responses were observed in comparing the activities of invertase, amylase, and cellulase. The effects of atrazine and cyanazine on invertase and amylase were inconsistent in that both depressive and stimulative activities were observed. Cellulase activity was diminished, but only at extremely high concentrations of atrazine; e.g., with 7100 mg/kg atrazine, the residual cellulolytic activity after 30 days of incubation was reduced to 10% of the control [67]. Concentrations of pesticides that are more than 1000-fold higher than those recommended are irrelevant for agricultural management practices, but such studies may be useful for assessing the environmental risk of pesticides in case of accidental spills.

The formulated herbicides pendimethalin and difenzoquat, both at 0.5 mg/kg, and thiobencarb at 2.5 mg/kg had no adverse effects on dehydrogenases in different soils. The fungicides folpet and captafol, at 1.0 mg/kg, inhibited dehydrogenases slightly. However, a 10-fold increase in the dosage of folpet and captafol reduced the activity significantly, but it recovered to the control level after 21 days. No effects on microbial phosphatase activity were reported and the mineralization of cellulose was only temporarily depressed by 10 mg/kg captafol, which was associated with inhibition of soil fungi [68]. The reversible effects of the pesticides on the oxidoreductive and cellulolytic activities might be explained by ready recovery of the affected microorganisms as a result of fast degradation or immobilization of the pesticides under the experimental conditions. On the other hand, it is possible that the impairment of one segment of the microbial community was compensated by increased activity of other genotypes with a resulting unchanged overall microbial activity. Whatever mechanism is responsible for the observed effects in this study, both reflect the high degree of soil homeostasis of the microbiological activity.

Sixteen different herbicides, either alone or in combinations with each other, were applied to a loamy sand (1.2% organic matter, pH 6.1) and a silty clay (3.4% organic matter, pH 5.9) as formulations at recommended field rates, i.e., between 0.3 and 11.8 kg/ha. The herbicides tested included S-ethyldipropylthiocarbamate (EPTC), dinoseb, trifluralin, bensulide, chlorthaldimethyl (DCPA), dinitramine, diuron, fluometuron, linuron, nitralin, profluralin, prometryne, alachlor, chloramben, metribuzin, and naptalam. Regardless of the soil type, no inhibition of dehydrogenase was reported. The respiratory activity and the ability to decompose alfalfa and to solubilize $Ca_3(PO_4)_2$ were also unaffected. Most of the herbicides, however, increased by as much as a factor of 3 the oxidative activity responsible for the transformation of sulfur to SO_4^{2-}. This might be explained by direct stimulation of thiobacilli or by reduction of their microbial competitors. Algal populations were severely depressed by dinoseb, i.e., by 90% in the loamy sand [69]. This shows that herbicides might affect specific functional groups of the soil microflora without changing the overall activity of the microbiota probably as a result of stimulation of resistant microorganisms. Therefore, attempts to investigate side effects of pesticides on the soil microbiota by measuring a single activity parameter often lead to misleading results.

The effects on dehydrogenase activity of formulated hexachlorocyclohexane (HCH), carbaryl, benomyl, and atrazine in an alluvial soil (0.7% organic matter, pH 6.4) were tested under flooding conditions; i.e., after pesticide application the soil was flooded with standing water. At 1 mg/kg, no inhibiton occurred with any of the pesticides, but at a 10-fold rate with HCH, dehydrogenase activity dropped temporarily to about 30% but recovered after 20 days. An even more drastic reduction was observed with benomyl at 10 mg/kg, with only 10% residual activity detected initially and still less than 50% after 20 days of incubation. Carbaryl at the same concentrations showed no effect. At a dose of 100 mg/kg, HCH and benomyl essentially abolished dehydrogenase activity with no recovery after 20 days, whereas carbaryl reduced the activity by only a factor of 2 throughout the testing period. Only atrazine induced no depression at all concentrations. In a different experiment, the same soil was flooded with standing water for 15 days before treatment with pesticides. Under such conditions HCH and benomyl at 100 mg/kg induced only negligible loss in activity [70], which shows that the experimental conditions have a major influence on the activities of soil enzymes. Absence of any information on the persistence of the chemicals under the experimental conditions is an imperfection of many reports on side-effect testing of pesticides. Clearly, 20 days of monitoring microbial activities will be considerably shorter than the half-lives of some of the pesticides tested in this study.

Three soils, a loamy sand (0.8% organic carbon, pH 5.6), a silty clay (1.4% organic carbon, pH 7.3), and a loam (13.5% organic carbon, pH 7.5), were treated with a formulated mixture of propham and medinoterb acetate at recommended field doses and at 10- and 50-fold doses. In all soils, dehydrogenase activity was inhibited by the basic dosage, i.e., by propham at 2.9 to 9.1 mg/kg and by medinoterb acetate at 1.5 to 4.5 mg/kg, for at least 1 month and for longer at the higher doses. The 50-fold dose reduced the enzyme activity significantly, reaching residual activities of less than 10% after 60 days. The effects on various microorganisms were reversible. Algae, including cyanobacteria, were inhibited by the lowest dosage, whereas aerobic bacteria (total) were stimulated, probably due to killed nonresistant organisms serving as a nutrient source for the surviving species [71].

In a sandy loam soil (4.0% organic carbon, pH 5.1) with and without lucerne meal amendment, glyphosate, paraquat, trifluralin, and atrazine as formulations were applied at normal field rates, i.e., at 5.4, 2.2, 3.4, and 2.2 kg/ha, respectively. Although the amendment of organic matter stimulated the activity of both dehydrogenases and phosphatases, no effect on the activities was observed as the result of pesticide treatment during the 8-week experiment. When fourfold doses of these pesticides were applied, only glyphosate at 21.6 kg/ha stimulated temporarily dehydrogenase activity due to the microbial utilization of either the chemical itself as carbon source or of microorganisms killed by the chemical. Relating the dehydrogenase activity to its natural variability, the chemically induced change became insignificant. The activities of phosphatase and urease were unchanged by all chemical applications [72].

Application of 3,6-dichloropicolinic acid at 0.1 kg/ha to a sandy loam with 2.8% organic carbon and a pH of 7.0 failed to inhibit dehydrogenase activities. The capacity of the soil to decompose straw was also not affected [73].

In a long-term field experiment, several soil sites with a crop rotation of sugar beet, winter wheat, and winter barley were successively treated with a pesticide mixture. One series of treatments contained chlortoluron, 2.5 kg/ha; mecoprop, 2.2 kg/ha; carbendazim, 0.2 kg/ha; triadimefon, 0.1 kg/ha; and oxydemetonmethyl, 0.2 kg/ha. The second series, used on a different soil site, comprised methabenzthiazuron, metoxuron, mecoprop, tridemorph, thiophanate-methyl, and dimethoate at 2.8, 3.2, 2.2, 0.6, 0.4, and 0.2 kg/ha, respectively. The soils used were a sandy loam type containing 1.7 to 2.8% organic carbon and with a pH of 6.9 to 7.0. In all the experiments, the reduction in dehydrogenase activity was less than 20%, and in the third year there was a stimulation in activity. Generally, strong dependences of the climatic and location characteristics on activity were reported. The decomposition of organic matter was not affected by the treatments [74]. Similar experiments using the same soil sites were conducted for 5 years with the following pesticide formulations, applied successively at recommended field rates: series 1 contained di-allate, chloridazon, aldicarb, oxydemeton-methyl, parathion, and pirimicarb; series 2 contained cycloate, metamitron, aldicarb, oxydemetonmethyl, parathion, and pirimicarb. Only slight, sometimes antagonistic but always transient, effects on dehydrogenase activity and on straw decomposition occurred [75]. It has been reported that the combination of pesticides may induce additive or even synergetic inhibitions of microbial activities [76]. However, there was no indication of such effects in this study.

Application of the formulated fungicides mancozeb, captan-folpet-folcidin, quintozene, and benomyl at recommended doses to an alluvial soil with a carbon content of 6.6% and a pH of 5.2 induced only reversible effects on dehydrogenase activity. Thus, up to

20% inhibition occurred during the first 6 weeks with benomyl or captan-folpet-folcidin, and thereafter all pesticides stimulated dehydrogenase activity by up to 40% until the control level was reached after about 20 weeks. The stimulation of dehydrogenase activity indicating the detoxification of the added biocides was correlated only partially with the respiratory activity of the soil. Urease activity was stimulated up to 20% during the first 4 weeks by quintozene but was not affected by the other pesticides. However, a pronounced and irreversible inhibition of xylanase activity was caused by quintozene: in line with the long persistence of quintozene in soil, the residual xylanase activity was only about 80% of the control after 24 weeks of incubation. The lag phase of about 3 weeks before xylanase activity became reduced suggests that quintozene interfered with xylanase-producing microorganisms rather than with the enzyme directly. Furthermore, quintozene diffusion in soil is limited and slow because of its high hydrophobicity [77]. The distinct inhibition of xylanase activity by quintozene shows that the quantification of enzymatic activities will provide specific functional information that would be masked by the use of parameters reflecting the soil's overall microbiological activity, such as soil respiration measurements. Ideally, specific functional tests and assays of the overall activity should be applied synergetically.

The effect of the herbicide dalapon at 2.6, 26.6, 266, and 2660 mg/kg on the activity of the soil microflora was tested in a loamy sand (2.4% organic carbon, pH 6.4) and a sandy loam (1.7% organic carbon, pH 7.6). The pesticide was readily degraded in both soils: no residues were detected with the two lowest concentrations after 2 and 4 weeks, respectively. However, with application at 2660 mg/kg, degradation was retarded, and 1215 mg/kg of the parent compound was analyzed in the loamy sand and 190 mg/kg in the sandy loam after 32 weeks. Dehydrogenase activity was inhibited at the two lowest concentrations, but the effects were statistically not significant. However, the activity was inhibited for 32 weeks at 266 and 2660 mg/kg, i.e., at much higher concentrations than those used for normal agricultural management practices. Phosphatase activity was slightly reduced at the normal field rate (2.6 mg/kg), but the effect disappeared after 2 weeks. Bacterial numbers fluctuated randomly, and no permanent effect of dalapon even at high concentrations was observed. Soil respiration was depressed for about 4 weeks at the lowest concentration of the pesticide, followed by small intermittent stimulations. At the highest concentrations, evolution of CO_2 and soil nitrification were severely inhibited. The reduced dehydrogenase and respiration activity in the loamy sand at high pesticide concentrations and, in contrast, the initial increased respiration concomitant with decreased dehydrogenase activity in the sandy loam are difficult to interpret. Even more complex is the evaluation of the influence and significance of the observed effects for the soil ecosystem as a whole. One simple approach is to compare chemically induced effects on microbial activities with the changes in activity due to natural climatic and topographic variations. When doing so, reversible pesticide effects that are smaller than the background fluctuation must be considered insignificant. The conclusion of this study is that dalapon used at the recommended concentration will not impair soil fertility, as all effects were reversible due to the ready microbial degradation of the pesticide and of its major metabolites [78].

The application to a clay loam soil (1.0% organic carbon, pH of 7.2) of chlorfenvinphos, chlorpyrifos, diazinon, ethion, ethoprophos, fensulfothion, fonofos, leptophos, malathion, parathion, phorate, thionazin, triazophos, trichloronate, terbufos, chlordane, dieldrin, gamma-HCH, carbofuran, oxamyl, permethrin, captan, maneb, thiram, and 2,4-D, all at 5 and 10 mg/kg, and of dichloropropane-dichloropropene, methyl isothiocyanate, and nitrapyrin, at usual field rates, resulted in no inhibition of dehydrogenase activity within 7 days, although several of the pesticides stimulated the activity (Table 1), probably as a result of an increase in resistant genotypes that catabolically utilized the affected microorganisms. Phosphatase activity was stimulated by malathion, parathion, chlordane, dieldrin, and thiram. Dichloropropane-dichloropropene and methyl isothiocyanate applied at high doses, 300 and 80 mg/kg, respectively, suppressed phosphatase activities by about 30% after 7 days of incubation. Only trichloronate reduced the activity of urease significantly, whereas several chemicals stimulated the activity both at low and high applications. Thiram reduced bacterial numbers, but they were stimulated by the other pesticides [79]. Unlike the protocol in this study, monitoring of microbial activities should be continued until the chemicals and their significant metabolites are degraded if there is any indication of impairment of the microbiota, such as the reduction of urease activity by trichloronate.

When the same pesticides were applied to an organic soil with 26.8% organic carbon and a pH of 7.2, dehydrogenase activity was reduced after 1 week by terbufos, triazophos, trichloronate, dichloropropane-dichloropropene, and nitrapyrin, with the last causing a 60% reduction. It is surprising that dehydrogenase activity was inhibited by some pesticides in the soil with a much higher organic matter content than the clay loam, since usually a higher sorption will reduce the interfering effects of chemicals added to soil. After 2 weeks, several of the pesticides stimulated dehydrogenase activity (Table 1). No close correlation between changes in the activities of dehydrogenase with those of phosphatases and urease was observed. With the exception of parathion, triazophos, carbofuran, oxamyl, permethrin, and fonofos, the low pesticide treatments reduced the phosphatase activity by up to a factor of 2, but the authors did not indicate how long these inhibitions lasted. On the contrary, chlordane, at 5 mg/kg, stimulated phosphatase activity about 1.5-fold. Urease activity was inhibited by all chemicals applied at 10 mg/kg (especially by thiram: more than 40%), but the activity recovered within 2 weeks. Stimulated urease activity after 2 weeks was observed with most of the pesticides applied at the low doses [80]. The results of this study reveal that chemicals with a broad range of structures and biological activities may exert similar effects on overall microbial metabolism resulting from temporary stimulation or repression of metabolic enzymes. It is therefore not possible to predict the effects on soil enzymes based only on knowledge of the chemical and physicochemical properties of the pesticides.

In an alluvial soil, 4.1% organic matter, pH 5.6, the formulated insecticides gamma-HCH, endrin, parathion, methidathion, omethoate, and propoxur at recommended field rates had different reversible effects on dehydrogenases. Propoxur and endrin induced stimulated activities of up to 40% after 3 and 14 weeks, respectively. Gamma-HCH, methidathion, and parathion reduced the activity of dehydrogenase for about the first 6 weeks. In all experiments, activities like those in the control soil were observed after about 20 weeks of incubation. The responses followed closely the respiratory activity of the soil, as measured by release of CO_2. Xylanase activity was 10 to 15% reduced by parathion throughout 24 weeks, but the responses to the other pesticides were either inconsistent or close to the xylanase activity in the control soil. Urease was 40 to 60% less active in presence of propoxur, gamma-HCH, and endrin during the 6 initial weeks but then recovered to the level of activity in the control soil. The other pesticides had little influence on urease activity [81]. Knowledge of the soil microflora and its relationship with soil fertility is insufficient to assess the significance of the different effects on the various enzyme activities caused by the chemicals. The only obvious conclusion to be drawn is that such effects reflect the disturbance of microbiological equilibria that usually recover after the chemical and active metabolites have been degraded.

In a field and laboratory experiment in which two different soils were used (a loamy sand with 0.6% organic carbon, pH 5.8, and a sandy loam with 2.5% organic carbon, pH 7.0), the effect of dinoseb acetate, at 1.7 kg/ha, on various microbiological activities was significantly influenced by the soil type but less by temperature and soil humidity. In the heavier soil, stronger adsorption and, thus, lower availability of the chemical accounted for the lower inhibitory effect on the activity of dehydrogenase, short- and long-term respiration, ATP content, nitrogen transformation,

and straw decomposition. Besides, differences in the composition of
the biocoenosis and the amount of nutrients in the soils may be responsible for the influence of the soil types. The inhibitions observed under laboratory conditions pertained for several months
after application of the chemical, although most of it had been degraded during incubation. One reason for the duration of the pesticide effects may be that a change in soil biomass under laboratory
conditions may only slowly recover from the chemical stress, while
in the field easier recovery is possible, e.g., as a result of im-
migration of microorganisms from deeper soil layers [82].

When dinoseb, paraquat, 2,4-D, 2,4,5-T, simazine, and chloroxuron were each applied as formulations at recommended doses to
a soil with 4.1% organic matter and a pH of 5.2, there was an initial decrease in dehydrogenase activity (up to 30% with paraquat
and 2,4-D) after 2 weeks, followed by an increase in activity after
10 weeks. The control level was reached after about 20 weeks.
Different response patterns were observed for the production of
CO_2 and the activities of xylanase and urease. The CO_2 production was enhanced for 6 weeks after application and was decreased
and again stimulated thereafter until it became similar to the respiration in the control soil after 20 weeks. The increased production of CO_2 after treatment might be seen as a stress reaction by
adaptation of the soil microorganisms. At a certain stress intensity nonresistant species became inhibited with concurrent decreased respiration, followed by stimulation due to the growth of
selected resistant organisms which were supported by the nutrient
supply of the killed species. All herbicides inhibited xylanase,
with dinoseb inducing the most pronounced reduction of up to 25%
during about 12 weeks, and the activity level in the control soil
was reached only after about 24 weeks. The stimulation of urease
activity by dinoseb and chloroxuron and the ~15% depression by
2,4,5-T and simazine were prevalent for about the first 6 weeks
and diminished thereafter. The increased urease activity after
application of chloroxuron might be interpreted in terms of utilization of the urea-based pesticide, which could act as a growth substrate for urease-producing microorganisms [83].

In a loamy soil with a sand content of 6.2% (1.4% organic carbon, pH 6.8) simazine, atrazine, propazine, and prometryne at
conventional field rates stimulated dehydrogenase activity; e.g.,
with simazine at 1 kg/ha the activity was about 140% of the control activity. On the other hand, aminotriazole, fenuron, defen-
uron, lenacil, chlorpropham, and propham inhibited the activity,
the last by almost 50%. These effects paralleled the influence of
the chemicals on populations of soil fungi but not of bacteria.
However, in a soil with 64.6% sand content (0.7% organic carbon, pH 5.9) all pesticides had an inhibitory effect on dehydrogenase activity, with proximpham, propham, and atrazine at 10,

10, and 2 kg/ha, respectively, reducing the activity by more than 50% 10 days after application. The more pronounced inhibitions in the sandy soil with its limited nutrient supply may be explained by the induction of significantly higher numbers of bacteria by the pesticides and concurrent reduction of other metabolic activities. Besides, the bioavailability of the pesticide in such soils is higher than in heavier soils. Soil respiration was not correlated with dehydrogenase activities [84].

The effect of cycloate (2.9 kg/ha) alone and in combination with the phospholipid phosphatidylcholine (1.6 kg/ha) was tested in a sandy loam (0.9% organic carbon, pH 7.1) and in a silt loam soil (2.6% organic carbon, pH 7.3). Phospholipids, which can be applied in combination with some agrochemicals to increase their biological activity and, thus, reduce their application rates, rarely have inhibitory effects on enzyme activities and serve as a phosphorus source for the soil microbiota. Dehydrogenase activity was decreased by both treatments during the first week and then returned to that of the control soils, but the effects were more pronounced in the soil with a lower organic carbon content. The inhibitory effect was still observed after 20 weeks of incubation when a 10-fold concentration of the chemicals was applied. Long-term respiration and N mineralization were stimulated by the treatments [85]. The combination of dinoseb acetate (2 kg/ha) with phosphatidylcholine (1.6 kg/ha) caused a significant reduction in dehydrogenase activity that was still 40% reduced in the sandy loam and 20% lower in the silt loam, respectively, after 16 weeks of incubation. Concomitantly, the soil's short-term respiration was inhibited and nitrogen mineralization was enhanced [86].

Four insecticides (tefluthrin, trimethacarb, and two experimental phosphorothioates), applied at 10 mg/kg, were tested for their effects on some biological parameters in a sandy loam (1.7% organic matter, pH 7.6) and in an organic soil (41.3% organic matter, pH 6.8). In the sandy loam, the activity of dehydrogenases was stimulated by all pesticides during 2 weeks. In the organic soil, activity was initially reduced by tefluthrin and unaffected by the other chemicals, but after 2 weeks the activity was slightly increased with each pesticide. The activity of amylase was unchanged in sandy soil but stimulated in the organic soil. Phosphatase activity was depressed by all chemicals in the organic soil, by tefluthrin more than twofold, but no information about the duration of the inhibition was provided. In contrast, phosphatase activity was stimulated in the sandy soil. Invertase activity was slightly reduced in the sandy loam during the first few days but recovered readily. The activity of urease was quite similar to that of the nontreated control soil, but it was slightly increased by all pesticides in the heavier soil 14 days after application. In line with the higher content of organic matter in the organic soil, all enzymes investigated consistently exhibited higher activities than in the sandy loam soil [87].

In a spill simulation study, very high concentrations of formulated alachlor, metolachlor, atrazine, and trifluralin were added to a silty clay loam containing 3.1% organic carbon, pH 5.6. Alachlor was applied either alone or in combination with the other pesticides, each at 10,000 mg/kg. Dehydrogenase activity was literally abolished and did not recover within 125 days. Similarly, fungal populations were reduced after 1 day, became undetectable after 7 days, and did not recover within the testing period of 90 days. On the other hand, carboxylic esterase activity was only temporarily reduced, similar to bacterial numbers, and recovered to a normal level after 7 days. Dehydrogenase and esterase activities are both taken as index of the microbial biomass in soil. The distinct differences in activities in response to the chemicals may be explained by shifts in microbial composition in favor of organisms less sensitive to high concentrations of herbicides and with high esterolytic but lower oxidoreductive potential. When added at usual doses (10 mg/kg), none of the pesticides inhibited dehydrogenase and esterase activities. Alachlor and the other herbicides at 10 mg/kg were rapidly degraded within 30 days, followed by a slower decline to about 10 to 20% of the initial dose over about 300 days. However, at 10,000 mg/kg, degradation was negligible over the same period. When a pesticide is utilizable as a nutrient source for the soil microbiota, large concentrations under favorable conditions may stimulate microorganisms to degrade the pesticide rapidly. All four herbicides tested in this study are biodegraded by cometabolism in soil; i.e., none of the chemicals serves as an energy and nutrient source for the microbiota, and, thus, at high concentrations they may persist for long periods because microbial activity is inhibited. Enhanced degradation of such pesticides might, however, be brought about by the addition of readily oxidizable organic matter by overall stimulation of microbial activity [88].

Linuron (10 mg/kg) stimulated the activities of dehydrogenase and protease in peat soil but reduced that of phosphatase. When the soil was sterilized by gamma irradiation, the extracellular activities of catalase, proteases, and phosphatases were not affected by the herbicide. After 5 months of incubations in native, irradiated, and autoclaved soil, 33, 44, and 78% of linuron added initially could be extracted as the nondegraded parent compound. Degradation of substrates in soil comprises the activities of microbiota and of accumulated enzymes as well as abiotic transformation mechanisms. Assuming no activity of living organisms but retained activity of accumulated enzymes in the irradiated soil, and no microbial and enzymatic activity in the autoclaved soil, the authors estimated by simple arithmetic that 22% of the chemical was degraded abiologically, 11% was degraded by live microbiota, and 34% was metabolized by soil accumulated enzymes. In the same study, TCA and dalapon inhibited the activity of humus-containing preparations of amylase and protease significantly less than that of further purified enzymes, which reflects the usual stabilization of extracellular enzymes in immobilized form. Phosphatase activity, however, was inhibited in both the purified and the humin-bound fraction [89].

Catalase

Hydrogen peroxide is formed during respiration as a by-product and is decomposed by catalase to water and oxygen. Catalase activity is measured by volumetric or titrimetric quantification of the oxygen liberated from added H_2O_2 or, alternatively, by the recovery of the substrate itself. However, H_2O_2 is also degraded by inorganic catalytic mechanisms. In experiments with streptomycin-inhibited catalase-producing soil microorganisms, the accumulated catalase activity was distinguished from the catalytic activity by the microorganisms [1].

As discussed in the section on dehydrogenases, the activity of accumulated catalase in gamma-irradiated peat soil was not affected by the herbicide linuron applied at 10 mg/kg [89].

The soil fumigant methyl bromide and the fumigant precursor sodium azide were added to pine nursery beds at application rates of 650 and 22–134 kg/ha, respectively. Microbial catalase activity increased steadily during the 1-year field experiment, but the changes in activity were similar to those in an untreated field plot. In contrast, methyl bromide reduced saccharase activity to about two-thirds of that in the control soil during the first 3 months, but then the enzyme activity returned to the level of the control. Amylase activity was not affected by methyl bromide during the first 3 months but was significantly and continuously suppressed for the next 9 months. The activities of saccharase and amylase were not affected by sodium azide. The bacterial populations were stimulated by both treatments for 3 months before they returned to the control level [90].

Peroxidases

This class of enzymes oxidizes substrates in the presence of H_2O_2 and is involved in the degradation of lignin and phenols [91]; see the chapter by Bumpus in this volume. In soil, the activity can be assayed by the oxidation of *o*- or *p*-dianisidine in the presence of H_2O_2 [92,93].

Paraquat, applied in a 6-week field test at 1.1 and 11.0 kg/ha, reduced the activity of peroxidase and concomitantly that of phosphatases in a calcareous loam with a pH of 8.0. The inhibition was independent of the application of either the active ingredient or the formulated product and, besides, was not dependent on the concentration of paraquat. Bacterial and fungal populations were reduced at 11.0 kg/ha of the active ingredient but not by the formulation. The populations of actinomycetes were not affected. Most of the effects at 11.0 kg/ha were significantly different from those in the untreated soil but not from those at the low concentration (1.1 kg/ha) of paraquat. Since paraquat ions kill plants within 1 day to 1 week after application, the appearance of effects from week 3 onward seems to indicate that the denudation of the vegetative cover by the chemical treatment and the resultant increase in oxidable plant matter, i.e., the nutritional status of the soil, had a more prounounced effect on the tested biological soil parameters than the chemical itself, which by then would have been adsorbed on clay minerals [94].

Phenoloxidases

Phenols are generated in soil by degradation of lignin or by microbial synthesis [95] and they may represent intermediate metabolites of pesticides or originate from other pollution sources. Their oxidation by phenoloxidases or laccases has been demonstrated for monophenols, o- and p-diphenols such as catechol and hydroquinone, and benzenetriols such as pyrogallol [3]. The resultant phenolic hydroxyl radicals may be incorporated into humic polymers to form bound soil residues [56].

Activity assays may be based on the quantification of oxygen consumption in the oxidation of phenolic substrates catalyzed by phenoloxidases or on the spectrophotometric determination of substrate turnover or product formation [96]. Some substrates, such as D,L-3,4-dihydroxyphenylalanine (DOPA), however, are also oxidized by inorganic catalysis in soil. A phenoloxidase assay has been developed that is based on the quantification of the substrate guaiacol, 2-methoxy-phenol, that is separated from its oxidized reaction products by means of high-performance liquid chromatography [96]. Guaiacol occurs naturally in soil.

The herbicide phenmedipham applied to a silt soil (2.9% organic carbon, pH 5.7) at a concentration of 10 mg/kg slightly increased the microbial activities of urease and invertase for 45 days, but it did not affect those of diphenoloxidases, proteases, cellulase, β-glucosidase, acid and alkaline phosphatases, and pyrophosphatase. Although the numbers of actinomycetes and fungi were increased, bacterial counts remained unchanged [97].

B. Ester Hydrolyases

Carboxylic Ester Hydrolases

The carboxylic ester hydrolase activity of soil, measured by spectrophotometric quantitation of fluorescein released from fluorescein diacetate [98], is indicative of the soil's overall microbiological activity. The hydrolysis of p-nitrophenyl acetate and subsequent analysis of the released p-nitrophenolate anion can be used to assay the aryl esterase activity of soils [99]. The activity of glycerol ester hydrolase, commonly called lipase, has been assayed by hydrolysis of 4-methyl umbelliferone butyrate to butyric acid and the fluorescent 4-methyl umbelliferone [100]. An improved method is based on the hydrolysis of 4-methyl umbelliferone nonoate, which shortens the incubation times considerably [101]. The microbial and accumulated lipase activity may also be assayed by titration of butyric acid released from tributyrin [102] or by spectrophotometric quantification of p-nitrophenolate liberated from p-nitrophenyl palmitate [65].

As discussed in the section on dehydrogenases, alachlor alone or in mixture with atrazine, metolachlor, and trifluralin, each at 10,000 mg/kg to simulate accidental spills, had very little effect on carboxylic esterase activity of a silty clay loam. In fact, only when all the pesticides were applied together at such high concentrations, the esterase activity decreased for the first 3 weeks and then recovered to the level of the control soil, similar to the number of bacteria that apparently adapted in favor of species resistant to the pesticide wastes [88].

Formulated fenitrothion, chlorothalonil, paraquat, and trichlorphon were applied to a sandy loam soil at the recommended doses and at fivefold doses in a field test. The microbial aryl esterase activity was only slightly reduced at both doses during the testing period of 30 days. Acid phosphatase activity of the microbiota was not significantly changed, whereas aryl acylamidase activity was continuously inhibited by up to 30% with fenitrothion under laboratory conditions and 40% with trichlorfon during 15 days. The authors speculated that the effects on aryl acylamidase were only temporary because of the ready degradation of the phosphorothioate and phosphonate in soil. However, no experimental evidence (i.e., analysis of the pesticide residues) for the hypothesis was provided [99].

The overall lipolytic activity of different strains of soil actinomycetes in vitro did not change significantly when formulated metobromuron was added at concentrations equivalent to 10 to 1000 mg/kg, based on the weight of the test media. This may be explained by the dose-dependent inhibition of growth of the microorganisms, and the concurrent stimulation of the lipolytic, as well as the amylolytic and proteolytic, activities of the remaining genotypes, so that the integral activities remained constant. The authors speculated that the observed enzymatic responses reflect unspecific detoxification activities [103]. The in vitro approach to assessing the side effects of pesticides on specific soil microorganisms has the disadvantage that biodegradative changes of the tested pesticides caused by other microorganisms present in soil are neglected. Besides, the adsorptive capacity of the soil matrix may lower the effects of the parent compound and of metabolic derivatives significantly. Therefore, such tests give only an indication of potential interferences with the soil microbiota.

Phosphatases

Phosphatases represent a broad range of intracellular as well as soil-accumulated activities that catalyze the hydrolysis of both the esters and anhydrides of phosphoric acid [104]. They include the following:

Phosphoric monoester hydrolyases, i.e., acid and alkaline phosphatases, which are named according to their pH optima and prevail in acidic and alkaline soils, respectively
Phosphoric diester hydrolases
Triphosphoric monoester hydrolases
Enzymes hydrolyzing phosphoryl-containing anhydrides
Enzymes hydrolyzing P—N bonds

Phosphatases are not related to microbial numbers and the respiratory activity of soil and are only sometimes related to the available P content of soil [1]. The relatively unspecific monoesterases are assayed by analysis of either the release of orthophosphate or, advantageously, of various organic groups from orthophosphoric molecules, e.g., phenol, p-nitrophenyl, and β-naphthylphosphates [105–108]. Phosphodiesterase activity can be measured by incubation with bis-p-nitrophenyl phosphate [109], and the unclassified phosphotriesterases that hydrolyze compounds in which all the hydrogens of orthophosphoric acid are substituted can be assayed with tris-p-nitrophenyl phosphate as the substrate [110]. The release of phosphate from pyrophosphate, $P_2O_7^{4-}$, is catalyzed by inorganic pyrophosphatase. The available phosphate formed can be quantified spectrophotometrically after the addition of molybdate as a heteropoly blue complex [111].

As discussed above, pendimethalin, difenzoquat, thiobencarb, folpet, and captafol at doses of 0.5–2.5 mg/kg [68], as well as glyphosate, paraquat, trifluralin, and atrazine at 5.4, 2.2, 3.4, and 2.2 kg/ha, respectively [72], did not change the phosphatase activity of soil. Dalapon at 2.6 mg/kg inhibited phosphatase activity for 2 weeks but the activity recovered afterward [78]. Malathion, parathion, chlordane, dieldrin, and thiram at 5 mg/kg stimulated the activity of phosphatase in a clay loam soil, whereas dichloropropane-dichloropropene and methyl isothiocyanate at 300 and 80 mg/kg, respectively, inhibited phosphatase [79]. In an organic soil, however, most of the pesticides reduced phosphatase activity except chlordane, which stimulated the activity at 5 mg/kg [80]. Tefluthrin, trimethacarb, and two experimental phosphorothioates at 10 mg/kg induced a reduction of activity in an organic soil but stimulation in a sandy soil [87]. Linuron at 10 mg/kg inhibited phosphatase in a native soil but did not affect the activity in gamma-sterilized soil. TCA and dalapon inhibited phosphatase both in purified preparations of the enzyme and in those containing humic material [89]. Paraquat at 1.1 kg/ha reduced phosphatase activity [94], while phenmedipham at 10 mg/kg [97] and the pesticides fenitrothion, chlorothalonil, paraquat, and trichlorphon [99] had no effect on the activity.

The inhibition of phosphodiesterase activity by orthophosphate, the natural product of organic phosphorus mineralization, has been studied, as this anion is added to soil in considerable amounts as a fertilizer. In two different clay loams with organic carbon contents ranging from 2.7 to 5.3% and pH from 6.2 to 7.0, 5 mM orthophosphate reduced the affinity of the enzyme for bis-p-nitrophenyl phosphate but did not change the rate of hydrolysis. Thus, orthophosphate is a competitive inhibitor of phosphodiesterases in soil [109]. p-Nitrophenyl phosphates are artificial substrates of phosphatases. Therefore, the significance of the results thus obtained for the interactions of phosphatases with natural substrates remains uncertain, but this disadvantage is accepted by many investigators, as a result of the sensitive, fast, and reproducible colorimetric analysis of p-nitrophenolate.

In a sandy loam (7.1% organic matter content, pH 5.0), the herbicide barban was applied as active ingredient and as a formulation at 200 mg/kg (the recommended field rate is 0.5 to 1 mg/kg). Phosphatase activity was slightly reduced by up to 8% during the first 40 days, but for the remainder of the experimental 220 days activities were stimulated by the active ingredient to 27% above those in the control soil. The formulated product had a lower stimulatory effect as a result of the slight inhibition of phosphatase activity by the formulation solvent alone, which was added in a separate experiment. An initial increase in soil respiration by the formulated product was accompanied by bacterial proliferation, apparently in response to the microbial utilization of the organic carbon of the formulation solvent. However, long-term and glucose-induced respriation was inhibited during the later stages of incubation, despite larger bacterial numbers than in the control soil, which indicates that the herbicide affected the respiratory activity directly. The microbial utilization of sugars, such as glucose, mannose, galactose, cellobiose, xylose, and arabinose, was inhibited in barban-treated soil, suggesting a pesticide-induced inhibition of the metabolism of carbohydrates. It may be that the increased phosphatase activity in barban-treated soil causes dephosphorylation of the sugars, thus inhibiting the primary step in their degradation [112].

The insecticide cartap hydrochloride was applied at 10, 100 and 1000 mg/kg to a flooded and nonflooded silty clay soil (1.6% organic carbon, pH 5.9). At the lowest concentration, phosphatase activity in the nonflooded soil was suppressed by about 10%, but it recovered after 1 week. The inhibition was dose dependent, but even at 1000 mg/kg the activity was similar to that in the control soil after 2 weeks of incubation. In the flooded soil, 10% suppression of phosphatases occurred only at the highest concentration of the pesticide and lasted about 60 days. Cellulase activity was increased by flooding the soil, but it was unaffected by the insecticide at all concentrations. Similarly, saccharase activity remained unchanged by insecticide treatment. In soil without standing water, proteases were less active throughout the 60 days after addition of 100 and 1000 mg/kg cartap, with the activities being about 80 and 60%, respectively, of that in the control soil. However, in flooded soil the inhibition was reversible, and all treated soils eventually had similar or higher activities than in the control soil. The decrease of protease activity correlated well with the decrease in the number of proteolytic microorganisms, particularly of fungi. Likewise, substrate-induced respiration under both flooding conditions was inhibited for 30 days by the highest cartap concentration, but it eventually approached the control level. In soil treated with 100 and 1000 mg/kg of the pesticide, NH_4^+-N accumulated and NO_2^--N and NO_3^--N diminished as cartap inhibited slightly nitrification processes. It was presumed that NH_4^+-N was derived by degradation of the hydrolyzed carbamoyl groups of cartap-HCl and of soil organic matter. The half-life of cartap in this soil was determined to be about 3 days [113].

Formulations of the nematicides oxamyl and fenamiphos were applied at 12.7 and 18.6 kg/ha, respectively, to a fine silty montmorillonitic soil (6.5% organic carbon content, pH 6.1) under grass-clover pasture in the field, and several enzyme activities and other biological parameters were assayed for 5 months. Phosphatase, invertase, and amylase activities were not significantly changed compared to nontreated soil, whereas cellulase activity increased up to 50%, particularly after treatment with fenamiphos, probably as a result of initial scorching and subsequent production of herbage. Urease activity was reduced up to 20% under both treatments, especially with the phosphoramidate fenamiphos, but after 5 months the effect was no longer statistically significant. The most obvious change in activity was observed with sulfatase, which was continuously suppressed by oxamyl and fenamiphos. After almost half a year, the residual activity was still about 20 and 30%, respectively, lower than in the control soil. The pesticides were degraded under the experimental conditions, with half-lives of 15 to 20 days for oxamyl and 50 to 100 days for fenamiphos. Neither nematicide had any significant influence on the rates of CO_2 and net mineral-N production [114]. In a similar experiment, but under laboratory conditions, both pesticides were applied at the recommended field rates, as well as at higher doses: oxamyl at 25 and 635 mg/kg and fenamiphos at 37 and 930 mg/kg. At the high application rates, CO_2 production was significantly reduced and N mineralization increased, with the latter occurring even at the lower rate. Oxamyl and fenamiphos had no deleterious influence on the activities of invertase, urease, and phosphatase. At the increased rate, both pesticides reduced amylase and cellulase activity by 24 and 44%, respectively, even 62 days after treatment. Sulfatase was also inhibited 17% by oxamyl after 15 days and 17% by fenamiphos after 62 days. Generally, the observed effects were less marked in the field than under laboratory conditions [115]. The inhibition of sulfatase activity must be taken as a serious indication of apparently irreversible distortion of a part of the microbial community that would have remained undetected if other biological activity parameters reflecting the soil's overall microbiological activity had been used.

Nitrofen, fluchloralin, methabenzthiazuron, metoxuron, 2,4-D, and isoproturon were applied as formulated products in the field to a sandy loam soil (pH 6.8) at the usual field rates, i.e., between 1.0 and 2.5 kg/ha. Two days after application the activity of acid phosphatase was depressed by the pesticides by 60 to 74% except for nitrofen, which reduced the activity by only 13%. Ten days after treatment, the activity was still 30 to 40% lower than in the control soil, but with minor effects of nitrofen. At harvest time, 110 days after application, activities had almost recovered. A similar response occurred with the activity of alkaline phosphatase. Most of the agrochemicals had a less marked inhibitory effect on urease activity. However, 2,4-D suppressed urease by 45% after 2 days and still by 13% after 110 days. Negligible effects on urease activity were observed with methabenzthiazuron and metoxuron, both urea-based herbicides [116].

Alachlor, metolachlor, and atrazine with active ingredients at 2 and 20 mg/kg were tested for their interference with various phosphatase activities in a clay loam, enriched or not enriched with maize residues. The organic carbon content of the soils was 1.07 and 2.03% and the pH 8.1 and 7.9, respectively. The organic enrichment of the soil promoted the proliferation of phosphatase-producing microorganisms, as seen by the increased activities of acid and alkaline phosphatases, phosphodiesterase, and phosphotriesterase. Both phosphomonoesterases were initially stimulated by all herbicides, probably due to the death of microorganisms unable to survive the herbicidal treatment and a temporary release of such enzymes after cell lysis. The activity then decreased to the level of the control soil after about 60 days, which may be explained by proteolysis of nonstabilized extracellular phosphatases. In the

unenriched soil, phosphodiesterase activity was slightly stimulated
by the acetanilide herbicides, but it was depressed by atrazine at
2 and 20 mg/kg by about 20% after 30 days incubation. After 60
days, the activity still remained 15% lower than in the control soil
in presence of atrazine at both doses. In the enriched soil with
its higher overall activity, however, no influence of any of the
pesticides was apparent. On the contrary, phosphotriesterase activity showed only minor changes, even at the higher doses, in the
unenriched soils but a marked depression by all chemicals at 20 mg/
kg in the maize-treated soil. Even after 60 days, the activity was
about 15% lower than in the control soil. After 60 days of incubation, 92, 66, and 80% of alachlor, atrazine and metolachlor, respectively, were metabolized under the experimental conditions [117].

Loamy soil sites in the field (1.3 to 4.8% organic matter, pH 5.3
to 7.7) that showed enhanced biodegradation of organophosphorodi-
thioate insecticides, as the result of pretreatments, were compared
to untreated sites in terms of their phosphatase activities. The
biological activities of the individual soils were normalized on the
basis of their dehydrogenase activities. Forty percent of all
cropped soils pretreated with either chlorpyrifos, terbufos, or
fenofos and two-thirds of the soils pretreated specifically with
fonofos had higher activities of acid phosphatase, although the
structure of the insecticides suggests that phosphotriesterases
were predominantly involved in their biodegradation. In the pretreated soils, fonofos and terbufos at 3.5 mg/kg were degraded by
85 and 74%, while in an untreated control soil, the pesticides were
less efficiently metabolized by 45 and 59%, respectively, 2 weeks
after application. The increased acid phosphatase activity was
most closely related to the insecticide application history rather
than to the pH, cation exchange capacity, or type of soil. The
correlation of the increased enzyme activity with enhanced biodegradation of terbufos or fonofos was verified by laboratory experiments [118]. Acid phosphatase activity may, thus, be used
for the prediction of enhanced biodegradation [59] in soils prone
to phosphorodithioate insecticide failure.

Sulfatases

This class of enzymes includes different accumulated alkyl and
aryl sulfatase activities that are responsible for the mineralization of sulfur in soil [119,120]. Aryl sulfatases catalyze the hydrolysis of aryl sulfates, $ROSO_3^-$, to the corresponding alcohol
and sulfate [104]. The assay can be based on the use of *p*-ni-
trophenyl sulfate as the substrate and quantification of the released *p*-nitrophenol [121]. Aryl sulfatase activity has been related to the organic matter content of soils [122,123].

The distinct inhibition of sulfatases by oxamyl and fenamiphos at 12.7 and 18.6 kg/ha, respectively, during about 6 months in a field experiment and under laboratory conditions was discussed in the section on phosphatases [114,115].

Aryl sulfatase is inhibited by several transition metal ions, such as Ag(I) or Hg(II), and is also affected by molybdate, tungstate, arsenate, and orthophosphate by competitive inhibition. The last, added for fertilization to a clay loam (2.7% organic carbon, pH 6.2) and a loam soil (2.9% organic carbon, pH 6.5), increased the Michaelis constant, K_M, for p-nitrophenyl sulfate by a factor of 1.5 without affecting the turnover rate, k_{cat}, at a concentration of 2400 mg/kg [124]. Thus, the activity of aryl sulfatase expressed by the ratio of turnover rate and Michaelis constant, k_{cat}/K_M, is inhibited by orthophosphate. Since orthophosphate is the end product of P mineralization in soil, the often observed decrease of aryl sulfatase activity in soils after a few weeks of storage may be due to inhibition of the enzyme by soluble native orthophosphate in soils.

C. Glycoside Hydrolases

Xylanase

Xylanase catalyzes the hydrolysis of β-1,4-xylans present as cell wall constituents of all land plants and, thus, takes part in the turnover of the biomass carbon in nature [125]. The D-xylose formed can be quantified by the addition of a 3,5-dinitrosalicylic acid reagent and subsequent photometric analysis [83]. The activity in toluene-treated soil reveals that the enzyme is, in part, extracellular. As discussed in the section on dehydrogenases, xylanase activity was irreversibly inhibited by the fungicide quintozene [77] and for about 24 weeks reduced by parathion [81], dinoseb, paraquat, 2,4-D, 2,4,5-T, simazine, and chloroxuron [83].

β-Glucosidase

Intra- and extracellular β-glucosidase is involved in the hydrolysis of maltose and cellobiose, the last having often been used as substrate for the assay of the enzyme's activity. The glucose formed can be quantified by exploiting its reducing properties, e.g., by the Fehling test, iodometric titration, or the Nelson-Somogyi method [126], or by use of a glucose oxidase—peroxidase reagent [127]. As an alternative, the hydrolysis of p-nitrophenyl-β-D-glucoside and subsequent colorimetric analysis of p-nitrophenol has been used to assay the activity [4]. As discussed in the section on phenoloxidases, 10 mg/kg phenmedipham did not affect the activity of β-glucosidase in soil [97].

β-1,3-Glucan Glucanohydrolase

This enzyme (trivial name, laminarinase or glucanase) is responsible for the transformation of laminarin, a water-soluble ubiquitous β-glucan present as a cell wall component in algae, bacteria, fungi, and higher plants and, thus, is found in soils [128]. Laminarinase has both exo- and endohydrolytic 1,3-β-glucanase and β-glucosidase activities. It accumulates in soil, and it can be assayed by analysis of the glucose that results from the hydrolysis of laminarin [129].

In a silt loam soil (2.2% organic carbon, pH 5.4) the effects of the herbicides 2,4-D, diallate, glyphosate, and benzoylprop-ethyl and of the insecticide malathion on the activities of laminarinase and urease were investigated. Each component was applied as formulated product at five times the recommended field rate, i.e., at 4.9, 11.8, 11.5, 7.0, and 14.8 mg a.i./kg, respectively. None of the pesticides affected the activity of laminarinase and of urease. Even when applied at 1000 mg/kg, glyphosate did not affect glucanase activity, but a gradual decrease in activity occurred with the other chemicals. After 111 days of incubation, the residual glucanase activity was 45 to 60% of that in the untreated soil with no apparent recovery, probably as a result of irreversible inhibition. The observed lag phase of about 10 days before glucanase activity decreased indicates either that the active inhibitory molecules were metabolites rather than the parent compound or that the diffusion of the only slightly water-soluble molecules to the enzyme in sufficient concentration was rate limiting. On the other hand, 2,4-D, at concentrations above 100 mg/kg, activated glucanase for about 50 days before the activity reached the control level. The glucanase stimulation by 2,4-D was not observed in the presence of sodium azide, which suggested that the enhanced enzyme activities resulted from stimulation of 2,4-D—resistant microbiota. Urease activity was not affected by benzoylprop-ethyl, diallate, and glyphosphate at 1000 mg/kg but was inhibited by 2,4-D at concentrations between 250 and 1000 mg/kg. At the highest concentration, only 30% residual urease activity was present after 100 days. Inasmuch as 2,4-D failed to inhibit urease isolated from jack bean, the observed reduction in urease activity in soil indicates an indirect response to the herbicide due to the intoxication of ureolytic microorganisms [130].

β-Fructofuranosidase

This enzyme (often named invertase, sucrase, or saccharase) catalyzes the hydrolysis of sucrose to glucose and fructose, which can be quantified as reducing sugars for the determination of activity [131]. Invertase activity in soil is associated with the organic matter content and microbial activity parameters, but it is also thought to accumulate extracellularly.

As discussed above, both stimulated and depressed activities of invertase were observed after applying atrazine and cyanazine at concentrations above 7100 mg/kg [67]. The activity of invertase was slightly reduced by tefluthrin, trimethacarb, and two experimental phosphorothioates, each at 10 mg/kg, but recovered after a few days [87]. Unchanged invertase activity was observed on applying 22—134 kg/ha of the fumigant precursor sodium azide, but reversible inhibition of invertase occurred with the fumigant methyl bromide at 650 kg/ha [90]. Phenmedipham, 10 mg/kg, slightly increased invertase [97], but cartap-HCl at 10—1000 mg/kg [113] and oxamyl and fenamiphos at 12.7 and 18.6 kg/ha, respectively, had no effect [114,115].

In a soil with a high organic matter content of 49% and a pH of 6.7, various pesticides were evaluated for their influence on invertase and amylase activities: chlorfenvinphos, chlorpyrifos, diazinon, ethion, ethoprophos, fensulfothion, fonofos, leptophos, malathion, parathion, phorate, thionazin, triazophos, trichloronate, terbufos, chlordane, dieldrin, gamma-HCH, carbofuran, oxamyl, permethrin, captan, maneb, thiram, 2,4-D, atrazine, dalapon, dicamba, nitrofen, picloram, glyphosate, and simazine, all at 5 and 10 mg/kg. Other pesticides applied at usual field rates were dichloropropane-dichloropropene, methyl isothiocyanate, and nitrapyrin. As measured for an incubation period of only 3 days, none of these agrochemicals reduced invertase activity, and most of them were stimulative (Table 1). Especially with triazophos, a phosphorothioate triazole, the activity was increased 10-fold. Stimulated activities were also observed for amylase, with chlorpyrifos and phorate inducing the strongest activation. However, the duration of the effects after the first 3 days was not described, nor was any information on the persistance of the chemicals in soil provided. With the exception of nitrapyrin, all chemicals reduced the level of ATP in the soil for only about 1 day, which might reflect the reduction of pesticide-sensitive microorganisms. However, no increased ATP levels were detected 3 days after application of the pesticide, indicating homeostatic recovery of the soil's overall activity [132]. Again, this study reveals that many pesticide-induced effects may be detected only by monitoring specific functional activities of the microbial community, whereas parameters such as the concentration of ATP may mask such effects.

Amylases

Both α- and β-amylases hydrolyze starch and related oligo- and polysaccharides that contain D-glucose units. Both enzymes accumulate in soil, with β-amylase activity prevailing. Assays are based on the hydrolysis of soluble starch and subsequent analysis of the reducing sugar formed [131,133].

As discussed above, inconsistent results, i.e., increased and depressed amylase activities, were reported after application of atrazine and cyanazine at concentrations above 7100 mg/kg [67]. Amylase was not affected by tefluthrin, trimethacarb, and two experimental phosphorothioates at 10 mg/kg in a sandy soil, but the activity was enhanced by the pesticides if applied to an organic soil [87]. The activity was irreversibly inhibited by the soil fumigant methyl bromide at 650 kg/ha and not affected by sodium azide at 22–134 kg/ha [90]. TCA and dalapon inhibited amylase isolated from soil more than in humin fractions containing preparations of the enzyme [89]. Unchanged amylase activity was observed after applying metobromuron at 10–1000 mg/kg [103] as well as oxamyl and fenamiphos at 12.7 and 18.6 kg/ha, respectively [114,115]. A variety of agrochemicals discussed in the section on β-fructofuranosidase [132] stimulated the activity of amylase (Table 1).

Cellulase

Cellulase acts on the β-1,4-glucan bonds present in cellulose and, thus, is involved in one of the major processes of the natural carbon cycle. Activity assays have been based on the decomposition of cellophane disks, cellulose powder, and the hydrolysis of carboxymethyl cellulose [3]. However, the rate of cellulose degradation in natural organic litter is not exclusively related to extra- and intracellular cellulase activity but is also related to a variety of other hydrolytic as well as oxidative enzymes [134,135]. The effects of herbicides on cellulose decomposition has been reviewed elsewhere [136].

As discussed above, the cellulolytic activity of soil was inhibited by high concentrations of atrazine and cyanazine (7100 mg/kg) [67], unchanged by 10 mg/kg phenmedipham [97] and cartap-HCl at 10–1000 mg/kg [113], and increased by oxamyl and fenamiphos at 12.7 and 18.6 kg/ha in a field soil under grass-clover pasture. However, under laboratory conditions oxamyl and fenamiphos reduced cellulase activity [114,115].

An anthraquic fluvisol soil (1.0% organic carbon, pH 5.3) was incubated for 4 weeks with the following formulated pesticides: the fungicides trichlamide, i.e., [(RS)-N-(1-butoxy-2,2,2-tri-chloroethyl)salicylamide], chlorothalonil, quintozene, and hymexazol; the herbicides paraquat, thiobencarb, propanil, and butachlor; and the insecticide diazinon. Under flooded soil conditions, trichlamide, at 10 times the recommended field rate (i.e., 400 mg/kg), inhibited the cellulolytic activity completely, whereas at the

usual field rate the activity was about 50% of that in the control
soil. Under upland conditions, the inhibition was comparatively
weak, amounting to about 30%. It is well known that the microflora in flooded soil is markedly different from that in upland soil.
Since fungi are the major cellulose degraders in upland soil, the
high selectivity of trichlamide, which is used for the control of
Aphanomyces, may explain the relatively weak inhibition under
such conditions. In flooded soil, however, the fungicide seems to
inhibit the activity of anaerobic cellulolytic bacteria, such as *Clostridium*, which were the predominant cellulose degraders in the
flooded soil, although fungi and aerobic bacteria were also active
to some extent. A distinct depression was observed with chlorothalonil under all conditions tested, i.e., at the usual doses and
at 10 times the dose in both flooded and nonflooded soil, indicating
that various cellulolytic microorganisms are affected. Chlorothalonil
is known to interact with thiol functional groups of enzymes [137],
and it may be that the inhibition of a certain step in glycolysis involving the action of thiol-dependent enzymes is responsible for the
observed reduced cellulolytic activities. Only minor effects on cellulase activities were reported for the other pesticides [138].

D. Proteases

Proteolytic activities in soil have been assayed with a variety of substances, e.g., with proteins such as albumin, casein, or gelatin and
with peptides such as benzyloxycarbonyl phenylalanyl leucine [65,
102,139]. The amino compounds released can be quantified by the
Lowry method [140] or by use of ninhydrin [141].

As discussed above, protease activity remained unchanged after
application of phenmedipham at 10 mg/kg [97] and metobromuron at
10—1000 mg/kg [103]. Linuron at 10 mg/kg stimulated protease activity in a native soil but did not affect the activity in gamma-sterilized soil. In the same study, TCA and dalapon reduced the activity in purified protease preparations more than in humin containing
preparations of protease [89]. Cartap-HCl at 100—1000 mg/kg reduced protease activity and the activity did not recover within 60
days. However, the inhibition was reversible if the pesticide was
applied at the same concentrations to a flooded soil [113].

E. Amidohydrolases

L-Asparaginase

L-Asparaginase is involved in the mineralization of nitrogen by cata-
lyzing the metabolism of L-asparagine to ammonium and aspartate, with
an apparent Michaelis constant with L-asparagine of about 6 mM. The
activity has been correlated with soil organic carbon and nitrogen

contents. The enzyme expressed by *Escherichia coli* has a molecular mass of 130,000 daltons and contains four subunits linked by disulfide bonds. The hydrolysis of the substrate involves either an aspartyl intermediate before ammonium release or the simultaneous release of aspartic acid and ammonium [142]. The assay of L-asparagine hydrolysis is based on the analysis of ammonium, e.g., by steam distillation or by use of Conway diffusion dishes [143—145]. Extracellular L-asparaginase activity accumulates in soil.

To three soil samples (silt loam, 2.6% organic carbon, pH 5.6; clay loam, 3.2% organic carbon, pH 7.6; clay loam, 5.3% organic carbon, pH 7.0) the following pesticides were applied as formulations at 10 mg/kg: the herbicides atrazine, naptalam, chloramben, dicamba, cyanazine, 2,4-D, dinitramine, EPTC + R-25 788 (*N,N*-diallyl-2,2-dichloroacetamide), alachlor, paraquat, butylate, and trifluralin; the fungicides maneb and captan; and the insecticides diazinon and malathion. After 30 min of incubation, all pesticides affected asparaginase activity. For example, activity was reduced between 30 and 50% by alachlor, paraquat, butylate, and both insecticides—i.e., chemicals with very different chemical and physicochemical properties, which points to lack of specificity of such assays for the assessment of side effects of pesticides. However, the additional determination of other enzymatic and microbial activities may allow a discrimination of the chemicals based on their mode of inhibition. Trifluralin and captan had the least inhibitory effect, i.e., below 10%, and the other pesticides caused about 20% inhibition compared to the control soil. The duration of the effects was not described, although it would be of interest to know, as the inhibition of L-asparaginase could lead to a reduction in the amount of NH_4^+-N derived from soil organic matter [142].

L-Glutaminase

Intra- and extracellular L-glutaminase specifically catalyzes the hydrolysis of glutamine to glutamate and ammonium. As for L-asparaginase, ammonium is quantified to assay the enzyme and its kinetic parameters. The apparent Michaelis constant of the enzyme with the substrate glutamine in soil is about 20 mM. The enzyme is strongly inhibited by reagents that interact with sulfhydryl groups, such as mercury salts and other metal ions. Its activity in soil is highly correlated with the soil organic matter content [146].

Twelve different herbicides, two fungicides, and two insecticides were applied to various soils to test their effects on glutaminase activity. The experimental conditions and chemicals were identical to those used for the asparaginase study described above

[142]. The enzyme activity was moderately reduced under all conditions. Activities were suppressed between 15 and 19% by naptalam, dicamba, cyanazine, dinitramine, paraquat, and malathion, each at 10 mg/kg. Again, the inhibition was unspecific, i.e., independent of the chemical, biological and physicochemical properties of the pesticides, and no information on the persistence of the observed interactions is available because all activity tests were performed after only 30 min of incubation [147].

Acylamide Amidohydrolases

Acylamide amidohydrolases, commonly called amidases, catalyze the hydrolysis of amides to ammonia and the corresponding carboxylic acids. They are widely distributed in nature, accumulate in soil, and the activity is correlated with the soil organic matter and clay contents. Like other microbial enzyme activities in soil, amidases are induced by their substrates [148]. For the quantification of substrate turnover rates, the soil is incubated with a buffered formamide solution, followed by analysis of the released ammonium [149]. Amidase activity has also been assayed by hydrolysis of acetanilide and colorimetric quantification of the liberated aniline after diazotization with N-(1-naphthyl)-ethylene diamine [99]. The lack of inhibition by metal ions suggests that amidase does not contain thiolate ligands at the active site participating in catalytic events, although thiol groups seem essential for stabilization of the active amidase conformation [150].

Irreversible inhibition of aryl acylamidase by fenitrothion and trichlorphon has been mentioned in the section on aryl ester hydrolases [99].

Only minor inhibition in amidase activity was observed after a 30-min incubation with some formulated pesticides applied at 10 mg/kg to a silty, a loamy, and a montmorillonitic soil with organic carbon contents of 2.6, 3.2, and 4.7% and pH values of 5.6, 7.0, and 7.6, respectively. The pesticides used were identical to those described in the section on L-asparaginase [142]. Alachlor, a branched amide; butylate, containing a linear amide; and the phosphorothioate diazinon inhibited amidase activity noncompetitively, as they reduced the maximum rate constant by 5, 17, and 12%, respectively, without changing the affinity of the enzyme for the substrate, formamide (K_M ca. 1.7×10^{-2} M). Because of the similarity in chemical structures of butylate and formamide, both containing a linear amide group, butylate would be expected to inhibit amidase competitively. It seems, however, that only molecules with terminal amide functional groups can compete for the active site, or that the substitution of the amide group in the thiocarbamate hinders the binding to the active site. Naptalam and malathion inhibited amidase activity by 6 and 7%, respectively, whereas the other pesticides caused even less reduction of activity [150].

Urease

Urease catalyzes the hydrolysis of urea to ammonium and carbon dioxide. Most of the ~7 million tons of urea produced industrially per year in the United States alone is used for fertilization of farmland. Other sources of urea in soil are mammalian excrements and degradation products of proteins and nucleic acids. Inasmuch as the ammonium formed represents a bioavailable form of nitrogen for plant uptake, this ubiquitous activity has a primary role in the cycling of nitrogen. Urease activity in soil originates predominantly from microorganisms and is correlated with the soil organic matter content [151]. Urease activity accumulates in soil to a significant extent [2].

Jack bean urease is a hexameric protein with a molecular mass of 590,000 daltons and contains two Ni(II) ions per subunit. A reaction mechanism has been proposed for urea hydrolysis in which one Ni^{2+} ion polarizes the urea carbonyl to facilitate the nucleophilic attack by an OH^- anion that is activated by the second Ni^{2+} center [152]. The formation of a carbamoyl-enzyme intermediate has been suggested, based on kinetic analyses [153], but the reaction could alternatively proceed via the formation of a carbamato complex. Bacterial ureases are also multimeric, being composed of two to six subunits, and they are of lower molecular mass than jack bean urease. They appear also to contain Ni^{2+} ions in varying stoichiometries [154]. Urease activities are inhibited by phosphoryl and thiophosphoryl amides acting as transition state analogs and by phenols, thiols, quinones, and thio reactive reagents [155], the last suggesting that ureases possess one or more thiolate ligands essential for activity. Urease inhibitors are deliberately added to agricultural soils to control the hydrolysis of urea and to retard ammonia loss by volatilization [156—160].

Kinetic parameters of urease activity have been obtained by quantification of the released ammonium [4,144,161—163] or carbon dioxide [164] and by colorimetric analysis of the remaining urea in a diacetylmonoxim assay [4,165].

As discussed above, urease activity was unchanged by glyphosate, paraquat, trifluralin, and atrazine at 5.4, 2.2, 3.4, and 2.2 kg/ha, respectively [72], and by mancozeb, captanfolpet-folcidin, and benomyl at recommended doses. In the same study, quintozene stimulated urease activity for 4 weeks before activity returned to a normal level [77]. Trichloronate at 5 mg/kg reduced urease activity, but it was stimulated by various other pesticides (Table 1) [79,80]. Propoxur, gamma-HCH, and endrin at recommended field rates inhibited urease during about 6 weeks but the activity recovered afterward. Parathion, methidathion, and omethoate had no effect on urease activity [81]. Urease activity was increased by application of recommended doses of dinoseb and chloroxuron, depressed by 2,4,5-T and simazine, and unaffected by paraquat and 2,4-D [83]. Slightly increased activities were observed with the pesticides tefluthrin and trimethacarb and two experimental phosphorothioates (10 mg/kg) [87]. Phenmedipham at 10 mg/kg did not change the urease activity of soil [97]. Oxamyl and fenamiphos at 12.7 and 18.6 kg/ha, respectively, reduced urease activity under field conditions, but after 5 months activity was like that in a control soil. Under laboratory conditions no effect was observed [114,115]. Nitrofen, fluchloralin, 2,4-D, and isoproturon at 1.0–2.5 kg/ha reduced the activity of urease, whereas methabenzthiazuron and metoxuron had a less pronounced inhibitory effect [116]. No effect was observed with recommended dosages of 2,4-D diallate, glyphosate, benzoylprop-ethyl, and malathion, but the activity was irreversibly lower in presence of 250–1000 mg/kg 2,4-D [130].

The addition of the phosphorothioates fenitrothion, malathion, and phorate, at elevated doses between 50 and 1000 mg/kg, to a sandy clay loam (1.9% organic carbon, pH 5.8) and a silt loam (2.2% organic carbon, pH 5.4) strongly inhibited urease activity, especially at the highest doses, throughout 60 days, i.e., about 40% with fenitrothion and by more than 50% with the others. Urease activity was inhibited by malathion at 1000 mg/kg continuously during 60 days of incubation. After that time, the recovery of malathion was less than 10 mg/kg; that is, it was metabolized by more than 99%. This suggested that the inhibition of activity was mediated by one or several metabolites of malathion rather than by the parent compound. The inhibitory effects of all three insecticides applied at 50 mg/kg were below 10% and reversible. In a separate experiment, isolated jack bean urease was inhibited by the pesticides applied at 1000 mg/kg. This enzyme was analyzed to mimic soil extracellular unbound urease, which was estimated by the authors to contribute 10 to 20% to the overall urease activity of soil. A corresponding immediate 10–20% decline in urease activity in soil treated with the phosphorothioates suggested that the reduced activities reflect in part a direct interaction of the chemicals with extracellular urease and also the interference of the chemicals with urease-producing microorganisms, since the overall long-term inhibition was above 20% [166].

The urea-based herbicides fenuron, monuron, diuron, siduron, linuron, and neburon were evaluated for their effects on urease activity in three soils (silt loam, 3.4% organic matter, pH 8.1; loam, 15.8% organic matter, pH 6.2; silt loam, 16.5% organic matter, pH 5.1). The pesticides, which were increasingly adsorbed by soil in the order in which they are listed, were applied at 4 to 20 mg/kg, except neburon at 1 to 6 mg/kg. Urease activity was determined 1 day after application of the chemicals. In contrast to jack bean urease, which was inhibited differently by each of the chemicals, the activity of soil urease was reduced essentially to the same degree, independent of the respective structures and physicochemical properties of the agrochemicals. Thus, about 90% residual activity was determined at the lowest application rates and 60 to 70% at the highest concentrations with only minor influence of the soil types used. The inhibition was both competitive and noncompetitive; i.e., it was of a mixed type, in that both the affinity of the enzyme for urea and the turnover rate of the enzyme-substrate complex were apparently affected by the herbicides. The authors derived a kinetic model that enabled the determination of inhibition constants based on the adsorption behavior of the herbicides in soil [39,167]. Although the structural details of the active center of jack bean urease remain to be further defined—i.e., the nature of the ligands for the two essential Ni^{2+} ions or the steric and mechanistic interrelation of both metal centers—a model for the competitive binding mode of the herbicides listed above is suggested in Figure 1 based on the conclusions of Cervelli et al. [39] and Andrews et al. [152]. In such a complex, the urea carbonyl oxygen would serve as the sixth ligand of one of the Ni^{2+} ions, which are believed to possess octahedral binding symmetry. The substitution of both amide groups impedes the sequence of catalytic events specific for the hydrolysis of the natural substrate, which proceeds via a carbamoyl or a carbamato intermediate. However, spectroscopic evidence supporting this tentative structure is lacking.

Two glucose-amended soils, a sandy clay loam (1.2% organic carbon, pH 8.1) and a silty clay (4.2% organic carbon, pH 7.7), were treated with ammonium salts and nitrates to obtain information about the regulation of microbial urease production by inorganic nitrogen forms applied as fertilizers. Addition of 18 to 900 mg NH_4^+/kg and of 62 to 3100 mg NO_3^-/kg resulted in a progressive reduction in urease activity, even though the microbial respiration, i.e., the production of CO_2, was stimulated. To study the mode of urease inhibition, experiments were conducted using a glutamine synthetase inhibitor, to block microbial assimilation of inorganic N and an NO_3^- reductase inhibitor, to block the microbial reduction of NO_3^- to NH_4^+. The glutamine synthetease inhibitor relieved the NH_4^+ and NO_3^- repression of urease production, whereas the NO_3^- reductase inhibitor relieved only the NO_3^- repression of urease synthesis and not the repression caused by NH_4^+. The results suggest that urease-producing microbiota are not directly inhibited by NH_4^+ and NO_3^- but are inhibited by their early assimilation products such as L-amino acids. This conclusion was confirmed by the observation

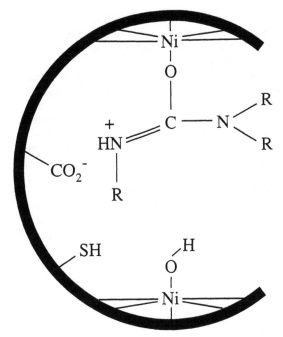

Figure 1   Tentative structural description of the active site in urease complexes with urea-substituted pesticides.   Adapted from Cervelli et al. [39] and Andrews et al. [152].

that urease activity was inhibited by the addition of the L-isomers of various amino acids but not the corresponding D-isomers [168].

III.   CONCLUSIONS

A.   Mode of Interactions Between Pesticides and Soil Enzymes

Obviously, there is little experimental evidence that agrochemicals properly applied will significantly and permanently interfere with the microbiological activity of soil.   A similar conclusion can be drawn from a recent publication in which about 3000 papers on the microbial degradation of pesticides and the side effects of pesticides on the soil microbiota have been reviewed [169].   However, at high concentrations that could result from accidental spills, leaks from dump sites, or use of persistent agrochemicals, severe impairments of the soil microbiota can occur. In such cases, extraneous enzymes in stabilized form may be used for the decontamination of polluted soil sites [170].

For a pesticide to impair the soil ecosystem significantly in either a reversible or irreversible mode, several criteria have to coincide (Figure 2). Of critical importance is the persistence of the chemical in soil, a parameter of minor concern for modern agrochemicals, as the result of environmental screening at an early stage of development and elimination of slowly degrading derivatives. Changes in enzyme activities lasting longer than the half-lives of agrochemicals under the experimental conditions have been presented, e.g., the inhibition of dehydrogenase activity by dinoseb acetate [82], the activity of acid phosphomonoesterase stimulated by fonofos and terbufos [118], the depression of sulfatase activity by oxamyl and fenamifos [114,115], and the reduced activities of phosphodiesterase

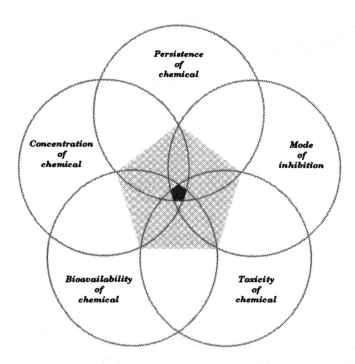

Figure 2 Potential interactions of pesticides with biological systems in soil, with the degree of detectable impairment as a measure of overlapping criteria.

in the presence of atrazine [117]. In such cases, it would be appropriate to confirm the results in field tests simulating agricultural practice conditions. Generally, for assessing the side effects of pesticides the duration of the experiments has to be adapted to the persistence of the parent chemicals and their major degradation products. Comprehensive information on the persistance of pesticides in soil under laboratory and field conditions is available [169], 171].

For any interaction between pesticides and soil microorganisms or accumulated soil enzymes to occur, pollutants have to remain bioavailable (Figure 2), a criterion that is directly related to the physicochemical properties of the chemicals and the soil and that is predominantly influenced by partition, adsorption, and immobilization phenomena. Furthermore, side effects are directly related to the application rate and the local concentration of a chemical, which is controlled by mass transport phenomena, e.g., by leaching, capillary flow, and volatilization, processes that will produce concentration gradients of a pesticide in soil. For example, dalapon, uniformly applied at 40 kg/ha to soil cores, gave concentrations in successive 1-cm soil layers of 180, 60, 50, 25, 8, and 0.8 mg/kg and between 6 and 10 cm less than 0.5 mg/kg [78].

Pesticides may affect soil enzyme activities either directly or indirectly. The prevalent mode of interaction is clearly of the indirect type, as indicated by most of the examples presented, but the distinction between indirect and direct effects is neither abrupt nor easy to determine, as soil enzymes have seldom been purified to the extent that kinetic and other mechanistic details of direct interactions of the protein molecules with pesticides could be elucidated. This is clearly a challenge for future research activities.

Direct inhibition of intra- and extracellular enzymes can occur by binding of the pesticides to the proteins in a reversible mode, i.e., competitively with the substrates by mimicking transition state analogs, or noncompetitively. Irreversible inhibition, on the other hand, is equivalent to the covalent binding of substrates or suicide substrates to functional protein groups that are essential for activity. Suicide substrates are compounds that become irreversible enzyme inhibitors after the parent substrate has been degraded by catalysis of the respective enzyme. Some examples of direct interactions of agrochemicals with accumulated soil enzymes, where the type of inhibition has been studied, are the inhibition of phosphodiesterase [109] and aryl sulfatase [124] by orthophosphate; the inhibition of amidase by alachlor, butylate and diazinon [150]; and the inhibition of urease by urea-based herbicides [39,167].

Indirect interactions of pesticides with soil enzymes, on the other hand, refer to changes in microbial activities and concurrently altered intra- and extracellular enzyme levels. Thus, the inhibition of photosynthesis in algae (e.g., by the bipyridilium herbicides, triazines, and uracils), the reduction in the biosynthesis of proteins and nucleotides, and the depression of cell division by antibiotics, benzimidazoles, and dinitroanilines are examples of this type of interference. Furthermore, cell membranes can be damaged, including disruption of membrane-enzyme complexes (e.g., by bipyridilium or thiocarbamate derivatives), either directly or via the formation of free radicals [7].

The effects of pesticides on microbiologically mediated activities of soil enzymes comprise a narrow range of general responses. A lag before changes in enzyme activities are detected may indicate that degradation products of the xenobiotics are the actual inhibitors or, in the case of hydrophobic molecules, that the rate of interference is controlled by their diffusion to the site of action. The disturbance of microbial equilibria in soil by the introduction of toxins can produce an environment with selective elimination of susceptible genotypes, concurrent with a temporary increase of exoenzyme activity, as the result of cell lysis. The released enzymes are either degraded in the soil solution or stabilized by immobilization on soil particles. Possible effects of the surviving microorganisms are the catabolic utilization of the killed species, the development of resistance to the pesticides, and the expression of detoxifying enzymes and stress proteins [172,173]. Stress proteins are cellular responses common to all living organisms upon exposure to chemical, physical, or biological irritations. For example, the stress protein metallothionein (MT) accumulates in earthworms that inhabit Cd-contaminated soils [174], and MT is also induced in rats that are treated with the herbicide paraquat [175]. Besides, it was documented that MT levels are enhanced by chemical and physical stress factors in various other animals, in plants as well as in eukaryotic and some prokaryotic microorganisms [176].

B. Significance of Soil Enzyme Assays

Predicting potential effects of pesticides on microbiological activities in soil based solely on the chemical, physicochemical, and biological properties of the pesticides and those of the soil is literally impossible, as many contradictory reports on the effects of the same chemical on the same enzyme indicate. This may be exemplified by atrazine and the activity of dehydrogenases. This herbicide was reported to be depressive at 5 mg/kg in a sandy soil [67], to have no effect between 1 and 100 mg/kg in a flooded alluvial soil [70], to stimulate activity at 2 kg/ha in a loam, and to inhibit at the same dose in a sandy soil [84]. There are numerous other examples of such seemingly inconsistent results, but these should not be interpreted as resulting from problems in assaying the

activity of enzymes per se; rather, they are caused primarily by
the variability of the experimental conditions. For example, the
effects of xenobiotics on soil organisms and enzymes are usually
lower in soils with high organic matter and clay content than in
lighter soils as the result of their adsorption [63,84,138,177].
Agricultural soil management techniques, such a monoculture vs.
crop rotations, liming, and fertilization; the methodology and type
of pesticide application; mechanical treatments, such as mulching,
ploughing, and waterlogging [94,102,117,178—181]; and climatic variations, especially precipitation and temperature [182,183], all influence the microbiological activities in the field. According to
Domsch et al. [184], fluctuations in bioactivities within two orders
of magnitude are not uncommon for microbes when coping with
natural environmental stress situations. As a consequence, reversible changes of enzyme activities caused by pesticides that
are smaller than those caused by natural conditions in an untreated control soil must be considered rather unimportant.
When studying the potential side effects of pesticides on the
soil microbiota it is, therefore, mandatory to report the results
against the background of natural fluctuations of the tested activities.

It is precarious to extrapolate results from laboratory studies
to conditions in the field, but this is not a problem peculiar to assays testing the effects of pesticides on enzyme activities in soil,
as it is relevant to all kinds of environmental monitoring. For
example, leaching studies of pesticides in the laboratory are usually
performed in glass or metal columns that are meticuously packed
with homogenized soil. A concentrated solution of the pesticide is
applied to the top soil layer, followed by addition of water to simulate artificial rainfall. Certainly, the subsequently analyzed distribution of the pesticide in the soil and in the leachate will provide
an indication of its potential mobility. However, the experiment
will only partially resemble the leaching behavior under field conditions, as the result of additional transport phenomena, such as
preferential flow, and the unpredictable variability of natural conditions, such as the heterogeneity of the soil in terms of humidity,
texture, and organic matter content, or the climate. In general,
significant effects indicated by laboratory data should be verified
on a case-by-case basis in field studies.

Mechanical treatment of soil, such as sieving or grinding, the
adjustment of pH, and the degree of aeration during the assays
may also have an influence on enzyme activities. Soil biomass is
usually decreased by prolonged storage of the soil samples (even
at 4°C) [185], by air drying, and by freeze-thaw events. In
addition, the biomass is usually related to a greater extent to
the soil moisture content than to the addition of pesticides at

recommended field rates [186,187]. For these reasons, to obtain relevant and reproducible data in soil enzyme assays it is indispensable to establish standardized conditions and adhere to the defined assay conditions precisely.

The inconsistent relation of enzyme activities to other parameters of metabolic activity in soil is sometimes mentioned as an argument against the significance of enzyme assays. However, activity tests, such as the measurement of long-term and substrate-induced respiration [188–190] or ammonification [191], calorimetry, or the quantification of ATP levels and adenylate energy charges [15], reflect the total microbial activity in soil, whereas enzymes catalyze specific reactions representing only minute parts of the soil's overall activity. Intrinsically, assays of the overall microbial and those of specific enzymatic activities will not necessarily produce the same results.

Parameters that reflect the overall metabolic activities in soil may mask the potentially negative effects of improperly used chemicals for several reasons. For example, the elimination of one microbial species is often compensated by the stimulation of others. Moreover, the microbial proliferation resulting from eutrophication with the debris of chemically killed weeds may mask reduced activities of certain segments of the microbiota. Such counterproductive effects are, in part, probably responsible for differences in observations between studies in the laboratory and in cropped fields [82,114,115]. In addition, the pesticides themselves may be utilized by the microorganisms as growth substrates and energy sources [192].

The microbiological activity of soil is far too complex to be described adequately by a single parameter. Thus, the assay of an enzyme activity alone is of little value for assessing the side effects of pesticides in soil, and it should be considered as a supplementary tool to obtain functional information on specific aspects of the bioactivity of soil. Some promising approaches with which to gain significant information by using enzyme methodologies have been discussed by Nannipieri et al. [15]; examples are the *biological fertility index* and the *average activity number* when referring to synergetic analyses of various oxidoreductases and hydrolases.

Summarizing, enzyme assays to test the effects of agrochemicals on nontarget sites, i.e., on microorganisms and accumulated enzymes in soil, have the following advantages and shortcomings.

Advantages	Shortcomings
Simple	Inconsistent relationship to other bioparameters
Fast	
Reproducible	Use of some unnatural substrates

Advantages	Shortcomings
Sensitive, as shown for phosphatases, peroxidases, and phenoloxidases [193]	Adsorption of substrates on soil
	Lack of knowledge of the inherent reactions
Represent in situ measurements	Results strongly dependent on experimental conditions
Provide specific functional information	Difficult to differentiate between microbial activities and those of unspecified exoenzymes
	Often nonspecific, as different chemicals may induce similar effects

According to a questionnaire on methods for evaluating environmental risks posed by xenobiotics, current knowledge is most advanced for aquatic ecosystems and there is still need to further characterize biological indicators for the soil environment [194]. Although it is difficult to define soil fertility in microbiological terms as a result of seasonal, climatic, and topographic activity changes, the holistic preservation of the productivity of the soil ecosystem, the characterization of biogeochemical processes, and how they are affected by agricultural management practices, e.g., by the utilization of pesticides, remain research topics of high priority. Clearly, additional basic research is indicated to analyze the implication of the manifold anthropogenic factors that affect the soil ecosystem.

Table 1 Effect of Agrochemicals on Enzyme Activities in Soil

CHEMICALS	DEHYDROGENASE	CATALASE	PEROXIDASE	PHENOLOXIDASE	CARBOXYL ESTERASE	PHOSPHATASE	SULFATASE	XYLANASE	β-GLUCOSIDASE	L-AMINOACID ARYLAMIDASE	INVERTASE	AMYLASE	CELLULASE	PROTEASE	L-GLUTAMAT ASPARAGINASE	L-ASPARAGINASE GLUTAMINASE	AMIDASE	UREASE	REFERENCES
Urea derivatives																			
Chloroxuron	0,+,0							+,-,0										+,-,+,-,0	[83]
Chlortoluron	-,+																	-	[74]
Diuron		0																	[39, 167]
																			[69]
Fenuron	-																		[39, 167]
																			[84]
Fluometuron	-,0																		[67]
	0																		[69]
Isoproturon						-,0												-,0	[116]
Linuron	+					-/0													[39, 167]
	0													+/0					[89]
																			[69]
Methabenzthiazuron	-,+																		[74]
Metobromuron						-,0												-,0	[116]
					0							0		0					[103]
Metoxuron	-,+																		[74]
						-,0												-,0	[116]
Monuron																			[39, 167]
Neburon																			[39, 167]
Siduron																		-	[39, 167]

Triazines & triazoles

CHEMICALS	DEHYDROGENASE	CATALASE	PEROXIDASE	PHENOLOXIDASE	CARBOX. ESTERASE	PHOSPHATASE	SULFATASE	XYLANASE	β-GLUCOSIDASE	L-GLUTAMINASE	INVERTASE	AMYLASE	CELLULASE	PROTEASE	L-ASPARAGINASE	L-GLUTAMIDASE	AMIDASE	UREASE	REFERENCES
Aminotriazole	-																		[84]
Atrazine	-,o										-,+	-,+	-						[67]
	o																		[70]
	o																	o	[72]
	+/-																		[84]
	o/-																		[88]
					o/-,o														[117]
					+,o/-														[132]
Cyanazine											o	o					o		[142, 147, 150]
										-	-,+	-,+	-		-		o		[67]
										-					-		o		[142, 147, 150]
Metamitron	o																		[75]
Metribuzin	o																		[69]
Prometryne	o																		[69]
	+/-																		[84]
Propazine	+/-																		[84]
Simazine	-,+,o							-,o										-,o	[83]
	+/-										o	o							[84]
																			[132]

Activity symbols: – (reduced), o (unchanged), + (stimulated). Activities separated by a comma (,) indicate changes during the observation period. Activities separated by a slash (/) refer to differences in experimental conditions (soil type, amendments, flooding, etc.). Details are discussed in the text.

Table 1 (Continued)

CHEMICALS	DEHYDROGENASE	CATALASE	PEROXIDASE	PHENOLOXIDASE	CARBOXYLESTERASE	PHOSPHATASE	SULFATASE	XYLANASE	β-GLUCOSIDASE	LAMINARINASE	INVERTASE	AMYLASE	CELLULASE	PROTEASE	L-ASPARAGINASE	L-GLUTAMINASE	AMIDASE	UREASE	REFERENCES
O- & S-carbamates																			
Aldicarb	o																		[75]
Barban	o																		[112]
Benomyl	-,+,o																	o	[70]
Carbaryl	o							+,-,o											[77]
Carbendazim	-,+																		[70], [74]
Carbofuran	o/o,+					o					o	o,+						+/o	[79, 80], [132]
Cartap-HCl	-					o/-,o					o		o	-/,o					[113]
Chlorpropham	o										o								[84]
Cycloate	-,o																		[75], [85]
Di-allate	o								o/-										[75]
EPTC	o																		[130]
EPTC + R-25 788	+,o																	o	[69]
Mancozeb	+					o/-					o	o				-	o		[142, 147, 150], [77]
Maneb						o/-		o										o/+	[79, 80], [132]
Oxamyl	o/o,+					o, o	-				o, o	o/-, o,+	+/-			-	o	o/+, -,o/o	[142, 147, 150], [79, 80], [114, 115], [132]

318

CHEMICALS	DEHYDROGENASE	CATALASE	PEROXIDASE	PHENOLOXIDASE	CARBOXYLESTERASE	PHOSPHATASE	SULFATASE	XYLANASE	β-GLUCOSIDASE	LAMINARINASE	INVERTASE	AMYLASE	CELLULASE	L-ASPARAGINASE	L-GLUTAMINASE	UREASE	REFERENCES
Phenmedipham				o		o			o		+		o			+	[97]
Pirimicarb	o																[75]
Propham	-																[84]
Propham (in mixture with medinoterb acetate)	-,o																[71]
Propoxur	+,o							o,+,o								-,o	[81]
Thiobencarb	o					o							o				[68] [138]
Thiophanate-methyl	-,+																[74]
Thiram	+/o					+/-					+,o					+	[79, 80] [132]
Phosphorus derivatives																	
Bensulide	o																[69]
Chlorfenvinphos	+/o,+					o/-						+				o/+	[79, 80] [132]
Chlorpyrifos	o/o,+					o/-, +					o	o,+				+	[79, 80] [118] [132]
Diazinon	o/o,+					o/-					o,+	+	o/-	-	-	+	[79, 80] [132] [138] [142, 147, 150]

Activity symbols: − (reduced), o (unchanged), + (stimulated). Activities separated by a comma (,) indicate changes during the observation period. Activities separated by a slash (/) refer to differences in experimental conditions (soil type, amendments, flooding, etc.). Details are discussed in the text.

319

Table 1 (Continued)

CHEMICALS	DEHYDROGENASE	CATALASE	PEROXIDASE	PHENOLOXIDASE	CARBOXYL ESTERASE	PHOSPHATAS	SULFATASE	XYLANASE	β-GLUCOSIDASE	LAMININARINASE	INVERTASE	AMYLASE	CELLULASE	PROTEASE	L-ASPARAGINASE	L-GLUTAMINASE	AMIDASE	UREASE	REFERENCES
Dimethoate	-,+																		[74]
Ethion	o/o,+					o/-					+,o	0,+						+,0/+	[79, 80] [132]
Ethoprophos	o/o,+					o/-					+,o	0,+						-,+/+	[79, 80] [132]
Experimental insecticides	+/o,+					+/-					+,o	0,+						0/+	[87]
Fenamiphos						o					-,o/o	0						-,0/o	[114, 115]
Fenitrothion					-	o					0	o/-	+/-				-		[99]
Fensulfothion	o					o/-	-				+,o							-,0 0/+	[166] [79, 80] [132]
Fonofos	o					o +					+,o	+						0	[79, 80] [118] [132]
Glyphosate	o					o				o	+,o	0,+						0 0	[72] [130] [132]
Leptophos	o/o,+					o/-					+,o	0						+/o	[79, 80] [132]
Malathion	o					+/-				o/-	0	0,+					-	+ -,0	[79, 80] [132] [130] [132] [166] [142, 147, 150]

320

CHEMICALS	DEHYDROGENASE	CATALASE	PEROXIDASE	PHENOLOXIDASE	CARBOXYLESTERASE	PHOSPHATASE	SULFATASE	XYLANASE	β,GLUCOSIDASE	LAMINARINASE	INVERTASE	AMYLASE	CELLULASE	PROTEASE	L,GLUTAMINASE	AMIDASE	UREASE	REFERENCES
Methidathion	-,+,0							+,0									+,0	[81]
Omethoate	0,+,0							0,+,0									0	[81]
Orthophosphate						-												[109]
Oxydemeton-methyl	-,+/o																	[74,75]
Parathion	o / o / -,0					+/o	-										+,o/+ , 0	[75] [79,80] [81] [132]
Phorate	o					o/-					o	0,+					0/+	[79,80] [132] [166]
Terbufos	+/-,+					o/- , +					+,o	0,+					-,0 , +	[79,80] [118] [132]
Thionazin	+/o					o/-					+,o	o					+,o/+	[79,80] [132]
Triazophos	o/-,+					o					+,o	+					+	[79,80] [132]
Trichloronate	+/-,+					o/-					+,o	o					-/+	[79,80] [132]
Trichlorphon	-					o												[99]

Activity symbols: – (reduced), o (unchanged), + (stimualted). Activities separated by a comma (,) indicate changes during the observation period. Activities separated by a slash (/) refer to differences in experimental conditions (soil type, amendments, flooding, etc.). Details are discussed in the text.

Table 1 (Continued)

CHEMICALS	DEHYDROGENASE	CATALASE	PEROXIDASE	PHENOLOXIDASE	CARBOX. ESTERASE	PHOSPHATASE	SULFATASE	XYLANASE	β-GLUCOSIDASE	LAMINARINASE	INVERTASE	AMYLASE	CELLULASE	PROTEASE	L-ASPARAGINASE	L-GLUTAMINASE	AMIDASE	UREASE	REFERENCES
Others																		
Alachlor	o / o/-				o/-,o										-	-	-		[69] [88] [117]
Ammonium salts						+,o													[142, 147, 150]
Benzoylprop-ethyl															-	-	-	-	[168]
Butachlor										o/-								o	[130]
Butylate	-,o												o						[138]
Captafol						o													[142, 147, 150]
Captan	+/o,+					o/-									-		-	o/+	[68] [79, 80] [132]
Captan-folpet-folcidin	-,+,o							-,o											[142, 147, 150] [77]
Chloramben	o											o					o	o	[69]
Chlordane	o/o,+					+					o	o,+						+	[142, 147, 150] [79, 80]
Chloridazon	o										o						o		[132] [75]
Chlorothalonil					-	o							-				o		[99] [138]

322

CHEMICALS	DEHYDROGENASE	CATALASE	PEROXIDASE	PHENOLOXIDASE	CARBOXYLESTERASE	PHOSPHATASE	SULFATASE	XYLANASE	β-GLUCOSIDASE	LAMINARINASE	INVERTASE	AMYLASE	CELLULASE	PROTEASE	L-ASPARAGINASE	L-GLUTAMINASE	AMIDASE	UREASE	REFERENCES
2,4-D	+/o,+					o/-												+	[79, 80]
	-,+,o							-,o										o	[83]
																		-	[116]
																		o/-	[130]
										o/+,o									[132]
											o	o							[142, 147, 150]
Dalapon	-					-													[89]
						-,o						-			-	-			[78]
											o						o		[132]
Dicamba											o	o							[132]
											o	o			-	-	o		[142, 147, 150]
3,6-Dichloropicolin. acid	o																		[73]
Dichloropropane, -ene	+/-,o					o/-												+	[79, 80]
											o	o							[132]
Dieldrin	o/o,+					+/-												+	[79, 80]
											o	o,+							[132]
Difenzoquat	o					o													[68]
Dinitramine	o														-	-	o		[69]
																			[142, 147, 150]
Dinoseb	o																	+,-,o	[69]
	-,+,-,+,o							-,o											[83]
Dinoseb acetate	-/o																		[82]
	-,o																		[86]

Activity symbols: – (reduced), o (unchanged), + (stimulated). Activities separated by a comma (,), indicate changes during the observation period. Activities separated by a slash (/) refer to differences in experimental conditions (soil type, amendments, flooding, etc.). Details are discussed in the text.

323

Table 1   (Continued)

CHEMICALS	DEHYDROGENASE	CATALASE	PEROXIDASE	PHENOLOXIDASE	CARBOXYLESTERASE	PHOSPHATASE	SULFATASE	XYLANASE	β-GLUCOSIDASE	LAMINARINASE	INVERTASE	AMYLASE	CELLULASE	PROTEASE	L-ASPARAGINASE	L-GLUTAMINASE	AMIDASE	UREASE	REFERENCES
Endrin	0,+,0							+,0										-,0	[81]
Fluchloralin	-,0					-,0												-,0	[116]
Folpet	0					0													[68]
Gamma-HCH	-,+,0					0/-		-,+,0										+/o, -,0	[79, 80] [81] [132]
HCH	0										o								[70]
Hymexazol													0						[138]
Lenacil	-																		[84]
Mecoprop	-,+																		[74]
Medinoterb acetate (in mixture with propham)	-,0																		[71]
Methyl bromide		o																	[90]
Methyl isothiocyanate	+/o					o/-					o,-	o,-						+	[79, 80] [132]
Metolachlor	o/-				o/-,o	+,0					o	o							[88]
Naptalam	o																		[117]
Nitralin	o														-	-	-		[69]
Nitrapyrin	0/-,o					0/-					o	o						+	[142, 147, 150] [69] [79, 80] [132]
Nitrates																		-	[168]

CHEMICALS	DEHYDROGENASE	CATALASE	PEROXIDASE	PHENOLOXIDASE	CARBOXYLESTERASE	PHOSPHATASE	SULFATASE	XYLANASE	β-GLUCOSIDASE	LAMINARINASE	INVERTASE	AMYLASE	CELLULASE	PROTEASE	L-ASPARAGINASE	L-GLUTAMINASE	AMIDASE	UREASE	REFERENCES
Nitrofen						-,o	-,o											-,o	[116]
Paraquat	o			-	-	o	o	-,o					o				o	o	[132] [72] [83] [94]
	-,+,o				-	-											o	-,o	[99] [138] [142, 147, 150]
Pendimethalin	o					o													[68]
Permethrin	o					o	o				+,o	0,+			-	-		+	[79, 80]
							o				o	o				-			
Picloram											+,o								[132]
Profluralin	o										o								[69]
Propanil	o												o/-						[69] [138]
Quintozene	+,o							-					-/o					+,o	[77] [138]
Sodium azide	o										o	o		-					[90]
2,4,5-T	o,+,o							-,o										-,o	[83]
TCA	-													-					[89]
Tefluthrin	+/-,+										-,o/o	+/+,o						o/+	[87]

Activity symbols: – (reduced), o (unchanged), + (stimulated). Activities separated by a comma (,) indicate changes during the observation period. Activities separated by a slash (/) refer to differences in experimental conditions (soil type, amendments, flooding, etc.). Details are discussed in the text.

Table 1 (Continued)

CHEMICALS	DEHYDROGENASE	CATALASE	PEROXIDASE	PHENOLOXIDASE	CARBOXYESTERASE	PHOSPHATASE	SULFATASE	XYLANASE	B-GLUCOSIDASE	LAMINARINASE	INVERTASE	AMYLASE	CELLULASE	PROTEASE	L-GLUTAMINASE	L-ASPARAGINASE	AMIDASE	UREASE	REFERENCES
Triadimefon	-,+																		[74]
Trichlamide																			[138]
Tridemorph	-,+																		[74]
Trifluralin	0,+																		[67]
	o																		[69]
	o																	o	[72]
	o/-				o/-,o														[88]
											-,o/o	+/+,o		-	-		o	o/+	[142, 147, 150]
Trimethacarb (experimental insecticide)	+/o,+					+/-													[87]

Activity symbols: − (reduced), o (unchanged), + (stimulated). Activities separated by a comma (,) indicate changes during the observation period. Activities separated by a slash (/) refer to differences in experimental conditions (soil type, amendments, flooding, etc.). Details are discussed in the text.

REFERENCES

1. Skujins, J. 1976. Extracellular enzymes in soil. Crit. Rev. Microbiol. 4:383–421.
2. Burns, R. G. 1978. Enzyme activity in soil: some theoretical and practical considerations, p. 295–340. *In* R. G. Burns (ed.), Soil Enzymes. Academic Press, New York.
3. Ladd, J. N. 1978. Origin and range of enzymes in soil, p. 51–96. *In* R. G. Burns (ed.), Soil enzymes. Academic Press, New York.
4. Tabatabai, M. A. 1982. Soil enzymes, p. 903–947. *In* A. L. Page, R. H. Miller, and D. R. Keeney (eds.), Methods of soil analysis, Part 2. American Society of Agronomy, Soil Science Society of America, Madison, WI.
5. Boyd, S. A., and M. M. Mortland. 1990. Enzyme interactions with clays and clay-organic matter complexes, p. 1–28. *In* J.-M. Bollag and G. Stotzky (eds.), Soil biochemistry, Vol. 6. Marcel Dekker, New York.
6. Kiss, S., M. Dragan-Bularda, and D. Radulescu. 1975. Biological significance of enzymes in soil. Adv. Agron. 27:25–87.
7. Cervelli, S., P. Nannipieri, and P. Sequi. 1978. Interactions between agrochemicals and soil enzymes, p. 251–293. *In* R. G. Burns (ed.), Soil enzymes. Academic Press, New York.
8. Wainwright, M. 1978. A review of the effects of pesticides on microbial activity in soils. J. Soil Sci. 29:287–298.
9. Anderson, J. R. 1978. Some methods for assessing pesticide effects on non-target soil microorganisms and their activities, p. 247–312. *In* I. R. Hill and S. J. L. Wright (eds.), Pesticide microbiology. Academic Press, London.
10. Simon-Sylvestre, G., and J. C. Fournier. 1979. Effects of pesticides on the soil microflora. Adv. Agron. 31:1–92.
11. Greaves, M. P., and M. P. Malkomes. 1980. Effects on soil microflora, p. 223–253. *In* R. J. Hance (ed.), Interactions between herbicides and the soil. Academic Press, London.
12. Stevenson, F. J. (ed.). 1986. Cycles of soil: carbon, nitrogen, phosphorus, sulphur, micronutrients. John Wiley & Sons, New York.
13. Edwards, C. A. 1989. Impact of herbicides on soil ecosystems. Crit. Rev. Plant Sci. 8:221–257.
14. Racke, K. D. 1990. Pesticides in the soil microbial ecosystem, p. 1–12. *In* K. D. Racke and J. R. Coats (eds.), Enhanced biodegradation of pesticides in the environment. American Chemical Society, Washington, DC.
15. Nannipieri, P., S. Grego, and B. Ceccanti. 1990. Ecological significance of the biological activity in soil, p. 293–355. *In* J.-M. Bollag and G. Stotzky (eds.), Soil biochemistry, Vol. 6. Marcel Dekker, New York.

16. Stotzky, G. 1986. Influence of soil mineral colloids on metabolic processes, growth, adhesion, and ecology of microbes and viruses, p. 305–428. *In* P. M. Huang and M. Schnitzer (eds.), Interactions of soil minerals with natural organics and microbes. Soil Science Society of America, Madison, WI.

17. McBride, M. B. 1987. Adsorption and oxidation of phenolic compounds by iron and manganese oxides. Soil Sci. Soc. Am. J. 51:1466–1472.

18. Rice, J. A., and P. MacCarthy. 1990. A model of humin. Environ. Sci. Technol. 24:1875–1877.

19. Stevenson, F. J. (ed.). 1982. Humus chemistry: genesis, composition, reactions. John Wiley & Sons, New York.

20. Aiken, G. R., D. M. McKnight, R. L. Wershaw, and P. MacCarthy (eds.). 1985. Humic substances in soil, sediment and water: geochemistry, isolation and characterization. John Wiley & Sons, New York.

21. Schulten, H.-R., B. Plage, and M. Schnitzer. 1991. A chemical structure for humic substances. Naturwissenschaften 78:311–312.

22. Bower, C. A., and J. O. Goertzen. 1959. Surface area of soils as determined by an equilibrium ethylene glycol method. Soil Sci. 87:289–292.

23. Chiou, C. T., J.-F. Lee, and S. A. Boyd. 1990. The surface area of organic matter. Environ. Sci. Technol. 24:1164–1166.

24. Pennell, K. D., and P. S. C. Rao. 1992. Comment on "the surface area of soil organic matter." Environ. Sci. Technol. 26:402–404.

25. Chiou, C. T., J.-F. Lee, and S. A. Boyd. 1992. Comment on "the surface area of soil organic matter." Environ. Sci. Technol. 26:404–406.

26. Burns, R. G. 1982. Enzyme activity in soil: location and a possible role in microbial ecology. Soil Biol. Biochem. 14:423–427.

27. Skujins, J. 1978. History of abiontic soil enzyme research, p. 1–49. *In* R. G. Burns (ed.), Soil enzymes. Academic Press, New York.

28. Frankenberger, W. T., Jr., and J. B. Johanson. 1986. Use of plasmolytic agents and antiseptics in soil enzyme assays. Soil Biol. Biochem. 18:209–213.

29. Kaplan, D. L., and R. Hartenstein. 1979. Problems with toluene and the determination of extracellular enzyme activity in soils. Soil Biol. Biochem. 11:335–338.

30. Anderson, J. P. E., and K. H. Domsch. 1975. Measurement of bacterial and fungal contributions to respiration of selected agricultural and forest soils. Can. J. Microbiol. 21:314–322.

31. Stotzky, G., and R. G. Burns. 1982. The soil environment: clay-humus-microbe interactions, p. 105–133. *In* R. G. Burns and J. H. Slater (eds.), Experimental microbial ecology. Blackwell, Oxford.

32. McLaren, A. D., and E. Packer. 1970. Some aspects of enzyme reactions in heterogeneous systems. Adv. Enzymol. 33:245–308.

33. Hattori, T., and R. Hattori. 1976. The physical environment in soil microbiology: an attempt to extend principles of microbiology to soil microorganisms. CRC Crit. Rev. Microbiol. 4:423–461.

34. Lai, C. M., and M. A. Tabatabai. 1992. Kinetic parameters of immobilized urease. Soil Biol. Biochem. 24:225–228.

35. Sinsabaugh, R. L., R. K. Antibus, and A. E. Linkins. 1991. An enzymic approach to the analysis of microbial activity during plant litter decomposition. Agric. Ecosyst. Environ. 34:43–54.

36. Sarkar, J. M., A. Leonowicz, and J.-M. Bollag. 1989. Immobilization of enzymes on clays and soils. Soil Biol. Biochem. 21:223–230.

37. Ladd, J. N., and J. H. A. Butler. 1975. Humus-enzyme systems and synthetic, organic polymer-enzyme analogs, p. 134–194. *In* E. A. Paul and A. D. MacLaren (eds.), Soil biochemistry, Vol. 4. Marcel Dekker, New York.

38. Burns, R. G. 1986. Interaction of enzymes with soil minerals and organic colloids, p. 429–451. *In* P. M. Huang and M. Schnitzer (eds.), Interactions of soil minerals with natural organics and microbes. Special Publication Number 17, Soil Science Society of America, Madison, WI.

39. Cervelli, S., P. Nannipieri, G. Giovannini, and A. Perna. 1977. Effect of soil on urease inhibition by substituted urea herbicides. Soil Biol. Biochem. 9:393–396.

40. Führ, F. 1987. Non-extractable pesticide residues in soil, p. 381–389. *In* R. Greenhalgh and T. R. Roberts (eds.), Pesticide science and biotechnology. Blackwell Scientific Publications, Oxford.

41. Anonymous. 1988. Erde, p. 493–498. *In* Brockhaus encyclopaedia, Vol. 6. F. A. Brockhaus, Mannheim.

42. Guth, J. A. 1980. The study of transformations, p. 123–157. *In* R. J. Hance (ed.), Interactions between herbicides and the soil. Academic Press, London.

43. Guth, J. A. 1981. Experimental approaches to studying the fate of pesticides in soil, p. 85–114. *In* D. H. Hutson and T. R. Roberts (eds.), Progress in pesticide biochemistry, Vol. 1. John Wiley & Sons, New York.

44. Briggs, G. G. 1990. Predicting the behaviour of pesticides in soil from their physical and chemical properties. Philos. Trans. R. Soc. London Ser. B 329:375–382.

45. Spencer, W. F., M. M. Cliath, W. A. Jury, and L.-Z. Zhang. 1988. Volatilization of organic chemicals from soil as related to their Henry's law constant. J. Environ. Qual. 17:504–509.

46. Ghodrati, M., and W. A. Jury. 1990. A field study using dyes to characterize preferential flow of water. Soil Sci. Soc. Am. J. 54:1558–1563.

47. Bollag, J.-M., and S.-Y. Liu. 1990. Biological transformation processes of pesticides, p. 169–211. *In* H. H. Cheng (ed.), Pesticides in the soil environment: Processes, impacts, and modeling. SSSA Book Ser. 2, Madison, WI.

48. Somasundaram, L., and J. R. Coats (eds.). 1991. Pesticide transformation products: fate significance in the environment. American Chemical Society, Washington, DC.

49. Kearney, P. C., J. S. Karns, and W. W. Mulbry. 1987. Engineering soil microorganisms for pesticide degradation, p. 591–596. *In* R. Greenhalgh and T. R. Roberts (eds.), Pesticide science and biotechnology. Blackwell Scientific Publications, Oxford.

50. Doyle, J. D., K. A. Short, G. Stotzky, R. J. King, R. J. Seidler, and R. H. Olsen. 1991. Ecologically significant effects of *Pseudomonas putida* PPO301(pRO103), genetically engineered to degrade 2,4-dichlorophenoxyacetate, on microbial populations and processes in soil. Can. J. Microbiol. 37: 682–702.

51. Short, K. A., J. D. Doyle, R. J. King, R. J. Seidler, G. Stotzky, and R. H. Olsen. 1991. Effects of 2,3-dichlorophenol, a metabolite of a genetically engineered bacterium, and 2,4-diclorophenoxyacetate on some microorganism-mediated ecological processes in soil. Appl. Environ. Microbiol. 57:412–418.

52. Ogram, A. V., R. E. Jessup, L. T. Ou, and P. S. C. Rao. 1985. Effects of sorption on biological degradation rates of 2,4-dichlorophenoxyacetic acid in soils. Appl. Environ. Microbiol. 49:582–587.

53. Speitel, G. E., C. J. Lu, M. Turakhia, and X.-J. Sho. 1988. Biodegradation of trace concentrations of substituted phenols in granular activated carbon columns. Environ. Sci. Technol. 22:68–74.

54. Horvath, R. S. 1972. Microbial cometabolism and the degradation of organic compounds in nature. Bacteriol. Rev. 36: 146–155.

55. Bollag, J.-M. 1982. Microbial metabolism of pesticides, p. 126–168. *In* J. C. Rosazza (ed.), Microbial transformations of bioactive compounds, Vol. 2. CRC Press, Boca Raton, FL.

56. Calderbank, A. 1989. The occurrence and significance of bound pesticide residues in soil. Rev. Environ. Contam. Toxicol. 108:71–103.

57. Dec, J., K. L. Shuttleworth, and J.-M. Bollag. 1990. Microbial release of 2,4-dichlorophenol bound to humic acid or incorporated during humification. J. Environ. Qual. 19:546–551.
58. Khan, S. U. 1991. Bound residues, p. 265–279. *In* R. Grover and A. J. Cessna (eds.), Environmental chemistry of herbicides, Vol. II. CRC Press, Boston.
59. Racke, K. D., and J. R. Coats (eds.). 1990. Enhanced biodegradation of pesticides in the environment. American Chemical Society, Washington, DC.
60. Racke, K. D. 1990. Implications of enhanced biodegradation for the use and study of pesticides in the soil environment, p. 269–282. *In* K. D. Racke and J. R. Coats (eds.), Enhanced biodegradation of pesticides in the environment. American Chemical Society, Washington, DC.
61. Worthing, C. R., and S. B. Walker. 1987. The pesticide manual, 8th ed. British Crop Protection Council, Thornton Heath, UK.
62. Gerber, H. R., J. P. E. Anderson, B. Bügel-Mogenson, D. Castle, K. H. Domsch, H.-P. Malkomes, L. Somerville, D. J. Arnold, H. van de Werf, R. Verbeken, and J. W. Vonk. 1991. Revision of recommended laboratory tests for assessing side-effects of pesticides on the soil microflora. Toxicol. Environ. Chem. 30:249–261.
63. Malkomes, H.-P. 1991. Existing alternative tests to measure side-effects of pesticides on soil microorganisms: dehydrogenase activity. Toxicol. Environ. Chem. 30:167–176.
64. von Mersi, W., and F. Schinner. 1991. An improved and accurate method for determining the dehydrogenase activity of soils with iodonitrotetrazolium chloride. Biol. Fertil. Soils 11: 216–220.
65. Dougherty, J. M., and G. R. Lanza. 1989. Anaerobic subsurface soil microcosms: methods to monitor effects of organic pollutants on indigenous microbial activity. Tox. Assess. 4:85–104.
66. Chander, K., and P. C. Brookes. 1991. Is the dehydrogenase assay invalid as a method to estimate microbial activity in copper-contaminated soils? Soil Biol. Biochem. 23:909–915.
67. El-Nawawy, A. S., A. T. El-Din, A. M. Komeil, M. A. S. Khalifa, S. T. El-Deeb, S. Abou Donia, and E. A. Kadous. 1977. Effect of several pesticides on the activity of soil enzymes. Med. Fac. Landbouw. Rijksuniv. Gent 42:901–909.
68. Atlas, R. M., D. Pramer, and R. Bartha. 1978. Assessment of pesticide effects on nontarget soil microorganisms. Soil Biol. Biochem. 10:231–239.
69. Lewis, J. A., G. C. Papavizas, and T. S. Hora. 1978. Effect of some herbicides on microbial activity in soil. Soil Biol. Biochem. 10:137–141.

70. Chendrayan, K., and N. Sethunathan. 1980. Effects of HCH,
     benomyl, and atrazine on the dehydrogenase activity in a
     flooded soil. Bull. Environ. Contam. Toxicol. 24:379–382.
71. Malkomes, H.-P. 1981. Effect of a sugar beet herbicide on
     the microflora of different soils in the laboratory. Zbl. Bakt.
     II. Abt. 136:451–460.
72. Davies, H. A., and M. P. Greaves. 1981. Effects of some
     herbicides on soil enzyme activities. Weed Res. 21:205–209.
73. Maas, G., and G. Heidler. 1981. Investigations on side-ef-
     fects of 3,6-dichloropicolinic acid–phytotoxicity, straw decom-
     position and dehydrogenase activity. Z. Pflkrankh. Pflschutz,
     Sonderheft IX:297–299.
74. Malkomes, H.-P. 1981. Effects of plant protection systems
     applied to winter cereals on biological activity in soils. Part I:
     Dehydrogenase activity and straw decomposition. Z. Pflkrankh.
     Pflschutz, Sonderheft IX:301–311.
75. Malkomes, H.-P. 1982. Effect of two pesticide treatment sys-
     tems for sugar beets on dehydrogenase activity and straw de-
     composition in soil. J. Plant Dis. Prot. 89:705–714.
76. Pasztor, Z., C. Dobolyi, and M. Kecskes. 1978. The effect
     of different pesticides on the mycoflora of an chernozem eco-
     system. Proc. Hung. Annu. Meet. Biochem. 18:21–22.
77. Mitterer, M., H. Bayer, and F. Schinner. 1981. The influ-
     ence of fungicides on microbial activity in soil. Z. Pflanzenern.
     Bodenkd. 144:463–471.
78. Greaves, M. P., H. A. Davies, J. A. P. Marsh, and G. I.
     Wingfield. 1981. Effects of pesticides on soil microflora using
     dalapon as an example. Arch. Environ. Contam. Toxicol. 10:
     437–449.
79. Tu, C. M. 1981. Effects of pesticides on activities of en-
     zymes and microorganisms in a clay soil. J. Environ. Sci.
     Health B16:179–191.
80. Tu, C. M. 1981. Effects of some pesticides on enzyme ac-
     tivities in an organic soil. Bull. Environ. Contam. Toxicol.
     27:109–114.
81. Bayer, H., M. Mitterer, and F. Schinner. 1982. The influ-
     ence of insecticides on microbial processes in A_h-materials of
     an agricultural soil. Pedobiologia 23:311–319.
82. Malkomes, H.-P. 1983. Testing and evaluating some methods
     to investigate side effects of environmental chemicals on soil
     microorganisms. Ecotoxicol. Environ. Safety 7:284–294.
83. Schinner, F., H. Bayer, and M. Mitterer. 1983. The influ-
     ence of herbicides on microbial activity in soil materials.
     Bodenkultur 34:22–30.
84. Hickisch, B., G. Machulla, and G. Mueller jun. 1984. Side
     effects of herbicides on soil organisms after one and several
     applications thereof. Zbl. Mikrobiol. 139:13–20.

85. Malkomes, H.-P. 1985. Effect of the herbicide Ro-Neet and its combination with a phospholipid on biological activities in soil under laboratory conditions. Zbl. Mikrobiol. 140:381–391.

86. Malkomes, H.-P. 1985. Influence of the herbicide dinoseb acetate and its combination with a phospholipid on microbial activities under laboratory and field conditions. J. Plant Dis. Prot. 92:489–501.

87. Tu, C. M. 1990. Effect of four experimental insecticides on enzyme activities and levels of adenosine triphosphate in mineral and organic soils. J. Environ. Sci. Health B25:787–800.

88. Dzantor, E. K., and A. S. Felsot. 1991. Microbial responses to large concentrations of herbicides in soil. Environ. Tox. Chem. 10:649–655.

89. Mas'ko, A. A., N. F. Lovchii, and L. A. Pototskaya. 1991. Stability of immobilized soil enzymes and their role in the degradation of herbicides. Vestsi Akad. Navuk BSSR, Ser. Biyal. Navuk 5:47–51.

90. Kelley, W. D., and R. Rodriguez-Kabana. 1979. Effects of sodium azide and methyl bromide on soil bacterial populations, enzymic activities and other biological variables. Pestic. Sci. 10:207–215.

91. Kirk, T. K., and R. L. Farrell. 1987. Enzymatic "combustion": the microbial degradation of lignin. Annu. Rev. Microbiol. 41:465–505.

92. Bartha, R., and L. M. Bordelau. 1969. Cell-free peroxidases in soil. Soil Biol. Biochem. 1:139–143.

93. Bordelau, L. M., and R. Bartha. 1969. Rapid technique for enumeration and isolation of peroxidase-producing microorganisms. Appl. Microbiol. 18:274–275.

94. Anderson, J. R. 1976. Effects of pure paraquat dichloride, "Gramoxone W," and formulation additives on soil microbiological activities. Zbl. Bakt. Abt. II 131:247–258.

95. Kirk, T. K., and M. Shimada. 1985. Lignin biodegradation: the microorganisms involved, and the physiology and biochemistry of degradation by white-rot fungi, p. 579–605. *In* T. Higuchi (ed.), Biosynthesis and biodegradation of wood components. Academic Press, New York.

96. Nannipieri, P., A. Gelsomino, and M. Felici. 1991. Method to determine guaiacol oxidase activity in soil. Soil Sci. Soc. Am. J. 55:1347–1352.

97. Bellinck, C., and J. Mayaudon. 1980. Influence of phenmedipham and its derivatives on the microflora and the biological and enzymological activities in a fresh soil. Rev. Ecol. Biol. Sol. 17:1–6.

98. Schnurer, J., and T. Rosswall. 1982. Fluorescein diacetate hydrolysis as a measure of total microbial activity in soil and litter. Appl. Environ. Microbiol. 43:1256–1261.

99. Nakamura, T., K. Mochida, Y. Ozoe, S. Ukawa, M. Sakai, and S. Mitsugi. 1990. Enzymological properties of three soil hydrolases and effects of several pesticides on their activities. J. Pestic. Sci. 15:593–598.

100. Pancholy, S. K., and J. Q. Lynd. 1972. Quantitative fluorescence analysis of soil lipase activity. Soil Biol. Biochem. 4:257–259.

101. Cooper, A. B., and H. W. Morgan. 1981. Improved fluorimetric method to assay for soil lipase activity. Soil Biol. Biochem. 13:307–311.

102. Hankin, L., D. E. Hill, and G. Stephens. 1982. Effect of mulches on bacterial populations and enzyme activity in soil and vegetable yields. Plant Soil 64;193–201.

103. Nowak, A., and W. Michalcewicz. 1988. Effect of the herbicide Patoran 50 WP on amylolytic, proteolytic, and lipolytic activity of mesophilic soil actinomycetes. Z. Pflkrankh. Pflschutz, Sonderheft XI:295–300.

104. Speir, T. W., and D. J. Ross. 1978. Soil phosphatase and sulfatase, p. 197–250. *In* R. G. Burns (ed.), Soil enzymes. Academic Press, New York.

105. Kroll, L., and M. Kramer. 1955. The influence of clay minerals on the activity of soil phosphatase. Naturwissenschaften 42:157–158.

106. Tabatabai, M. A., and J. M. Bremner. 1969. Use of *p*-nitrophenyl phosphate for assay of soil phosphatase activity. Soil Biol. Biochem. 1:301–307.

107. Pettit, N. M., L. J. Gregory, R. B. Freedman, and R. G. Burns. 1977. Differential stabilities of soil enzymes: assay and properties of phosphatase and arylsulphatase. Biochim. Biophys. Acta 485:357–366.

108. Ramirez-Martinez, J. R., and A. D. McLaren. 1966. Determination of soil phosphatase activity by a fluorimetric technique. Enzymologia 30:243–253.

109. Browman, M. G., and M. A. Tabatabai. 1978. Phosphodiesterase activity of soils. Soil Sci. Soc. Am. J. 42:284–290.

110. Eivazi, F., and M. A. Tabatabai. 1977. Phosphatases in soil. Soil Biol. Biochem. 9:167–172.

111. Dick, W. A., and M. A. Tabatabai. 1978. Inorganic pyrophosphatase activity of soils. Soil Biol. Biochem. 10:59–65.

112. Quilt, P., E. Grossbard, and S. J. L. Wright. 1979. Effects of the herbicide barban and its commercial formulation carbyne on soil micro-organisms. J. Appl. Bacteriol. 46:431–442.

113. Endo, T., T. Kusaka, N. Tan, and M. Sakai. 1982. Effects of the insecticide cartap hydrochloride on soil enzyme activities, respiration and nitrification. J. Pestic. Sci. 7:101–110.

114. Ross, D. J., T. W. Speir, J. C. Cowling, and K. N. Whale. 1984. Influence of field applications of oxamyl and fenamiphos on biochemical activities of soil under pasture. N.Z. J. Sci. 27:247–254.

115. Ross, D. J., and T. W. Speir. 1985. Changes in biochemical activities of soil incubated with the nematicides oxamyl and fenamiphos. Soil Biol. Biochem. 17:123–125.

116. Tarafdar, J. C. 1986. Effect of different herbicides on enzyme activity in controlling weeds in wheat crop. Pesticides 20:46–49.

117. Perucci, P., L. Scarponi, and M. Monotti. 1988. Interference with soil phosphatase activity by maize herbicidal treatment and incorporation of maize residues. Biol. Fertil. Soils 6:286–291.

118. Sikora, L. J., D. D. Kaufman, and L. C. Horng. 1990. Enzyme activity in soils showing enhanced degradation of organophosphate insecticides. Biol. Fertil. Soils 9:14–18.

119. Perucci, P., and L. Scarponi. 1984. Arylsulphatase activity in soil amended with crop residues: kinetic and thermodynamic parameters. Soil Biol. Biochem. 16:605–608.

120. Press, M. C., J. Henderson, and J. A. Lee. 1985. Arylsulphatase activity in peat in relation to acidic deposition. Soil Biol. Biochem. 17:99–103.

121. Tabatabai, M. A., and J. M. Bremner. 1970. Arylsulfatase activity of soils. Soil Sci. Soc. Am. Proc. 34:225–229.

122. Tabatabai, M. A., and J. M. Bremner. 1970. Factors affecting soil arylsulfatase activity. Soil Sci. Soc. Am. Proc. 34:427–429.

123. Dick, R. P., P. E. Rasmussen, and E. A. Kerle. 1988. Influence of long-term residue management on soil enzyme activities in relation to soil chemical properties of a wheat-fallow system. Biol. Fertil. Soils 6:159–164.

124. Al-Khafaji, A. A., and M. A. Tabatabai. 1979. Effect of trace elements on arylsulfatase activity in soils. Soil Sci. 127:129–133.

125. Wong, K. K. Y., L. U. L. Tan, and J. N. Saddler. 1988. Multiplicity of β-1,4-xylanase in microorganisms: functions and applications. Microbiol. Rev. 52:305–317.

126. Spiro, R. G. 1966. Analysis of sugars found in glucoproteins, p. 3–26. *In* E. F. Neufeld and V. Ginsburg (eds.), Methods in enzymology, Vol. 8. Academic Press, New York.

127. Lloyd, J. B., and W. J. Whelan. 1969. An improved method of enzymic determination of glucose in presence of maltose. Anal. Biochem. 30:467–470.

128. Wainwright, M. 1981. Assay and properties of alginate lyase and 1,3-β-glucanase in intertidal sands. Plant Soil 59:83–89.

129. Lethbridge, G., A. T. Bull, and R. G. Burns. 1978. Assay and properties of 1,3-β-glucanase in soil. Soil Biol. Biochem. 10:389–391.

130. Lethbridge, G., A. T. Bull, and R. G. Burns. 1981. Effect of pesticides on 1,3-β-glucanase and urease activities in soil in the presence and absence of fertilisers, lime and organic materials. Pestic. Sci. 12:147–155.

131. Ross, D. J. 1965. A seasonal study of oxygen uptake of some pasture soils and activities of enzymes hydrolysing sucrose and starch. J. Soil Sci. 16:73–85.

132. Tu, C. M. 1982. Influence of pesticides on activities of invertase, amylase and level of adenosine triphosphate in organic soil. Chemosphere 11:909–914.

133. Smith, B. W., and J. H. Roe. 1949. A photometric method for the determination of α-amylase in blood and urine with the use of starch iodine color. J. Biol. Chem. 179:53–59.

134. Eriksson, K.-E., and T. M. Wood. 1985. Biodegradation of cellulose, p. 469–503. *In* T. Higuchi (ed.), Biosynthesis and biodegradation of wood components. Academic Press, New York.

135. Ljungdahl, L. G., and K.-E. Eriksson. 1985. Ecology of microbial cellulose degradation. Adv. Microb. Ecol. 8:237–299.

136. Grossbard, E., and G. I. Wingfield. 1975. The effect of herbicides on cellulose decomposition, p. 236–256. *In* R. G. Board and D. W. Lovelock (eds.), Some methods for microbiological assays. Academic Press, London.

137. Vincent, P. G., and H. D. Sisler. 1968. Mechanism of antifungal action of 2,4,5,6-tetrachloroisophthalonitrile. Physiol. Plant 21:1249–1264.

138. Katayama, A., and S. Kuwatsuka. 1991. Effect of pesticides on cellulose degradation in soil under upland and flooded conditions. Soil Sci. Plant Nutr. 37:1–6.

139. Ladd, J. N., and J. H. A. Butler. 1972. Short-term assays of soil proteolytic enzyme activities using proteins and dipeptide derivatives as substrates. Soil Biol. Biochem. 4:19–30.

140. Lowry, O. H., M. J. Rosebrough, A. L. Farr, and R. J. Randall. 1951. Protein measurement with Folin-phenol reagent. J. Biol. Chem. 193:265–275.

141. Moore, S. 1968. Amino acid analysis: aqueous dimethyl sulfoxide as solvent for ninhydrin reaction. J. Biol. Chem. 243:6281–6283.

142. Frankenberger, W. T., Jr., and M. A. Tabatabai. 1991. Factors affecting L-asparaginase activity in soils. Biol. Fertil. Soils 11:1–5.

143. Frankenberger, W. T., Jr., and M. A. Tabatabai. 1991. L-Asparaginase activity of soils. Biol. Fertil. Soils 11:6–12.

144. Keeny, D. R., and D. W. Nelson. 1982. Nitrogen-inorganic forms, p. 643–698. *In* A. L. Page, R. H. Miller, and D. R. Keeney (eds.), Methods of soil analysis, Part 2, 2nd ed. American Society of Agronomy, Madison, WI.

145. McLaren, A. C., L. Reshetko, and W. Huber. 1957. Sterilization of soil by irradiation with an electron beam and some observations on soil enzyme activity. Soil Sci. 83:497–502.

146. Frankenberger, W. T., Jr., and M. A. Tabatabai. 1991. L-Glutaminase activity of soils. Soil Biol. Biochem. 23:869–874.

147. Frankenberger, W. T., Jr., and M. A. Tabatabai. 1991. Factors affecting L-glutaminase activity in soils. Soil Biol. Biochem. 23:875–879.

148. Frankenberger, W. T., Jr., and M. A. Tabatabai. 1981. Amidase activity in soils: III. Stability and distribution. Soil Sci. Soc. Am. J. 45:333–338.

149. Frankenberger, W. T., Jr., and M. A. Tabatabai. 1981. Amidase activity in soils: I. Method of assay. Soil Sci. Soc. Am. J. 44:282–287.

150. Frankenberger, W. T., Jr., and M. A. Tabatabai. 1981. Amidase activity in soils: IV. Effects of trace elements and pesticides. Soil Sci. Soc. Am. J. 45:1120–1124.

151. Beri, V., K. P. Goswami, and S. S. Brar. 1978. Urease activity and its Michaelis constant for soil systems. Plant Soil 49:105–115.

152. Andrews, R. K., R. L. Blakeley, and B. Zerner. 1984. Urea and urease, p. 245–283. *In* L. Eichhorn and L. G. Marzilli (eds.), Advances in inorganic biochemistry, Vol. 6. Elsevier Science Publishing, New York.

153. Reithel, F. J. 1971. Ureases, p. 1–21. *In* P. D. Boyer (ed.), The enzymes, Vol. 4. Academic Press, New York.

154. Hausinger, R. P. 1987. Nickel utilization by microorganisms. Microbiol. Rev. 51:22–42.

155. Mobley, H. L. T., and R. P. Hausinger. 1989. Microbial ureases: significance, regulation, amd molecular characterization. Microbiol. Rev. 53:85–108.

156. Gould, W. D., C. Hagedorn, and R. G. L. McCready. 1986. Urea transformations and fertilizer efficiency in soil. Adv. Agron. 40:209–238.

157. Beyrouty, C. A., L. E. Sommers, and D. W. Nelson. 1988. Ammonia volatilization from surface-applied urea as affected by several phosphoramide compounds. Soil Sci. Soc. Am. J. 52:1173–1178.

158. Bremner, J. M., and H. S. Chai. 1989. Effects of phosphoroamides on ammonia volatilization and nitrite accumulation in soils treated with urea. Biol. Fertil. Soils 8:227–230.

159. Creason, G. L., M. R. Schmitt, E. A. Douglass, and L. L. Hendrickson. 1990. Urease inhibitory activity associated with N-(n-butyl)-thiophosphoric triamide is due to formation of its oxon analog. Soil Biol. Biochem. 22:209–211.

160. Zhengping, W., O. van Cleemput, P. Demeyer, and L. Baert. 1991. Effect of urease inhibitors on urea hydrolysis and ammonia volatilization. Biol. Fertil. Soils 11:43–47.

161. Roberge, M. R., and R. Knowles. 1968. Urease activity in a black spruce humus sterilized by gamma radiation. Soil Sci. Soc. Am. Proc. 32:518–521.

162. Tabatabai, M. A., and J. M. Bremner. 1972. Assay of urease activity in soils. Soil Biol. Biochem. 4:479–487.

163. Bremner, J. M., and R. L. Mulvaney. 1978. Urease activity in soils, p. 149–196. In R. G. Burns (ed.), Soil enzymes. Academic Press, New York.

164. Skujins, J. J., and A. D. McLaren. 1969. Assay of urease activity using ^{14}C-urea in stored, geologically preserved and in irradiated soils. Soil Biol. Biochem. 1:89–99.

165. Mulvaney, R. L., and J. M. Bremner. 1979. A modified monooxime method for colorimetric determination of urea in soil extracts. Commun. Soil Sci. Plant Anal. 10:1163–1170.

166. Lethbridge, G., and R. G. Burns. 1976. Inhibition of soil urease by organophosphorus insecticides. Soil Biol. Biochem. 8:99–102.

167. Cervelli, S., P. Nannipieri, G. Giovannini, and A. Perna. 1976. Relationships between substituted urea herbicides and soil urease activity. Weed Res. 16:365–368.

168. McCarty, G. W., D. R. Shogren, and J. M. Bremner. 1992. Regulation of urease production in soil by microbial assimilation of nitrogen. Biol. Fertil. Soils 12:261–264.

169. Domsch, K. H. (ed.). 1992. Pestizide im Boden: Mikrobieller Abbau und Nebenwirkungen auf Mikroorganismen. VCH, Weinheim.

170. Nannipieri, P., and J.-M. Bollag. 1991. Use of enzymes to detoxify pesticide-contaminated soils and waters. J. Environ. Qual. 20:510–517.

171. Ware, G. W. (ed.). 1992. Reviews of environmental contamination and toxicology, Vol. 123. Springer-Verlag, New York.

172. Ho, T.-H. D., and M. M. Sachs. 1989. Stress-induced proteins: characterization and the regulation of their synthesis, p. 347–378. In A. Marcus (ed.), The biochemistry of plants, Vol. 15. Academic Press, San Diego, CA.

173. Ödberg-Ferragut, C., M. Espigares, and D. Dive. 1991. Stress protein synthesis, a potential toxicity marker in *Escherichia coli*. Ecotoxicol. Environ. Safety 21:275—282.

174. Morgan, J. E., C. G. Norey, A. J. Morgan, and J. Kay. 1989. A comparison of the cadmium-binding proteins isolated from the posterior alimentary canal of the earthworms *Dendrodrilus rubidus* and *Lumbricus rubellus*. Comp. Biochem. Physiol. C Comp. Pharmacol. Toxicol. 92C:15—21.

175. Sato, M., A. Ohtake, K. Takeda, H. Mizunuma, and Y. Nagai. 1989. Metallothionein-I accumulation in the rat lung following a single paraquat administration. Toxicol. Lett. 45:41—47.

176. Kägi, J. H. R., and A. Schäffer. 1989. Biochemistry of metallothionein. Biochemistry 27:8509—8515.

177. Strzelec, A. 1984. Effect of herbicides on biochemical transformation in soils. Rocz. Glebozn. 35:107—121.

178. Verstraete, W., and J. P. Voets. 1977. Soil microbial and biochemical characteristics in relation to soil management and fertility. Soil Biol. Biochem. 9:253—258.

179. Wingfield, G. I., H. A. Davies, and M. P. Greaves. 1977. The effect of soil treatment on the response of the soil microflora to the herbicide dalapon. J. Appl. Bacteriol. 43:39—46.

180. Rao, J. L. N., I. C. Pasalu, and V. R. Rao. 1983. Nitrogen fixation (C_2H_2 reduction) in the rice rhizosphere soil as influenced by pesticides and methods of their application. J. Agric. Sci. 100:637—642.

181. Pulford, I. D., and M. A. Tabatabai. 1988. Effect of waterlogging on enzyme activities in soils. Soil Biol. Biochem. 20:215—219.

182. Cook, K. A., and M. P. Greaves. 1987. Natural variability in microbial activities, p. 15—43. *In* L. Sommerville and M. P. Greaves (eds.), Pesticide effects on the soil microflora. Taylor & Francis, London.

183. Rastin, N., K. Rosenplänter, K. and A. Hüttermann. 1988. Seasonal variation of enzyme activity and their dependence on certain soil factors in a beech forest soil. Soil Biol. Biochem. 20:637—642.

184. Domsch, K. H., G. Jagnow, and T.-H. Anderson. 1983. An ecological concept for the assessment of side-effects of agrochemicals on soil microorganisms. Res. Rev. 86:65—105.

185. Ross, D. J. 1991. Microbial biomass in a stored soil: a comparison of different estimation procedures. Soil Biol. Biochem. 10:1005—1007.

186. Wardle, D. A., and D. Parkinson. 1990. Comparison of physiological techniques for estimating the response of the soil microbial biomass to soil moisture. Soil Biol. Biochem. 22:817—823.

187. Wardle, D. A., and D. Parkinson. 1991. Relative importance of the effect of 2,4-D, glyphosate, and environmental variables on the soil microbial biomass. Plant Soil 134:209–219.
188. Anderson, J. P. E., and K. H. Domsch. 1985. Maintenance carbon requirements of actively-metabolizing microbial populations under in situ conditions. Soil Biol. Biochem. 17:197–203.
189. Insam, H., and K. H. Domsch. 1988. Relationship between organic carbon and microbial biomass on chronosequences of reclamation sites. Microb. Ecol. 15:177–188.
190. Bauer, E., C. Pennerstorfer, P. Holubar, C. Plas, and R. Braun. 1991. Microbial activity measurements in soil–a comparison of methods. J. Microbiol. Methods 14:109–117.
191. Vonk, J. W. 1991. Testing of pesticides for side-effects on nitrogen conversions in soil. Toxicol. Environ. Chem. 30:241–248.
192. Hill, I. R., and S. J. L. Wright (eds.). 1978. Pesticide microbiology. Academic Press, London.
193. Domsch, K. H. 1991. Status and perspectives of side-effect testing. Toxicol. Environ. Chem. 30:147–152.
194. Peichl, L., and D. Reiml. 1990. Biological effect-test systems for the early recognition of unexpected environmental changes. Environ. Monit. Assess. 15:1–12.

9

Microbial Degradation of Chlorinated Biphenyls

DENNIS D. FOCHT *University of California, Riverside, California*

I. INTRODUCTION

Chlorinated aromatic hydrocarbons (CAHs), compounds composed solely of carbon, hydrogen, and chlorine, are the most persistent class of chemicals present in the environment. They have had a long history and diverse usage in agriculture and industry, although the addition of new organochlorine compounds to the market probably diminished after 1963, when patents filed for these compounds declined [1] (Figure 1). Despite laws banning or curtailing use and production, polychlorinated biphenyls (PCBs) and diphenyldichlorotrichloroethane (DDT:1,1,1-trichloro-2,2,2-bis[p-chlorophenyl]ethane) continue to be major environmental problems. Because of their chemically similar structure, the two compounds share many of the common problems pertinent to their biodegradation by microorganisms in vitro and in situ.

DDT was the first and most infamous environmental contaminant brought to public attention by the publication of Silent Spring in 1962 [2]. It was originally synthesized as the topic of a German Ph.D. thesis in 1874 by Zeidler, but it languished on the shelf for more than 60 years until it was discovered to have excellent insecticidal characteristics [3]. DDT was hailed as an excellent agent for combatting agricultural pests and for improving human health by controlling insect vectors of disease. Its major problems in the environment have been the indiscriminate destruction of beneficial insects and its accumulation and biomagnification in the food chain. DDE (1,1-bis-[p-chlorophenyl]-2,2-dichloroethylene), the primary metabolite of DDT, is a potent inducer of steroid hydroxylase and

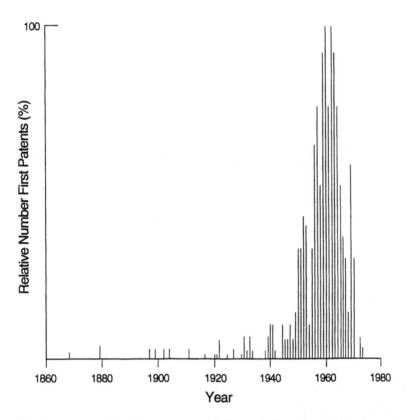

Figure 1   Relative percentage of first U.S. patents by year for organochlorine compounds.   Redrawn from Ref. 1 with permission.

is particularly disruptive of avian calcium metabolism, which leads
to the production of thin eggshell.   Ironically, this same metabolic
conversion involving the removal of a chlorine and hydrogen from
DDT by a hydrolytic dehalogenation is how insects detoxify the insecticide [4].   The usage of DDT was banned in the United States
in 1972.
     With the advent of the gas chromatograph and the developing
concern about pesticide residues, it became apparent that DDT-re-
lated compounds were being found in places not associated with
agricultural practices.   The realization that a more pervasive group
of chemicals, namely PCBs, was present in the environment was
first made by Jensen in 1966 and published 6 years later [3].   The
refining of better gas chromatographic methods, particularly the
development and interfacing of the quadrupole mass spectrometer,

enabled the two classes of compounds, DDT and PCBs, to be distinguished from one another.

Control of PCBs in the United States was mandated by Congress under the Toxic Substance Control Act (TSCA) of 1976. Between 1929 and 1977, about 0.5 million tons (metric) were produced in the United States. Because of their thermal stability and high viscosity, PCBs have been used in virtually all major industrial processes, for example, lubricants, stabilizers in printer's ink and paints, and heat transfer agents in electrical transformers. Their major usage in electrical equipment has been in transformers that raise or lower the voltage of a power line or capacitors, the bread box–sized can on utility poles that maintains constant voltage to homes, and electromagnets and voltage regulators in fluorescent fixtures. Monsanto Chemical Company, the sole producer of PCBs in the United States, restricted sales to closed-system applications, such as transformers and capacitors, since 1971 and discontinued production entirely in 1978, when dioctyl phthalate was substituted in the manufacture of capacitors. In 1977, manufacture and distribution in a nontotally closed system was banned. The U.S. Environmental Protection Agency made exceptions to the rule in 1979 and determined that 50 ppm or less did not come under the ban [5].

It has been estimated that industrial production of PCBs in the United States has amounted to more than 600,000 metric tons and that about 15% has entered aquatic and terrestrial ecosystems [6]. PCBs, because of their lipophilic nature and recalcitrance to biodegradation by soil and aquatic microbiota, readily bioaccumulate and have been found in places as remote as the Antarctic [7].

Incineration, the only proven technology for the destruction of PCBs, is very costly and extremely impractical for treatment of contaminated soil. The development of technology for the biodegradation of PCBs would provide a safe and cost-effective alternative to incineration. Whereas biodegradation of PCBs was thought to be extremely improbable a decade ago, recent work in independent laboratories in academia, government, and private industry has demonstrated the potential of biological treatment.

II. CHEMISTRY AND ENVIRONMENTAL SIGNIFICANCE

PCBs are prepared by direct chlorination of biphenyl, using anhydrous chlorine and iron filings or ferric chloride as a catalyst [8]. There are 209 possible congeners, ranging from the three monochloroisomers to decachlorobiphenyl. In reality, less than half of all possible combinations are present in commercial mixtures. In the United States and Great Britain, the trade name Aroclor has been used by the major manufacturer, Monsanto Chemical Company,

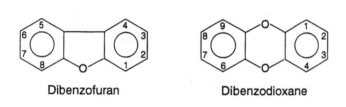

Biphenyl   1,1-Diphenylethane

Diphenylether

Dibenzofuran   Dibenzodioxane

Figure 2   Structures and numbering system for biphenyl and related diphenyl compounds.

and is affixed with a number that designates the extent of chlorination: Aroclor 1221 and Aroclor 1260 represent mixtures having average contributions by weight, of 21 and 60% chlorine, respectively. The first two digits, 12, refer to the total number of carbons representing the biphenyl nucleus (Figure 2). Aroclor 1016 and 1043 do not follow the normal rationale for nomenclature of PCBs. The composition and relative transformation rates of commercial Aroclors are presented in Table 1, from data of Tucker et al. [9]. Different commercial names have been given by manufacturers in other countries for PCB mixtures: Kanechlor (Mitusubishi-Monsanto, Japan), Phenoclor and Pyralene (Prodelec, France), Clophen (Beyer, Germany), and Fenclor (Caffaro, Italy).

The features of PCBs that make them so suitable as heat transmitters in facilities that require fire-retardant capabilities are high thermal and chemical stability, high electrical resistivity, high hydrophobic properties, and resistance to acids, alkali, and corrosive agents. Unfortunately, the same factor that increases the physical stability of the biphenyl nucleus (i.e., the degree of chlorination) also impedes biodegradation.

Table 1 Congener Mass Composition (%) of Aroclors and Relative
Transformation Rates (%) of Polychlorinated Biphenyls
with Biphenyl Transformation Being 100%

Chlorines per bi-phenyl	Aroclor				
	1221	1043	1016	1242	1254
0	11	—	—	—	—
1	51	22	1	1	—
2	32	72	20	16	—
3	4	6	57	49	1
4	2	—	21	25	21
5	—	—	1	8	48
6	—	—	—	1	23
7	—	—	—	—	6
Relative transformation rate (%)					
100	81	56	33	26	15

Source: Ref. 9.

Because of their low water solubility, PCBs are not normally a
problem in ground water contamination via normal transport processes. Solubilities range from 6 ppm for 2-chlorobiphenyl to 0.015
ppm for decachlorobiphenyl [8]. Their primary mode of dispersal
in rivers and oceans is direct discharge and transport by sediments or particulate matter. Concentrations of PCBs vary considerably in soil and tend to be log-normally distributed. It has been
estimated that the average concentration in soil throughout North
America is about 100 ppb, although more realistic estimates range
from a few ppb in rural areas to 100 ppm in urban areas [10].
Only 15% of soil samples from Indiana that received PCB-contaminated sludges were found to exceed the acceptable level of 50 ppm
set by the U.S. EPA [11]. Similar estimates have been found in
soils from Japan, where only 5% of the samples had concentrations
in excess of 100 ppm: these higher concentrations could be traced
to industries that used and disposed of PCBs [12]. Log-normal
distributions of concentrations of DDT and PCB have also been
reported in coastal sediments [13].

PCBs are much more toxic than DDT and have been shown to cause reproductive toxicity, birth defects, and behavioral changes in animals. Moreover, children born of mothers consuming large quantities of PCB-contaminated fish (5 kg fish/year, containing more than 2 ppm) have been shown to have low birth weights and learning and behavioral deficits [14]. In 1968, a Yusho (rice oil) poisoning episode occurred in Japan that affected 1057 people. Their symptoms of illness—chloroacne, headaches, nausea, diarrhea—were traced to a heat transfer agent (PCB) that was used during the manufacture of cooking oil and was present in 2000 to 3000 ppm quantities. The level of ingestion of PCBs was estimated to be about 0.5 to 2.0 g/person. In 1980, a second episode involving 1000 people was reported in Taiwan, also traced to cooking oil contaminated with PCBs [3,5].

Not all congeners of PCBs are toxic to higher life forms, although all have the potential of being bioaccumulative in the absence of microbial degradation. The nonortho isomers are the most toxic, presumably because ortho substitution prevents a coplanar configuration of the two rings. Chlorinated dioxins, which are by-products in the commercial synthesis of chlorinated aromatic compounds, are among the most toxic synthetic compounds known. 2,3,7,8-Tetrachlorodibenzodioxane (TCDD) is the most toxic dioxin congener. Thus, 3,3',4,4'-tetrachlorobiphenyl and other PCBs having a similar structure (e.g., 3,4,5,3',4',5'-hexachlorobiphenyl) are highly toxic PCB congeners [15,16]. In contrast, 2,4,5,4'-tetrachlorobiphenyl is sufficiently different in structure that it is not as toxic [17]. The assays for determining toxicity of PCB congeners are based on induction of arylhydrocarbon hydroxylase (AHH) and ethoxy resorufin O-deethylase (EROD). Sassa et al. [18] found excellent agreement between calculated coplanarity from ortho substitution and observed effects on liver cell metabolism, in which ortho congeners were least reactive.

Although TCDDs are much more toxic than PCBs, the much greater distribution of the latter raises the question of which of the two groups pose a potentially greater risk. Dewailly et al. [19] found that coplanar, non-ortho-chlorinated PCBs were excreted in human milk fat at concentrations of mg/g, whereas monoortho congeners were excreted at ng/g levels. They concluded that PCBs represent a higher risk than polychlorinated dibenzodioxanes (PCDDs) and polychlorinated dibenzofurans (PCDFs) because of their much greater concentration in vivo. Tanabe et al. [20] determined concentrations of PCBs in sediments and found levels of all eight congeners (3,3',4',4'; 3,4,5,3',4',5'; 3,4,5,3',4'; 2,3,4,3',4'; 2,4,5,3',4'; 2,4,3',4',5'; 2,3,4,5,4'; and 2,4,5,3',4',5') to be much lower than those of less toxic congeners. Because meta and para chlorines are removed more readily than ortho chlorines, they suggested that reductive dehalogenation in the environment accounted for the low level of these congeners. However, microbial metabolism of PCBs is much slower in the ocean than in sediments or soil, and the relative bioconcentration and metabolic capacity of terrestrial and marine mammals for these chemicals suggest that the toxic threat of coplanar PCBs increases from land to ocean, whereas the reverse is true for PCDDs and PCDFs. Thus, the bioaccumulation of 3,3',4,4',5 and 2,3,3',4,4' in carnivorous marine mammals is cause for concern [21].

III. METABOLISM AND MICROBIAL DIVERSITY

A. Putative Growth of PCB and DDT Utilizers

Bacteria that use PCBs (or DDT) as sole carbon sources probably do not exist in nature. Although there are several reports of direct isolation of PCB- or DDT-degrading microorganisms in the literature, these claims must be considered equivocal as they invariably lack the requisites for establishing utilization of the target substrate as a sole carbon source. These requisites are that (1) the culture must be available for validation by other scientists; (2) the composition of the growth media must be adequately described for others to reproduce; (3) utilization of the substrate must be verified when exogenous carbon sources (e.g., yeast extract, glucose, peptone) are present by comparisons of growth curves and yields (or maximal cell densities) in the absence of the CAH; (4) mineralization of the substrate as evidenced by chloride liberation or $^{14}CO_2$ production from labeled substrate must be demonstrated; and (5) purity of the substrate must be known (e.g., biphenyl is an impurity in some Aroclor mixtures).

Yeast extract (YE) at a concentration as low as 50 ppm results in a bacterial density of about 10^8/ml, which causes visible turbidity in a culture flask. The use of 0.1% YE in the presence of 0.01% concentrations of chlorobenzenes and chlorotoluenes [22] simply results in the selection of YE-utilizing bacteria (generally pseudomonads) that are tolerant to the CAH and does not ensure that CAH-utilizing bacteria will be isolated by enrichment culture. Growth on solid media, but not in liquid, is cause for concern because of impurities in the agar and because soil harbors a modest number of agar-degrading microorganisms (10^2 to 10^4/g). If PCB- or DDT-utilizing bacteria could be directly obtained by enrichment culture, it would follow that these compounds would be readily biodegradable in the environment from where they were isolated; this is clearly not the case.

As research on the metabolism and ecology of xenobiotic-metabolizing bacteria has progressed, it has become more apparent that recalcitrant compounds are not degraded by single organisms but by the action of microbial consortia. As a rule, compounds that

are very biodegradable generally can be utilized as sole carbon
sources by single species of microorganisms, which makes isolation by the enrichment culture procedure an easy task. When two
or more members are involved in the biodegradation of recalcitrant
compound, it is rarely possible to obtain each of them directly by
enrichment culture with the substrate. The rationale, methodology,
and limitations of the enrichment culture procedure are discussed
elsewhere (D. D. Focht, Methods of Soil Analysis: Biochemical and
Microbiological Properties, American Society of Agronomy, in press).

Before any further discussion of the microbial metabolism of recalcitrant compounds such as PCBs and DDT, it is important to define the term "biodegradation" clearly and to distinguish its proper
usage from more limited forms of metabolism. Biodegradation as defined herein refers to the complete mineralization (CO_2 production)
or metabolism of a substrate to innocuous products (biomass, humus, natural metabolites). Transformation refers to the conversion
of substrate to a product(s) that cannot be utilized for growth or
energy.

B. Cometabolism and Oxidation

Aerobic metabolism of CAHs occurs primarily by cometabolism, whereby an organism is capable of transforming a compound through one
or more steps of a catabolic pathway but is unable to carry out sufficient catabolic conversion to obtain energy for growth [23]. The
mechanism for this reaction is believed to be the result of low substrate specificity of enzymes, caused by evolutionary pressures to
utilize other substrates, which fortuitously metabolize the chlorinated
analog [24]. Thus, cultures isolated by enrichment or diphenylmethane cometabolized several of the corresponding chlorinated
analogs shown in Figure 3 [25] to their respective aromatic acids
via the meta-fission pathway [26], which was first described in its
entirety by Dagley et al. [27] for metabolism of catechol and methyl-
catechols.

The meta-fission pathway is the universal pathway involved in the
mineralization of all aromatic hydrocarbons including benzene, toluene, xylene, naphthalene, phenanthrene [28], diphenylmethane
[29], diphenylethylene [30], diphenylethane [31], and biphenyl
[32,33] by natural isolates obtained from enrichment culture. The
pathway was first proposed for oxidation of mono- and dichlorobi-
phenyls to the corresponding chlorobenzoates by Ahmed and Focht
[34] and later confirmed by Furukawa et al. [35,36]. The example
shown in Figure 4 is an illustration of how cometabolism of 4,4'-di-
chlorobiphenyl occurs concomitantly when cells are grown on biphenyl. Although washed cell suspensions can cometabolize PCBs
in the absence of biphenyl, the activity is rapidly lost and the rate

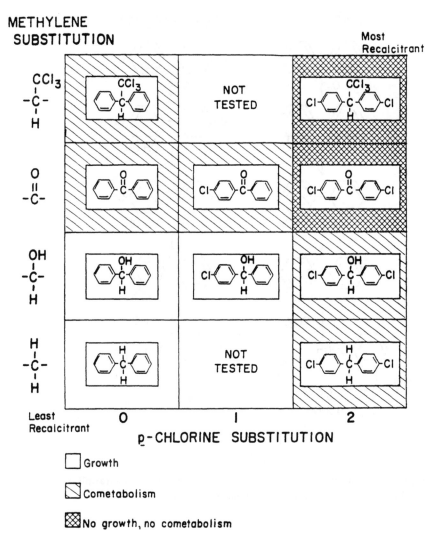

Figure 3   Aerobic cometabolism of DDT and related analogs by *Hydrogenomonas* sp. From Ref. 25, © AAAS.

and extent of cometabolism are less pronounced than when cells are actively growing [37–39].

The production of the corresponding benzoic acids from PCBs is analogous to other biphenyls joined together by alkyl substituents.

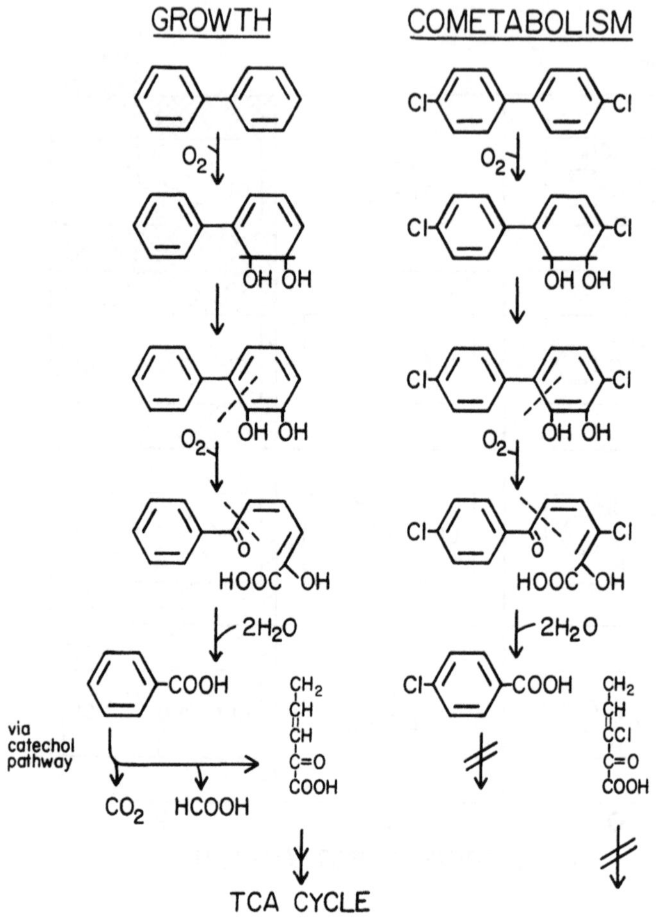

Figure 4 Bacterial metabolism of biphenyl and cometabolism of 4,4'-dichlorobiphenyl. Redrawn from Ref. 247.

The identical pathway for metabolism of diphenylmethane, 4,4'-dichlorodiphenylmethane, 2,2,2-trichlorodiphenylethane, 1,1-diphenylethylene, and 2,2-diphenylethane gives the corresponding aromatic acids, respectively: phenylacetic acid, 4-chlorophenylacetic acid, 2-phenyl-3,3,3-trichloropropionic acid, 2-phenylpropenoic acid, and 2-phenylpropanoic acid [26,30,31].

Bacterial oxidation of biphenyls specifically occurs at carbons 2 and 3 (Figure 4) prior to metabolism through the meta-fission pathway. However, PCB congers not containing unsubstituted 2,3 sites (e.g., 2,5-, 2,6-, 3,5-dichloro) on both rings were, nevertheless, cometabolized by *Acinetobacter* sp. P6, causing Furukawa et al. [36] to conclude that dioxygenation at 3,4-positions might occur. Carey and Harvey [40] found trace quantities of an unidentified lactone, which might have originated from 3,4-dioxygenation of ^{14}C-labeled 2,5,2',5'-tetrachlorobiphenyl and 2,5,2'-trichlorobiphenyl that were added to flasks containing nutrient-enriched seawater.

The most convincing evidence for 3,4-dioxygenation comes from studies with *Alcaligenes eutrophus* H850 [41,42] and *Pseudomonas* sp. LB400 [43]. Both strains were found to be superior to others in oxidizing PCB congeners containing no free 2,3 sites. Surprisingly, strain LB400 attacked neither 2,4,6,4'-tetrachloro-, 2,6,2',6'-tetrachloro-, nor 2,4,6,2',4'-pentachlorobiphenyl, yet it attacked corresponding congeners that were more chlorinated, namely 2,4,6,3',5'-pentachloro- and 2,4,6,2',4',6'-hexachlorobiphenyl, which neither *Acinetobacter* sp. P6 [37] nor *Corynebacterium* sp. MB1 [44,45] transformed. Inasmuch as strains LB400 and H850 also attack congeners with free 2,3 sites by normal 2,3-dioxygenation, it remains unclear whether a single dioxygenase, with broad substrate specificity, or two dioxygenases are involved. The latter possibility could be inferred from DNA hybridization data: strain LB400 hybridized only to strain H850 and to none of the strains having poor or no 3,4-dioxygenase activity [46].

Bedard [47] proposed another possible mode of attack of 2,5,2',5'-tetrachlorobiphenyl by *A. eutrophus* H850 and *Pseudomonas* sp. LB400 that involves a 2,3-dioxygenation that would cause spontaneous removal of the ortho chlorine as the result of chemical instability of the chlorohydrodiol. A similar mechanism was proposed by Springer and Rast [48] for the oxidative elimination of both ortho chlorines from 1,2,4,6-tetrachlorobenzene by *Pseudomonas* sp. 4CB1 to yield 3,4-dichlorocatechol. The elimination of one chlorine, during conversion of 1,2,4,5-tetrachlorobenzene to 3,4,6-trichlorocatechol, was reported by Sander et al. [248] in two other strains of *Pseudomonas*.

C. Metabolism of Hydroxylated PCBs

Not all PCB congeners that are cometabolized by biphenyl-utilizing bacteria are converted to chlorobenzoates. Chlorophenylphenols are commonly observed products of the metabolism of chlorinated biphenyls and are produced chemically during extraction of culture supernatants by acid catalysis of dihydrodiols [36,49]. Phenolic products are generally produced by metabolism of the more highly chlorinated PCB congeners. It is uncertain whether hydroxylated PCBs are transient intermediates or terminal products of cometabolism in these cultures. Chlorophenylphenols functioned as pseudosubstrates, by eliciting O_2 uptake, yet they were not transformed by biphenyl utilizers or by phenylphenol utilizers [50, 51].

Bacteria that grow on biphenyl are generally unable to utilize phenylphenols, and bacteria that utilize phenylphenols cannot utilize the former; this phenomenon is analogous to that of benzene and phenol utilizers. Biphenyls containing a hydroxyl at either the 2 or 3 position undergo monooxygenase attack to form 3-phenyl-catechol (2,3-dihydroxybiphenyl). However, hydroxyl substitution at the 4 position, even with cells grown on 3-hydroxybiphenyl, results in 2,3-dioxygenation of the unsubstituted ring to form 2,3,4'-trihydroxybiphenyl, which is metabolized through the meta-fission pathway to 4-hydroxybenzoate [50].

IV. ALTERNATIVE PATHWAYS

A. Phenyl Alkanoic Acids

Biodegradation of biphenyls and chlorinated biphenyls also results in the production of other metabolites, some of which cannot be explained by the meta-fission pathway. Omori et al. [52] purified three reducing enzymes from *Pseudomonas cruciviae* S93B1 that reduced the ring fission product, 2-hydroxy-6-oxo-6-phenyl-2,4-dienoic acid (OHPDA) [53], to 2,6,dioxo-6-phenylhexanoic acid. The ring fission products from 3-isopropylcatechol and 3-methyl-catechol were also substrates for the three enzymes, which were produced from cells grown in glucose nutrient broth medium.

Ahmad et al. [54] investigated several strains able to utilize 4-chlorobiphenyl for growth and found that six ring fission products, not common to the normal pathway, were produced by *Pseudomonas putida* DA2 (Figure 5). Two of these could be rationalized by the successive reduction of the ring fission product to the hex-4-enoic and hexanoic acids, respectively. Three other shorter-chained acids could be generated by an alpha-oxidation and subsequent hydrolytic decarboxylation or by beta-oxidation according to the scheme shown in Figure 5. However, the formation of 4-chlorocinnamic acid via the normal meta-fission pathway is difficult to envision without a dehydration reaction, which would be uncommon in an aerobic catabolic pathway. The same six metabolites were also detected with *Pseudomonas testosteroni* strains B-356 and B-206, *Achromobacter* sp. B-218, and *Bacillus brevis* [55]. Several of these metabolites were also reported by Omori et al. [56].

A picolinic acid derivative of the ring fission product, 2-(p-chlorophenyl)picolinic acid, was also identified [51]. However, products of meta ring fission are readily transformed chemically to picolinic acids in basic solutions containing ammonia [57]; thus, the origin of the picolinate derivative was unclear.

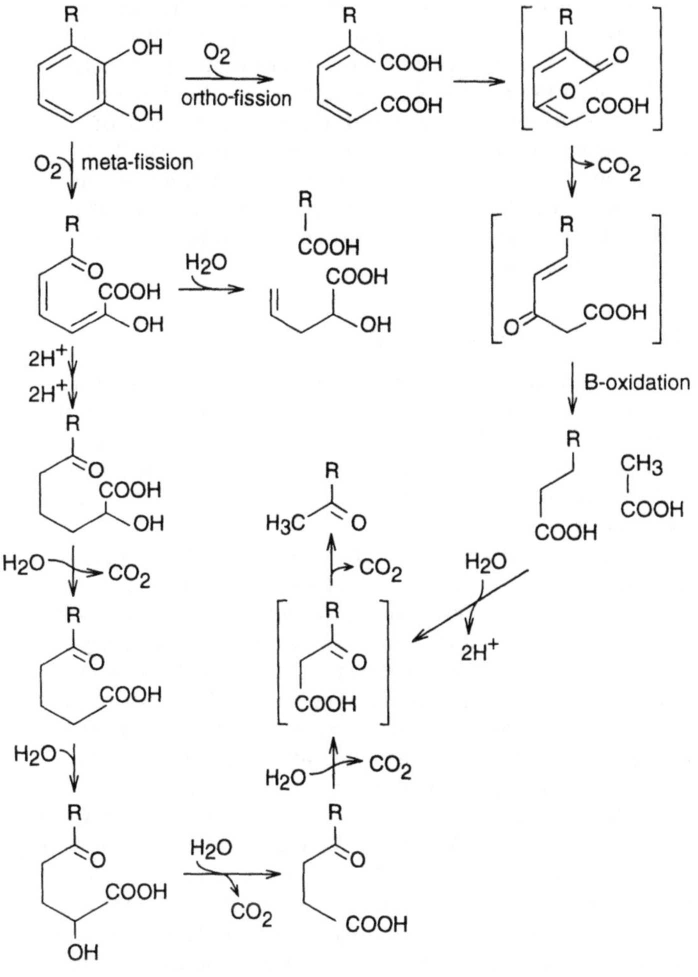

Figure 5   Alternative pathways for bacterial degradation of biphenyl and chlorobiphenyl compounds.   R = phenyl or chlorophenyl substituent.   From Refs. 49, 52–56.

Alcaligenes sp. BM-2 [49] was isolated from enrichment culture, supplemented with 0.1 g YE/L, with diphenylmethane.   Although the isolate was unable to grow on any of the halogenated biphenyl compounds tested, except 4-chlorobiphenyl and 1,2-dichlorodiphenyl-ethane, it cometabolized a diverse number of congeners containing chlorines on both rings.   Several mono- and dichlorobenzoates were produced from tetrachlorobiphenyl isomers.   Chlorinated cinnamic acids, having both aliphatic and aryl chlorines, and chlorinated phenylacetic acids were identified by gas chromatography—mass spectrometry (GC-MS). As neither of these products could have arisen through the standard meta-fission pathway, the authors concluded that they must have originated from meta fission of 4-phenylcatechols or from ortho fission of the ring (between 2,3- or 3,4-dihydroxyl—bearing carbons). Because dioxygenation at the 3,4 position occurs in the absence of free 2,3 sites [42,43], the ortho-fission method, proposed in Figure 5, would be the more plausible of the two hypotheses.

B. Diphenyl Ethers, Dibenzofurans, Dioxanes

Support for an ortho-fission pathway for biphenyls comes from studies with *P. crucivae* S93B1 that grew on 10 different biphenyl compounds and cometabolized bromo- and nitrobiphenyls to the corresponding aromatic acids [58]. Phenol and 2-phenoxymuconate were produced from diphenyl ether through the ortho-fission pathway shown in Figure 6.

Metabolism of diphenyl ethers is not confined to 2,3-dioxygenation and subsequent ring fission. A novel dioxygenase attack at the 1,2 positions has been reported in *Sphingomonas* sp. SS3 [59]. The resulting hemiacetal is unstable and spontaneously hydrolyzes to phenol and catechol. Nonspecific dioxygenase attack occurred with all F, Cl, or Br isomers of diphenyl ethers. The culture cometabolized dibenzodioxane by the same mechanism to 2,2',3'-trihydroxybiphenyl ether, which was dioxygenated and ruptured by ortho fission to 2-(2'-hydroxyphenoxy)-*cis,cis*-muconate.

Microbial metabolism of dibenzodioxane may have relevance to cometabolism of its highly toxic 2,3,7,8-tetrachloro analog, as nothing is known about the catabolic fate of TCDD. Gibson et al. [60] reported that dibenzodioxane was cometabolized to 1,2-dihydroxy dibenzodioxane by *Beijerinckia* sp. A study in which *Sphingomonas* sp. strain RW1 was isolated by enrichment culture clearly demonstrates that dibenzodioxane is biodegradable [61]. The degradation pathway was similar to that reported in *Sphingomonas* sp. SS3 [59] except that 2,2',3'-trihydroxydiphenyl ether was meta cleaved to catechol and a hypothetical 2-hydroxymuconate (Figure 6). Catechol was metabolized by both the ortho- and meta-fission routes, and the culture grew on a vast range of compounds including dibenzofuran, hydroxylated dibenzofurans, hydroxylated dioxanes, and hydroxylated biphenyls, but it did not grow on biphenyl, benzene, naphthalene, diphenyl ether, fluorene, xanthene, 4-hydroxybiphenyl, or phenol. Oxidation of dibenzodioxanes and dibenzofurans was inducible, whereas oxidation of 2,3- and 2,2',3'-hydroxybiphenyls was constitutive.

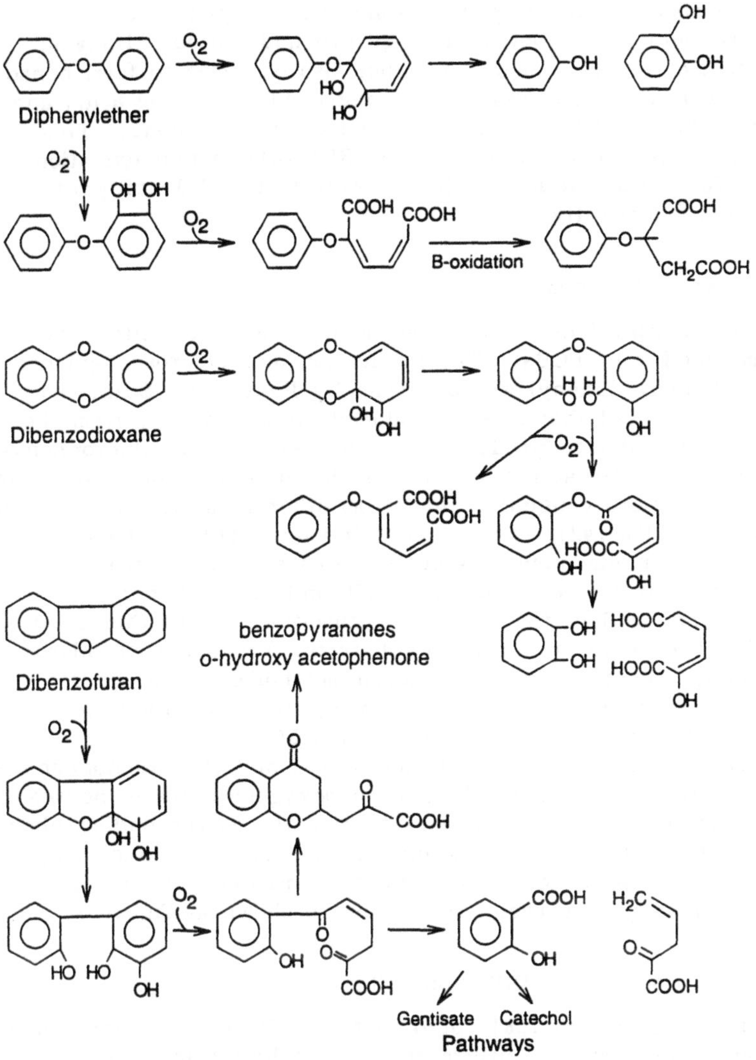

Figure 6 Microbial catabolism of diphenyl ether, dibenzodioxane, and dibenzofuran. From Refs. 58, 59, 61—64.

Dibenzofuran also undergoes dioxygenation at the tertiary carbon to form the diol (Figure 6), which is then ruptured to form 2,2',3'-trihydroxybiphenyl [62—64]. Further metabolism occurred through the meta-fission pathway with *Pseudomonas* sp. HH69 [63]. Hydrolysis

of the ring fission product yielded salicylate and gentisate, both of
which were metabolized slowly after growth had ceased. Internal
lactonization of the ring fission product also occurred to form benzo-
pyran-4-one acid, which was further degraded to shorter acids and
to methyl- and benzopyranone and *o*-hydroxyacetophenone. This
strain was similar to *Sphingomonas* sp. RW1 [61] in that oxidation
of dibenzofuran was inducible while oxidation of 2,2',3'-trihydroxy-
biphenyl was constitutive.

C. Chloroacetophenones

Under certain conditions, chlorinated acetophenones are produced
during metabolism of PCBs. The production of 2-chloroacetophenone
from 2,4'-dichlorobiphenyl was observed by Baxter and Sutherland
[65], who concluded that it was produced photochemically from the
yellow meta ring fission product, 2-hydroxy-6-oxo-6-(*o*-chlorophenyl)-
2,4-dienoic acid. Greater quantities of chloroacetophenones were produced from 3-chloro congeners than from other isomeric congeners in
A. eutrophus H850 [42]. Barton and Crawford [66] reported the
production of copious quantities of 4-chloroacetophenone from 4-
chlorobiphenyl by *Pseudomonas* sp. MB86 and found that it was a
toxic dead-end product that inhibited utilization of the other product, 4-chlorobenzoate, as a growth substrate. Production of chloroacetophenones strictly through bacterial metabolism could occur
through the catabolic pathway shown in Figure 5 by either the
ortho- or modified meta-fission pathway [49,54].

Bacterial metabolism of acetophenones [67] and chloroacetophenones
[68] occurs through a Bayer-Villiger monooxygenase attack between
the carbonyl and 1-aromatic carbons to give the corresponding phenylacetate ester, which is subsequently converted to the phenol by a
nonspecific esterase attack. Chlorophenols are monooxygenated to
catechols, which pass through the pathways shown in Figure 7.

D. Fungal Metabolism of DDT and PCBs

Gibson and Subramanian [28] noted that no eukaryotic organism had
ever been demonstrated to utilize an aromatic hydrocarbon as a
growth substrate: no exception to this rule has since been demonstrated. Nevertheless, fungal extracellular enzymes, particularly
those of *Phanerochaete chrysosporium*, have been the subject of attention because of their broad activity on a variety of xenobiotic
compounds as reviewed by Higson [69]. Although other species of
fungi have been studied with respect to "biodegradation" of DDT
and PCBs, the analytical procedures are generally based only on
the disappearance of substrate. The distinction between nonspecific adsorption of PCBs on glass surfaces and microorganisms and
actual metabolism of the substrate has been discussed by Unterman

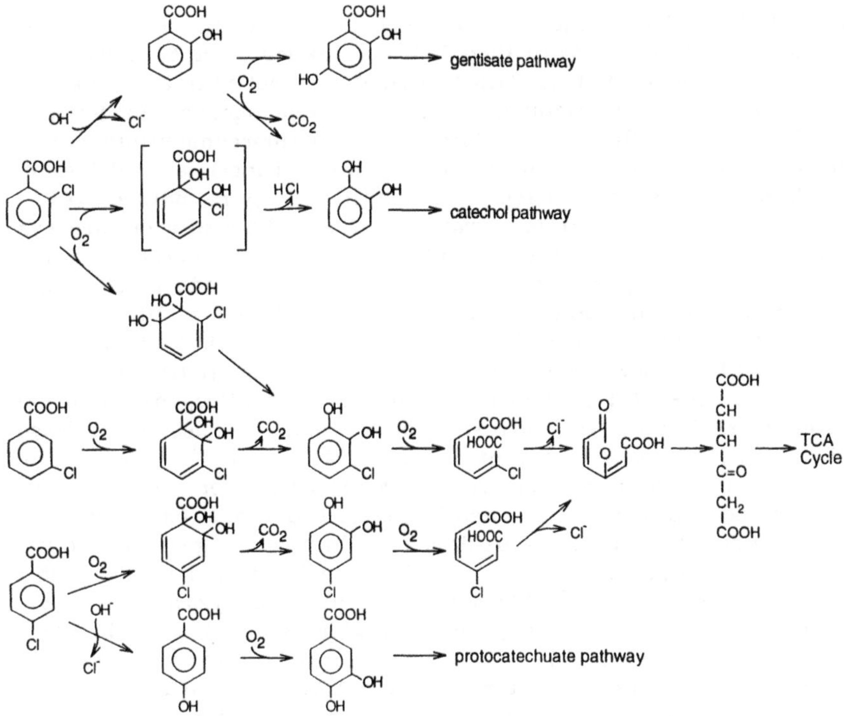

Figure 7 Bacterial metabolism of chlorobenzoates. From Refs. 112–120, 125–132, 137–143, 146.

et al. [70] with respect to PCBs. Studies with *P. chrysosporium* involving ^{14}C-labeled substrates with accompanying $^{14}CO_2$ evolution data have provided clear evidence of biodegradation.

P. chrysosporium, first isolated by Burdsall and Eslyn [71], had been studied for many years with regard to its ligninolytic activities before Bumpus et al. [72] demonstrated the partial mineralization (8%) of DDT in soil inoculated with it. The fungus is unable to utilize any CAH as a sole carbon source for growth, although in pure culture, mineralization can be extensive. Of six simple carbohydrates used to support growth of *P. chrysosporium*, 1% cellulose effected the most rapid and sustained mineralization of DDT: 14% in 30 days and 32% in 60 days [73]. The rate of DDT degradation followed a sharp transition from first-order kinetics between 0.01 and 100 ppm DDT and leveled off abruptly to zero-order kinetics between 100 and 1000 ppm DDT. Although degradation of DDT in soil was enhanced by inoculation with *P. chrysosporium* and addition of substrate, it was much lower than in culture, as no treatment resulted in more than 10% degradation after 60 days.

The pathways for degradation of DDT have not been elucidated, although several of the metabolites, common to bacterial metabolism (see Section VII,E), DDD (1,1-dichloro-2,2-bis[p-chlorophenyl]ethane), dicofol (1,1-dichloro-2,2-bis[p-chlorophenyl]ethanol), and DBP (p,p'-dichlorobenzophenone), have been identified in pure culture [74]. The appearance of DDD as a transient intermediate in a culture system that produces peroxidases and other ligninases indicates some degree of cellular partitioning for protection of the reduced electron carrier involved in reductive dehalogenation of DDT to DDD.

P. chrysosporium has also been shown to cause limited mineralization of PCBs at low concentrations. About 7% mineralization of ^{14}C-labeled Aroclor 1254 (0.3 ppm) was observed in nitrogen-limited cultures over 22 days, in contrast to other fungal species that released 1% or less [75]. Lower concentrations (0.04 ppm) of ^{14}C-labeled Aroclor 1242 and 1254 underwent 20 and 18% mineralization, respectively.

The major difference between bacterial and fungal oxidation of chlorobiphenyls is that the oxidation by bacteria is generally, but not always, induced by growth on the nonchlorinated analog. In contrast, degradation of chlorobiphenyls by *P. chrysosporium* is effected by extracellular nonspecific peroxidases (ligninases) in the presence of peroxide [76]. The generation of hydroxyl radicals, similar to the action of Fenton's reagent [77] may also have a role in breaking carbon-carbon bonds. The much lower mineralization rates that are observed in soil than in culture are probably caused by competition of soil organic matter as a substrate for the enzymes and catalytic reagents that are excreted by the fungus.

The failure of *P. chrysosporium* to oxidize hexachlorobenzene [78] may indicate that fission of the aromatic ring requires one or more unsubstituted sites. Oxidation and fission of the aromatic ring might be similar to the bacterial process. The wood-rotting fungus *Coriolus versicolor* transformed 2,2'-dihydroxy-3,3'-dimethoxy-5,5'-dicarboxy biphenyl (DDVA) via phenoxy radicals, which resulted in polymers, quinones, and ring fission. 2-Hydroxy-3-methoxy-5-carboxybenzoic acid was isolated and identified, confirming that the meta-fission pathway was used [79]. However, the mechanism was not the same as bacterial processes because the fungus did not grow on DDVA, nor was *meta*-protocatechuate 4,5-dioxygenase activity present.

V. THE BIPHENYL PATHWAY

A. Enzymology

There are four enzymes in the biphenyl pathway, namely biphenyl dioxygenase, dihydrodioldehydrogenase, 3-phenylcatechol dioxygenase, and 2-hydroxy-6-oxo-6-phenyl-2,4-dienoic acid (OHPDA) hydrolase. Oxidation of biphenyls follows the same mechanism as oxidation of all aromatic hydrocarbons, namely the direct incorporation of both atoms of molecular oxygen into the substrate to form the corresponding cis-dihydrodiol (Figure 4). The aromatic hydrocarbon dioxygenases, as characteristic of benzene [80,81], toluene [82–85], and naphthalene [86–89] dioxygenases, have been difficult to purify, but it is likely that biphenyl dioxygenase is similar to them in consisting of three subunits. Electron flow from NADH is transferred through an NAD reductase to a ferredoxin protein and finally to an iron-sulfur protein, which reduces the substrate for incorporation of dioxygen. More information on aromatic hydrocarbon oxidation can be found in a review by Sariaslani [90].

Dehydrogenases that act on biphenyldihydrodiols appear to have the broadest specificity of all enzymes in the biphenyl pathway. There is a strong sequence similarity (65%) between the dihydrodiol dehydrogenases of *Pseudomonas pseudoalcaligenes* and *P. putida* [91] and liver cell actIII and 17β-hydroxysteroid dehydrogenases [92]. The location of the NAD^+-binding site is believed to reside in the NH_2-terminal part of the dihydrodiol dehydrogenase, which contains the nucleotide binding domain, while the oxidoreductase functions for a diverse group of substrates, namely polyols, aromatic hydrocarbons, antibiotics, and steroids [92].

Catechol 2,3-dioxygenase (EC 1.13.11.2), generally referred to as *meta*-pyrocatechase, was well characterized with respect to activity on monoaromatic catechols many years ago [93]. The catechol dioxygenase involved in the fission of 3-phenylcatechol has been characterized in *P. pseudoalcaligenes* and *P. aeruginosa* [94]. Because *meta*-pyrocatechases characteristically have low substrate specificity, catechol is a substrate for 3-phenylcatechol dioxygenase. However, the relative affinity for the two catechols appears to be strain dependent. Kuhm et al. [95] found that *Pseudomonas paucimobilis* Q1, originally isolated on biphenyl, had almost three times higher activity of 3-phenylcatechol dioxygenase when grown on naphthalene than on biphenyl. The V_{max} for activity of 3-phenylcatechol dioxygenase from *P. paucimobilis* Q1 on catechol was about one-fifth of that for 3-phenylcatechol [94]. However, the purified dioxygenase (250,000 daltons) from *P. pseudoalcaligenes* KF 707 was specific only for 3-phenylcatechol and had no activity on catechol, 3-methyl-, 4-methyl-, or 4-phenylcatechol [96].

The similarity between OHPDA hydrolase and the hydrolase involved in the catechol pathway is uncertain. The five-carbon acid produced from the oxidation of catechol and methylcatechol is different from that produced from the oxidation of phenylcatechol. The former, 2-oxo-5-hydroxypentanoic acid, involves two hydrations [27], whereas the latter, 2-oxopen-4-enoic acid, involves only one [56]. Although the same suite of enzymes appears to be involved in the catabolism of biphenyl and 4-chlorobiphenyl, growth is always faster, cell yields are higher, and there is little or no accumulation of the yellow-colored ring fission product (OHPDA) with biphenyl compared to 4-chlorobiphenyl. High specificity of the hydrolase was observed with *Pseudomonas cepacia* MB2, which uniquely utilized 3-methyl-2-chlorocatechol via the meta-fission pathway. Although other methyl-chlorocatechol isomers were meta cleaved, the ring fission products were not further metabolized [97]. A naphthalene degrader, *Pseudomonas* sp. HV3, with low substrate specificity for the first three steps in the meta-fission pathway (see Figure 4) rapidly cometabolized biphenyl to a yellow-colored ring fission product, which was not further metabolized [38]. These observations indicate that the hydrolase has a much higher specificity than the other three enzymes of the biphenyl pathway.

B. Molecular Biology

The genes coding for the degradation of biphenyl and PCBs to benzoate and polychlorobenzoates, respectively, have been cloned. These genes (designated *bph*) have been identified in *Pseudomonas* sp. LB400 [98], *P. putida* OU83 [99–102], *P. testosteroni* [103], *P. pseudoalcaligenes* KF715 [104], and *P. putida* KF707 [105]. All *bph* genes characterized to date have been chromosomally located. Four *bph* genes and their products have been identified: *bphA* (biphenyl dioxygenase), *bphB* (dihydrodiol dehydrogenase), *bphC* (2,3-dihydroxybiphenyl dioxygenase), and *bphD* (OHPDA hydrolase). Whereas the *bphABC* genes are generally clustered, the location of the *bphD* gene varies from being closely associated with the *bphABC* genes [98] to being separated some distance from them [104,105].

The cloned *bph* genes have been utilized to assess diversity in bacterial populations. Furukawa et al. [106] reported that the *bph* genes in 6 of 15 soil bacteria examined were almost identical to those in *P. pseudoalcaligenes* KF707. Furthermore, the *bph* genes of all soil isolates were chromosomally coded. However, there is considerable genetic diversity in the *bph* genes [100,101]. Walia et al. [107] used two probes from recombinant plasmids of biphenyl-utilizing bacteria against isolates from garden top soil and chemically contaminated industrial soil. The plasmids, pAW313-BSS and pAW6194-NSS, coded for low and high catechol specificity of 3-phenylcatechol dioxygenase activity, respectively. Less than 1% of bacterial colonies isolated from the garden soil and over 80% of colonies isolated from the contaminated soil showed DNA homology to both probes. However, some of the putative biphenyl-degrading isolates, as detected by either probe, did not show clearing of biphenyl on agar places, and several biphenyl-clearing isolates did not hybridize with either probe. Some strains hybridized with one probe but not the other and vice versa. Generally, the DNA probes detected larger numbers of biphenyl utilizers than the plate method. The authors concluded that the genetic diversity among biphenyl and 4-chlorobiphenyl degraders in soil is very high.

Taira et al. [96] sequenced the *bphC* gene and found the 298 amino acid residues of the enzyme (33,000 daltons) to be coded by 900 base pairs. Homology (38%) of amino acids was low between *P. paucimobilis* Q1 [108] and *P. pseudoalcaligenes* KF 707 [105], although it was higher near the center of the gene (60%) when the nucleotide sequences were optimally aligned. In terms of codon usage, strain Q1 appeared closer to the catechol 2,3-dioxygenase that is coded by the *xylE* gene of TOL plasmid, pWWO [109], despite even less homology (24%). This extent of homology was hardly greater than that observed (23%) between the *bphC* gene of *P. paucimobilis* Q1 and the control, the gene which coded for carboxypeptidase G2 [110]. Clearly, comparisons of the DNA homology of the *bphC* gene between bacteria is elusive and will require further clarification of the nature and similarity of the active catalytic sites.

The cloned *bphABCD* genes have also been considered for bioremediation of PCBs by recombinant organisms. Relatively high levels of expression in *Escherichia coli* were attained when the *bphABCD* genes were cloned from *P. putida* LB400 into a pUC-18 vector [98]. Resting cells of the wild-type strain LB400 and the engineered *E. coli* carrying the genes transformed similar amounts of Aroclor 1242. The recombinant organism had two possible advantages over the wild type, which could result in enhanced degradation of PCBs. First, the product of the *bph* gene in the recombinant was constitutive and not dependent on the presence of biphenyl. Second, the number of *bph* genes in a cell could be increased by cloning into a vector that was maintained at high copy numbers.

One of the major problems in biodegradation research is obtaining the necessary substrates for isolation of organisms or for transformation studies. Cloning of the *bph* genes may have practical usage in generating metabolites that could be used as substrates for isolation of other organisms by enrichment culture. Studies of 3-phenylcatechol dioxygenase became possible only when the substrate became available commercially a few years ago. However, chlorinated dihydrodiols, catechols, and ring fission products are not commercially available, so it is difficult to undertake enzymatic studies comparing differences between chlorinated and nonchlorinated metabolites. In order to generate the chlorocatechols for future studies addressing the similarity of the substrates and products in the biphenyl/4-chlorobiphenyl pathway,

Khan and Walia [111] produced 1,2-dihydroxy-4'-chlorobiphenyl from 4-chlorobiphenyl by using a strain of *E. coli* that contained a chimeric plasmid pAW6194-T17, specifying *A* and *B* genes but lacking the *C* gene. Cloning specific genes and combinations thereof may be a viable procedure for producing sufficient quantities of desired metabolites that would otherwise be prohibitive by chemical synthesis.

VI. CHLOROBENZOATE METABOLISM

A. Metabolism and Enzymology

Chlorobenzoates, the major and most transformed products of aerobic PCB metabolism, represent the stage at which biodegradation occurs. The subject of dehalogenation of aromatic compounds has been reviewed by Reineke and Knackmuss [112], who have categorized dehalogenation reactions by four different mechanisms: (1) reductive displacement with a hydrogen, (2) displacement with a hydroxyl, (3) oxygenolytic halogen-carbon bond cleavage, and (4) chloride elimination from aliphatic compounds.

Reductive dehalogenations are known to occur under anaerobic conditions. This subject has been reviewed by Mohn and Tiedje [113] and thus will not be covered here. However, there have been a few reports of reductive dehalogenations occurring with obligate aerobic bacteria. van den Tweel et al. [114] identified 4-chlorobenzoate as the major product formed from 2,4-dichlorobenzoate during aerobic growth of *Alcaligenes denitrificans* NTB-1. 4-Chlorobenzoate was further metabolized by the hydrolytic dehalogenation pathway shown in Figure 7 (p. 357). Growth on 2,3-dichlorobenzoate by *Pseudomonas aeruginosa* JB2 resulted in the formation of 4-chlorocatechol [115]. 2,3-Dichlorophenol, the chemical decomposition product of the carboxyl-dihydrodiol, was also identified, which led the authors to conclude that the reductive elimination of chloride occurred during decarboxylation and reduction of the diol.

Displacement of chlorine by a hydroxyl ion apparently involves a nucleophile attack. This reaction is most common with 4-chlorobenzoate, although it has been reported with orthohalobenzoates in *P. aeruginosa* [116]. Studies with ^{18}O on the conversion of 4-chlorobenzoate to 4-hydroxybenzoate show that the origin of OH^- is from water and not dioxygen in *Arthrobacter* sp. [117] and *Pseudomonas* sp. CBS3 [118]. Although the reaction occurs in the presence or absence of O_2, it is faster in its absence [119, 120]. The dehalogenase in *Pseudomonas* sp. CBS3 consists of three components: component I, II, and III, which have respective molecular masses of 3000, 86,000 and 92,000 daltons [121]. Only the first component is unstable. As the enzyme requires neither O_2 nor NADH, the authors concluded that an electron transport chain was not required. Copley and Crooks [112] found that the purified enzyme from *Acinetobacter* sp. 4-CB1 rapidly hydrolyzed synthetic 4-chlorobenzoyl coenzyme A (CoA) to 4-hydroxybenzoate and that metabolism of 4-chlorobenzoate required both ATP and acetyl CoA. ATP-CoA coupled dehalogenation of 4-chlorobenzoate has also been shown in *Pseudomonas* sp. CBS3 [123], and energy-dependent uptake was shown in a coryneform bacterium [124], which metabolized 4-chlorobenzoate through the hydrolytic pathway.

Dioxygenolytic removal of chlorine is generally restricted to ortho-chlorinated benzoates. However, dioxygenation of 3-chloro-, 4-chloro-, and 3,5-dichlorobenzoate at positions not containing a chlorine substituent has been reported [125,126]. The reaction (Figure 7) involves an attack identical to that of the normal benzoate dioxygenase, described many years ago [127], at the 1,2 positions to give a 1-carboxy-1,2-dihydroxy-cyclohexa-3,5-diene. With an ortho-chloro substituent, the instability of two electron-withdrawing groups on the same 2-carbon causes spontaneous loss of chloride and CO_2, with subsequent formation of catechol [128–131]. A novel 2,3-dioxygenase attack of 2-chlorobenzoate (in contrast to the normal 1,2 attack) was suggested in *Pseudomonas* sp. 2CBA to account for the formation of 2,3-dihydroxybenzoate [132].

Removal of chlorines from aliphatic products occurs after fission of the aromatic ring by catechol 1,2-dioxygenase, also referred to as *ortho*-pyrocatechase (EC 1.13.11.1), in which the catechol intermediate is ruptured between the two hydroxyl-bearing carbons to form *cis,cis*-muconate. The chlorine atom is spontaneously eliminated during lactonization of the ring fission product when it is on the same lactone carbon (as from 4-chlorocatechol) or adjacent carbon (as from 3-chlorocatechol). All bacteria that utilize chlorinated aromatic acids utilize the ortho-fission pathway with 3-substituted catechols because meta fission of 3-chlorocatechol produces an acyl chloride which reacts with and irreversibly denatures the catechol 2,3-dioxygenase [112,133,134].

Although the normal catechol and chlorocatechol pathways appear similar, different enzymes are induced by the growth substrate. *ortho*-Pyrocatechase type I has high specificity for catechol and is induced by growth on benzoate. Its isoenzyme, *ortho*-pyrocatechase II, has broad specificity and high activity on chlorocatechols and is induced by 3-chlorobenzoate [135,136]. The lactonization of *cis,cis*-muconate also involves two isoenzymes: cycloisomerase type I is highly specific for *cis,cis*-muconate, whereas cycloisomerase type II has high activity on *cis,cis*-muconate, 2-chloro-, and 3-chloromuconates [137]. The most obvious difference between the normal catechol and chlorocatechol pathways is the formation of the lactone. Whereas *cis,cis*-muconate yields

4-carboxymethylbut-2-en-4-olide, both chloromuconates give rise to the corresponding hydrogenated lactone, 4-carboxymethylenebut-2-en-4-olide, as a result of HCl elimination during lactonization or isomerization. The dienelactone dehydrogenases of both pathways have no cross reactivity with the other substrates, nor are they isoenzymes [112]. Reduction of maleylacetate produces β-oxoadipate [137], which converges on the standard ortho-fission pathway to yield succinate and acetate [138].

Metabolism of di- and trichlorinated benzoates proceeds through combinations of the pathways shown for monochlorobenzoates (Figure 7; Table 2). 3,5-Dichlorobenzoate proceeds through the same pathway as 3-chlorobenzoate to 3,5-dichlorocatechol [112,139], which is also a central intermediate in the metabolism of the herbicide 2,4-D [140,141]. The first chlorine is eliminated by lactonization of 2,4-dichloro-cis,cis-muconate, and the hydrolysis of the lactone gives β-chloromaleylacetate, which is converted to chlorosuccinate, presumably through beta-oxidation, and then to succinate by *Arthrobacter* sp. [142]. Metabolism by *Pseudomonas* sp. B13 is different in that 3-oxoadipate is formed directly from β-chloromaleylacetate through three steps involving reduction (NADH coupled), dehydrodehalogenation, and subsequent reduction (NADH coupled) [112,137,143].

The dioxygenolytic pathway appears to have the broadest specificity with regard to all o-chlorobenzoates, as noted by the diversity of substrates utilized for growth by *A. denitrificans* [144], *P. aeruginosa* JB2 [115], and *P. putida* P111 [126]. *P. putida* P111 differs from *P. aeruginosa* and *A. denitrificans* in its ability to utilize 4-chlorobenzoate and 3,5-dichlorobenzoate, although the latter cannot be utilized as a sole carbon source. In contrast to growth on o-chlorobenzoates, growth on 3- or 4-chlorobenzoate requires induction of a functional dihydrodiol dehydrogenase to form 3- or 4-chlorocatechol, respectively. In the presence of either monochlorobenzoate, stoichiometrically greater cell yields and chloride release occurred when 3,5-dichlorobenzoate was added. In contrast, cells grown on o-chlorobenzoates did not induce for a diol dehydrogenase, and resting cells incubated with 3,5-dichlorobenzoate accumulated the dihydrodiol. Thus, *P. putida* P111 was unable to grow on 3,5-dichlorobenzoate as a sole carbon source because the substrate did not induce for a functional dihydrodiol dehydrogenase. A similar phenomenon was noted by Timmis et al. [145] with a recombinant strain of *Pseudomonas* sp. B13 that received *xyl* genes for broadening its dioxygenase activity. That organism also was able to metabolize 3,5-dichlorobenzoate only in the presence of 3- or 4-chlorobenzoate because transcription of the permease, coded by the *xylS* gene, was initiated only with the two monochlorobenzoates.

Table 2 Biodegradation of Polychlorinated Benzoates[a]

Substrate[b]	Products Prior to Ring Fission	Dehalogenation
2,3-DCBa →	2,3-Dichloro-1-carboxy-1,6-diol →	
	4-chlorocatechol →	Reductive (2)
	via chlorocatechol pathway (Figure 7)	Aliphatic (3)
2,4-DCBa →	4-CBa →	Reductive (2)
	via chlorobenzoate pathway (Figure 7)	Hydrolytic (3)
2,5-DCBa →	4-Chlorocatechol →	Oxygenolytic (2)
	via chlorocatechol pathway (Figure 7)	Aliphatic (3)
3,4-DCBa →	3-Chloro-4-hydroxybenzoate →	Hydrolytic (4)
	carboxy-3,4-benzoquinone →	Hydrolytic (3)
	via protocatechuate pathway	
3,5-DCBa →	3,5-Dichloro-1-carboxy-1,2-diol →	
	3,5-dichlorocatechol →	
	via chlorocatechol pathway (Figure 7)	Aliphatic (3,5)
2,3,5-TCBa →	3,5-Dichlorocatechol →	Oxygenolytic (2)
	via chlorocatechol pathway (Figure 7)	Aliphatic (3,5)
2,3,6-TCBa →	2,5-DCBa →	Reductive (2)
	4-chlorocatechol →	Oxygenolytic (6)
	via chlorocatechol pathway (Figure 7)	Aliphatic (3)

[a]Products are listed in sequential order of transformation, with the mechanism of dehalogenation presented in the next column. Numbers in parentheses refer to the original position of the chlorine substituent on the aromatic ring.
[b]DCBa, dichlorobenzoate; TCBA, trichlorobenzoate.

   3,5-Dichlorobenzoate was also a potent inhibitor of *o*-chlorobenzoate metabolism in *P. putida* P111 at concentrations as low as 50 μM, and growth did not occur on any *o*-chlorobenzoate in its presence [126]. Although the organism had broad activity on many polychlorinated benzoates, a single congener (3,5-dichlorobenzoate) did not induce biodegradation and inhibited metabolism of all *o*-chlorobenzoate congeners. The question of how biodegradation of specific compounds, such as PCBs, is effected when a multitude of xenobiotic compounds are present in the environment is worth considering here. The aerobic cometabolism of PCB congeners will clearly lead to a chemical diversity of chlorobenzoates. Broadening the range of compounds that a single organism can metabolize may have its limits.

Another example of a bacterium having the ability to mineralize a compound, yet being unable to utilize it as a growth substrate, is *Acinetobacter* sp. 4-CB1, which dehalogenates 3,4-dichlorobenzoate to 4-hydroxy-3-chlorobenzoate (4-H-3-CB) [146]. The product (4-H-3-CB) was used as a sole carbon source for growth and was dehalogenated to 4-carboxy-1,2-benzoquinone (4-CBQ) prior to ring fission [119,146]. Both products, 4-H-3-CB and 4-CBQ, were inhibitory to one or all of the steps in the metabolism of 4-chlorobenzoate, the normal growth substrate from which the strain was isolated [147]. The culture was unable to grow on 3,4-dichlorobenzoate because the compound did not induce for dehalogenation. However, growth on 4-chlorobenzoate or 4-H-3-CB induced for dehalogenation of 3,4-dichlorobenzoate.

No organism has been isolated that is able to utilize 2,6-dichlorobenzoate. However, Gerritse and Gottschal [148] have developed a novel reactor system for the complete mineralization of the herbicide 2,3,6-trichlorobenzoate, which involved reductive dehalogenation to 2,5-dichlorobenzoate by unidentified anaerobic bacteria in the center of Perlite granules. Complete mineralization of the herbicide occurred when the reactor was inoculated with *P. aeruginosa* JB2, whereas 2,5-dichlorobenzoate accumulated in the reactor not inoculated with strain JB2.

B. Molecular Biology

Three chlorocatechol (*clc*) genes, which code for the modified ortho-fission pathway, have been identified [149]: *clcA* (*meta*-pyrocatechase II); *clcB* (muconate cycloisomerase II); and *clcD* (dienelactone hydrolase). The *clc* genes were shown to be plasmid coded in *Pseudomonas* sp. B13, *P. putida*, and *Alcaligenes eutrophus* [150—155] and organized in clusters. These clusters varied in size from 4.2 kb on pAC27 [149] to approximately 5 kb on pJP4 [156]. The organization on the third plasmid, pWR1, has yet to be clearly delineated, but the *clc*-encoded activities were reported to be contained within a 10-kb fragment [155]. The plasmid coding and clustering of the *clc* genes is similar to what has been described for catabolic pathways of other xenobiotic chemicals [157,158].

DNA:DNA hybridization experiments have demonstrated homology between the *clc* gene clusters carried on pAC27 and pJP4 [149], and restriction mapping showed a high degree of similarity between pAC27 and pWR1 [159]. Thus, it can be assumed that the three plasmids are all highly homologous to each other. The *clc* genes are positively regulated and are amplified when induced through a gene dosage effect. Amplification of the structural genes, which are necessary for high-level expression, occurs only under selection pressure. Succinate-grown cells did not oxidize 3-chlorobenzoate, and benzoate-grown cells released only about 25% of the chloride normally observed with cells grown on 3-chlorobenzoate [149]. The high degree of similarity observed in the *clc* genes is of particular interest considering that the plasmids came from bacteria that were isolated from widely separated points on the globe (pAC27, United states; pWR1, West Germany, pJP4, Australia).

The similarity observed in the *clc* genes, however, is accompanied by divergence. For example, Wyndham and Straus [160] found no homology between the *clc* cluster from pAC27 and from pBR60, a plasmid isolated from *Alcaligenes* sp. BR60, a 3-chlorobenzoate degrader. Stem-loop structures have been identified in the *clc* gene cluster from pJP4 which were not detected in pAC27 [149]. Also, transfer of pAC27, but not of pJP4, to a new *Pseudomonas* host readily conferred the ability to grow on 3-chlorobenzoate [161]. Only after extended selection were 3-chlorobenzoate-degrading transconjugants obtained from transfers of pJP4. The authors considered that pJP4 behaved as a cryptic operon that was activated occasionally through insertion, deletion-fusion, or other types of illegitimate recombination through step-loop structures [149].

The genes coding for the chlorobenzoate dioxygenase and the chlorocarboxydihydrodiol dehydrogenase (see Figure 7) are believed to be derived from those of the ortho-fission pathway for benzoate degradation [162]. The *benABC* and *benD* genes, which code for the respective benzoate dioxygenase, carboxydiol dehydrogenase, orthopyrocatechase, and *cis*, *cis*-muconate isomerase, have been cloned from *Acinetobacter calcoaceticus* in *E. coli* [163]. Neither organism, however, could utilize chlorobenzoates, which would indicate that the *benAB* genes are different from the corresponding AB genes for chlorobenzoate oxidation or that the two sets of genes are regulated differently. The existence of two separate dioxygenases for degradation of chlorobenzoates in a single bacterial strain has been suggested [115,128]. Recent studies with mutants of *P. putida* P111 unable to grow on orthochlorobenzoates, yet able to grow on benzoate or 3- or 4-chlorobenzoate, were cured of plasmids (V. Brenner, B. S. Hernandez, and D. D. Focht, Appl. Environ. Microbiol., in press). The authors suggested that the genes coding for *o*-chlorobenzoate catabolism in strain P111 resided on a 75-kb plasmid, whereas those coding for benzoate or 3- or 4-chlorobenzoate were located on the chromosome.

C. Construction of Catabolic Pathways

The idea of combining the genes for the metabolism of chlorobiphenyl and chlorobenzoate metabolism in a single bacterium was first suggested by Furukawa and Chakrabarty [164]. With few exceptions, most bacteria that grow on monochlorobiphenyls as sole carbon sources are unable to utilize the chlorobenzoates that are produced and, hence, do not dehalogenate and completely mineralize the substrate. Exceptions to this rule have been reported with 4-chlorobenzoate and 4-chlorobiphenyl [66,165]. *Pseudomonas* sp. JB1 grew on 2- and 4-chlorobiphenyls and cometabolized several mono- and dichlorobenzoates, but it did not dehalogenate or grow on chlorobenzoates [166]. Bedard and Haberl [167] reported that resting cell suspensions of the two efficacious PCB cometabolizers, *A. eutrophus* H850 and *Pseudomonas* sp. LB400, also completely metabolized 3-chlorobenzoate, which was produced from metabolism of 2,3'-dichlorobiphenyl. No products could be identified, nor was dehalogenation or growth on chlorobenzoates evident.

Construction of complementary catabolic pathways (e.g., PCBs/chlorobenzoates) involves the assemblage of the genes from two or more organisms into a single one to bring about complete mineralization of the target compound. Four distinct advantages of this approach in contrast to a coculture system (e.g., PCB cometabolizers/chlorobenzoate utilizers) have been offered: (1) no cosubstrate analog (e.g., biphenyl) would be required for growth and enzyme induction, (2) there would be no competition for the growth substrate (PCB) because indigenous bacteria could not utilize it, (3) diffusion and availability of cometabolic products (i.e., chlorobenzoates) would not be limited, and (4) only the growth rate of the recombinant, rather than of two or more cultures, would have to be managed [168]. The problems in optimization of consortia are discussed in the next section.

Strategies for construction of recombinant bacteria are founded upon three concepts: (1) enzymes with narrow specificity create "bottlenecks" in pathways through which "foreign" substrates cannot pass; (2) substrates are metabolized by the "wrong" enzyme to suicidal or dead-end products; (3) the natural pathway is discontinuous. Examples of the first type are the narrow substrate range of some chlorobenzoate dioxygenases. Examples of the second type are the misrouting of chlorocatechols through the *meta*-pyrocatechase pathway, which causes suicidal denaturation of the enzyme by the production of acyl halides [112,133,134]. An example of the third type is the PCB/chlorobenzoate pathway, which requires more than one organism. The key is to identify the point(s) at which bottlenecks or wrong "choices" occur and overcome these by recruitment of the appropriate broad-spectrum enzymes from other organisms.

Reineke and Knackmuss [169] were the first to extend the chlorobenzoate utilization spectrum of *Pseudomonas* sp. B13 from 3-chlorobenzoate to include 4-chloro- and 3,5-dichlorobenzoate. The bottleneck was identified as benzoate dioxygenase, which was active only on benzoate and 3-chlorobenzoate. A broad-spectrum benzoate dioxygenase, which acted on methyl- and chlorobenzoates, was identified in *Pseudomonas* sp. mt-2. Because strain mt-2 metabolized all substrates by meta fission, chlorobenzoates could not be utilized for growth because the chlorocatechol products denatured the enzyme. The genes, carried on pWR101, a TOL-like plasmid [170], were transferred by conjugation from strain mt-2 to strain B13. The resulting recombinant oxidized both 4- and 3,5-dichlorobenzoate to 4-chloro- and 3,5-dichlorocatechol, respectively, which were then metabolized through the modified ortho-fission pathway, rather than through the suicidal meta-fission pathway.

Similar matings have been conducted to broaden or reroute pathways for the complete degradation of chlorobenzoates [171,172], chlorophenols [173], and chlorobenzenes [134,174]. Thus, the utility of recombinants for the biodegradation of these chlorinated compounds has been demonstrated.

Strains that mineralize 3-chlorobiphenyl have been constructed by mating biphenyl utilizers with 3-chlorobenzoate utilizers. *P. putida* BN10 (biphenyl-degrader) and *Pseudomonas* sp. B13 (chlorobenzoate degrader) were mated to give hybrid strains able to grow on 3-chlorobiphenyl [175]. Some hybrid strains, such as BN210, resulted from genes being transferred from the chlorobenzoate utilizer, whereas others, such as B131, originated from transfers of genes to the chlorobenzoate utilizer. In contrast to wild-type strains, which grow on 3-chlorobiphenyl and accumulate 3-chlorobenzoate, chloride release was 90% stoichiometric.

An intergeneric mating between *Acinobacter* sp. P6, a biphenyl utilizer, and *Pseudomonas* sp. HF1, a 3-chlorobenzoate utilizer, resulted in a hybrid strain able to mineralize 3-chlorobiphenyl [176, 177]. The hybrid *Pseudomonas* sp. CB15, which was genetically and phenotypically more similar to the chlorobenzoate-utilizing parent, grew on 3-chlorobiphenyl and 3-chlorobenzoate but would not grow on 3,3'-dichlorobiphenyl. As neither 3,3'-dichlorobiphenyl nor the products of its cometabolism were toxic to the organism when it was growing on 3-chlorobiphenyl, the authors suggested that the lower respiration rate on 3,3'-dichlorobiphenyl, compared to that on 3-chlorobiphenyl, did not provide sufficient energy for growth.

Hickey et al. (W. J. Hickey, V. Brenner, and D. D. Focht, FEMS Microbiol. Ecol., in press) reported growth and mineralization of 2,5-dichlorobiphenyl and 2-chlorobiphenyl in a hybrid strain, *Pseudomonas* sp. UCR2, which was formed by transfer of the *bph* genes from *Arthrobacter* sp. B1Barc to *P. aeruginosa* JB2, the 2,5-dichlorobenzoate utilizer. The hybrid strain cometabolized 2,2'- and 2,3'-dichlorobiphenyl, but the rate of O_2 consumption on these substrates was about 10% of the rate of consumption on 2,5-dichlorobenzoate.

A somewhat surprising result was noted when *P. putida* JHR, which utilized all monochlorobiphenals, but was unable to metabolize the corresponding chlorobenzoates, was mated with *P. cepacia* JH230, which utilized 2-chlorobenzoate [178]. The resulting hybrid, strain JHR2, utilized and dehalogenated 3- and 4-chlorobiphenyl and utilized all monochlorobenzoates, but it could not utilize 2-chlorobiphenyl. The addition of 2-chlorobiphenyl inhibited growth on biphenyl or on 3- or 4-chlorobiphenyl, but not on chlorobenzoates. When strain JHR2 was grown on 4-chlorobenzoate and 2-chlorobiphenyl, a greater increase in cell yield was noted in comparison to cells grown only on 4-chlorobenzoate. Transfer to media containing 2-chlorobiphenyl resulted in poor growth after 30 days, whereupon successive transfers into fresh media resulted in rapid growth and utilization of 2-chlorobiphenyl (3 mM) in 48 h.

The construction of a complementary catabolic pathway for PCBs results in hybrid strains that cleave chlorophenylcatechols by meta fission (Figure 4) and that cleave chlorobenzoates by ortho fission (Figure 7). This is a violation of the Reineke-Knackmuss concept [112,133,134], because *meta*-pyrocatechases are irreversibly inactivated by chlorocatechols. Hybrid strains able to mineralize chlorobiphenyls, therefore, must have some mechanism that prevents metabolism of chlorocatechols by the 3-phenylcatechol dioxygenase. Although 3-chlorocatechol was a potent inhibitor of the 3-phenylcatechol dioxygenase in the hybrid strain, *Pseudomonas* sp. CB15, it was not metabolized by cell-free extract to an acyl halide [176]. 3-Chlorocatechol accumulated in cultures growing on 3-chlorobiphenyl, which resulted in the formation of black-colored polymers. A similar phenomenon was observed in pure cultures able to cometabolize both 3-chlorobiphenyl and 3-chlorobenzoate [179,180], in sewage [181] containing 3-chlorobiphenyl, and in cocultures containing 3-chlorobiphenyl [178].

The incompatibilities in the complementary catabolic pathway of 3-chlorobiphenyl for *Pseudomonas* sp. CB15 are shown in Figure 8. Three separate modes of inhibition regarding the two ring fission reactions were noted [176]. The first, and most predictable, was the inhibition by 3-chlorocatechol of meta-fission dioxygenase activity of 3-(*m*-chlorophenyl)catechol. The second was the competitive inhibition by 3-phenylcatechol of ortho-fission dioxygenase activity of 3-chlorocatechol. The third was the inhibition caused by high substrate concentrations of 3-phenylcatechol on the meta-fission dioxygenase.

What is not understood about all of these hybrid strains that mineralize chlorobiphenyls is why they are incapable of utilizing chlorinated biphenyls with chlorines on both rings, despite their ability to cometabolize them. The restricted ability of constructed strains to grow only on chlorobiphenyls with a nonchlorinated ring

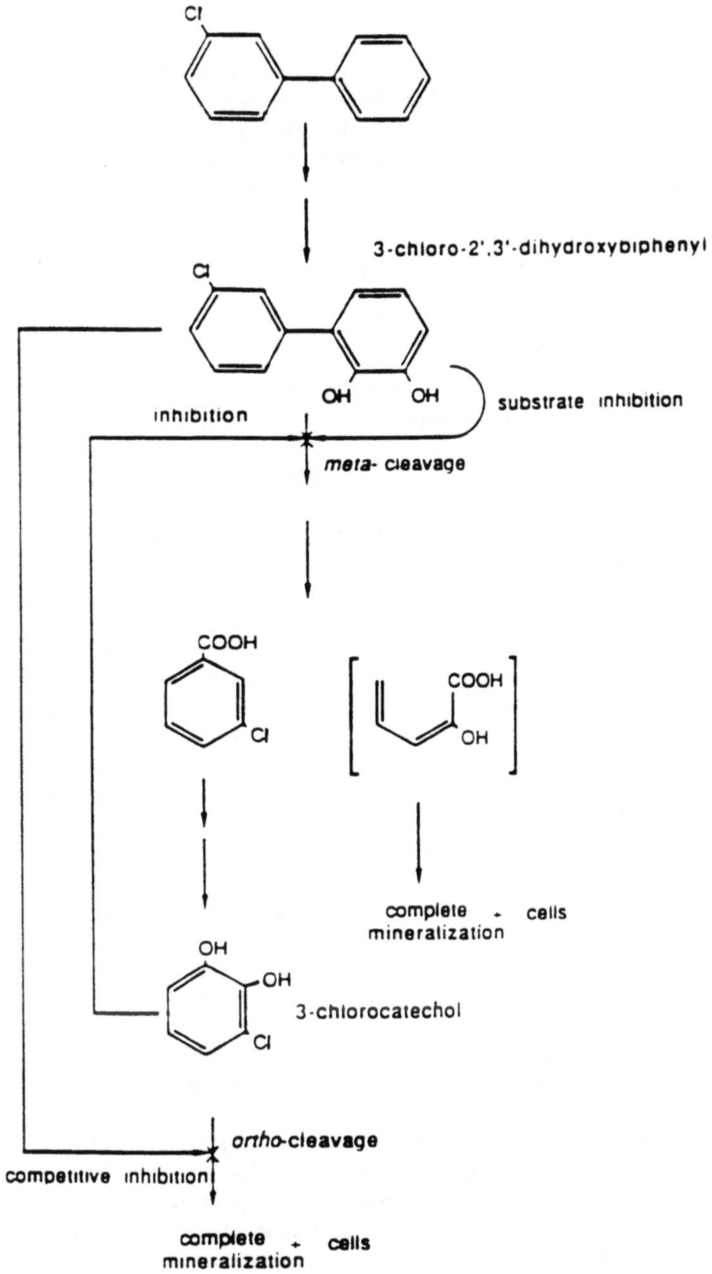

Figure 8 Inhibitory effects of catecholic substrates and products on the catechol dioxygenases during the mineralization of 3-chlorobiphenyl by *Pseudomonas* sp. CB15, a hybrid strain derived from biphenyl- and 3-chlorobenzoate—utilizing parental strains [176].

indicates that neither set of biphenyl or chlorobenzoate genes en-
codes for the enzymes required for dehalogenation of the five-carbon chloraliphatic acids, that the hydrolase is highly substrate
specific, or both. Clearly, resolution of this point is essential for
the construction of effective PCB-utilizing recombinants.

VII. BIODEGRADATION BY MICROBIAL CONSORTIA

A. Axenic Cocultures of Biphenyl and Chlorobenzoate Utilizers

Aerobic biodegradation of PCBs is brought about by at least two
known members of a catabolic consortium, namely biphenyl utilizers
and chlorobenzoate utilizers. There is probably a third group of
microorganisms, as yet unidentified, that dehalogenate the chlorinated aliphatic acids that are produced from cometabolism of PCBs
containing chlorines in both rings. In the example shown in Figure 4, hydrolysis of the ring fission product yields 4-chlorobenzo-
ate and a hypothetical chlorinated five-carbon aliphatic acid, which
has never been identified or isolated. It is unclear whether the
product accumulates in axenic cocultures of biphenyl utilizers
and chlorobenzoate utilizers or whether it is transformed, through
the meta-fission pathway, to more toxic products, such as chloro-
pyruvate. If a third member of the consortia were needed for
biodegradation of the chloraliphatic acid, it would be necessary
to identify the compound and to produce quantities sufficient for
isolation of bacteria from nature by enrichment culture.

The effectiveness of consortia may be limited by the ability to
maintain the individual members at levels of optimal activity, as
discussed in the previous section. The biodegradation of PCBs
is not a case of syntrophism, in which each member benefits from
the metabolism of the other, but of commensalism. The chlorobenzoate utilizer is the commensal and benefits from the production
of chlorobenzoates through cometabolism of PCBs by the biphenyl
utilizer. The biphenyl utilizer obtains no benefit from the cometabolism of PCBs and is, therefore, dependent on biphenyl for its
growth and maintenance.

A classic example of a cometabolizing/commensalistic consortium
is the biodegradation of cyclohexane. Despite the biodegradable
nature of the compound, it does not appear to select for microorganisms able to utilize it as a growth substrate because the initial transformation is cometabolic [182]. Complete catabolism of
cyclohexane involves two different types of microorganisms: those
that grow on n-alkanes and cometabolize cyclohexane to cyclohexanol and those that grow on cyclohexanol [183–185]. From both
examples, PCB and cyclohexane, clearly a cometabolic/commensalis-
tic consortia cannot function without exogenous substrate.

There have been numerous studies of the transformation of PCBs by pure cultures and in soil and other nonaxenic systems, but there have been comparatively few studies with defined axenic cocultures. Furukawa and Chakrabarty [164] were the first to report total co-culture degradation of 4-chloro- and 3,5-dichlorobiphenyl by axenic cultures of *Acinetobacter* sp. P6 and *Pseudomonas* sp. An axenic coculture of two species of *Acinetobacter* has been shown to produce inorganic chloride from 4,4'-dichlorobiphenyl and 3,3',4,4'-tetrachlorobiphenyl [119,186]. However, stoichiometric quantities of chloride were not produced in either case, and less chloride was released from the tetrachlorobiphenyl. Nevertheless, cometabolism of the tetrachloro congener is particularly important, because it is the most toxic of all PCB congeners, as discussed in Section II.

Although biodegradation of chlorobiphenyls by consortia should lead to complete mineralization and stoichiometric release of chloride, this rarely occurs. Fava and Marchetti [187] noted the formation of black-colored products from cocultures of *Pseudomonas* spp. grown on 3-chlorobiphenyl. The addition of benzoate enhanced biodegradation and reduced color formation, whereas the addition of 3-chlorobenzoate had the opposite effect. Havel and Reineke [178] compared the biodegradation of the three isomeric monochlorobiphenyls in separate cocultures containing the same biphenyl utilizer, *P. putida* JHR, which grew on all monochlorobiphenyls but did not metabolize the corresponding chlorobenzoates. *Pseudomonas* sp. B13 grew on 3-chlorobenzoate, *P. cepacia* JH230 grew on 2-chlorobenzoate, and *Pseudomonas* sp. WR216 grew on 4-chlorobenzoate. Black-colored products, which inhibited growth, were produced from cocultures grown on 2- and 3-chlorobiphenyl but not from 4-chlorobenzoate, and the respective chloride releases were 20, 50, and 80% of stoichiometry.

Although the addition of biphenyl maintains the growth of the cometabolizing bacterium, stability of the cometabolic/commensalistic consortium also requires a sufficient rate of cometabolism and availability of product (e.g., chlorobenzoate) to the commensal. Transformation and biodegradation of 4,4'-dichlorobiphenyl by a coculture containing two species of *Acinetobacter* were greater in batch culture [119] than in continuous culture [186] because of the difficulty in maintaining the commensal, which utilized 4-chlorobenzoate. Addition of benzoate, in an attempt to provide exogenous substrate for the commensal, exacerbated the problem because the cometabolizer was more competitive in utilizing benzoate for growth.

The role of a contributing member may not always be known. For example, Pettigrew et al. [188] established a three-membered bacterial consortium that grew on and mineralized 200 ppm 4-chlorobiphenyl or 4,4'-dichlorobiphenyl. Growth yields and dehalogenation were slightly lower with 4,4'-dichlorobiphenyl (68%) than with 4-chlorobiphenyl (87%).   The consortium contained three isolates: *P. testosteroni* LPS10A, which metabolized 4-chlorobiphenyl to 4-chlorobenzoate; *Arthrobacter* sp. LPS10B, which mineralized 4-chlorobenzoate; and *P. putida* LPS10C, whose role could not be determined.   None of these isolates contained plasmid pSS50, which codes for 4-chlorobenzoate metabolism in 4-chlorobiphenyl degraders [165,189,190].   In chemostat experiments with biphenyl and 4-chlorobiphenyl as primary carbon sources, the consortium outcompeted bacteria harboring pSS50 plasmids.   Normally, an organism able to utilize a compound for growth should have a distinct advantage over a consortium, for reasons discussed previously (Section VI,C).   However, 4-chlorobiphenyl was a poor growth substrate for *Alcaligenes* sp. A5, the organism in which this plasmid was originally discovered [165].   As 4-chlorobiphenyl utilizers normally have poor hydrolase activity (Sections V,A, VI,C), the "cryptic" member of the consortium might have facilitated hydrolysis of the ring fission product.

B.   Aerobic Biodegradation of PCBs in Soils

Previous studies [191,192] have shown that biodegradation of PCBs in soil could be enhanced by the addition of biphenyl and *Acinetobacter* sp. P6, a PCB-cometabolizing strain.   Analog enrichment with biphenyl resulted in more than 45% of the added [^{14}C]Aroclor 1242 being mineralized ($^{14}CO_2$ + soil carbonates) in 49 days in contrast to less than 2% in the absence of biphenyl (Figure 9).   Although addition of the PCB-cometabolizing inoculant enhanced the transformation of the more highly chlorinated congeners of Aroclor 1242, there was little difference in the amount mineralized between the inoculated and uninoculated soil.   Addition of biphenyl was the single important factor affecting mineralization of Aroclor 1242 in both studies.   Inoculation in the absence of biphenyl was ineffective.

During mineralization of PCBs in soil, $^{14}CO_2$ production from [^{14}C]Aroclor 1242 always followed the production of unlabeled CO_2 from biphenyl (Figure 9).   This lag is expected on kinetic grounds as chlorobenzoates are transient intermediates in the mineralization of PCBs.   The length of the lag phase was directly proportional to the initial concentration of biphenyl utilizers that were inoculated or indigenous to soil.   Therefore, the rate-limiting step in the biodegradation of PCBs would appear to be the initial oxidation, rather than the metabolism of chlorinated cometabolic products.   This concept was supported by a good fit of the data [192] to a first-order sequential reaction sequence at high inocula concentrations (10^9/g soil), at which the rate was independent of cell density.

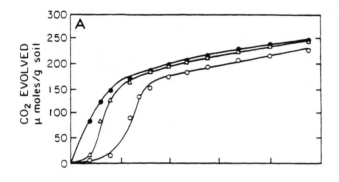

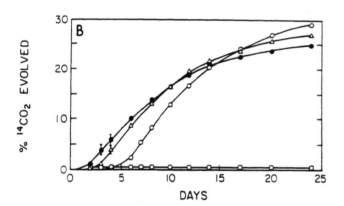

Figure 9   Biodegradation of biphenyl (A) and [^{14}C]Aroclor 1242 (B) in soil [192].   Biphenyl was added to all treatments except those indicated by squares in (B).   Soils were inoculated with *Acinetobacter* sp. P6 at 10^9 (filled circles) and 10^5 (triangles) cells/g or left uninoculated (open circles).

Although PCB-cometabolizing bacteria can be routinely isolated by enrichment on biphenyl [34,37,43,193], comparatively little is known about the presence of chlorobenzoate-degrading bacteria in soil. Pertsova et al. [194] were unable to isolate any 3-chlorobenzoate degraders from a soil column that was perfused with a solution of 3-chlorobenzoate over a 20-day period.   Moreover, there was no disappearance of 3-chlorobenzoate from the uninoculated soil, in contrast

to soils inoculated with chlorobenzoate-degrading pseudomonads.
Focht and Shelton [195] also found that 3-chlorobenzoate was recalcitrant in soil except when inoculated with a 3-chlorobenzoate
utilizer. Inasmuch as dichloro- and trichlorobenzoate degraders
are even less common, there is ample evidence to suggest that
chlorobenzoate degraders are not ubiquitous in soil.

W. J. Hickey, D. Searles, and D. D. Focht (Appl. Environ. Microbiol., in press) compared the efficacy of PCB biodegradation in
soil as affected by bacterial inoculants, specifically biphenyl and
chlorobenzoate degraders, and found the two chlorobenzoate degraders, *P. putida* P111 and *P. aeruginosa* JB2, to effect the greatest disappearance and mineralization of Aroclor 1242. Although indigenous populations of biphenyl utilizers were comparable to those
in an earlier study [192], 2-chlorobenzoate utilizers were not detected from the uninoculated soil at any time during the study.
Soil inoculated with the 2-chlorobenzoate utilizers, however, maintained populations between 10^7 and 10^9/g. Subsequent experimen-
tation showed that the isolates able to utilize both substrates increased substantially from 0 to 44% over 48 days in the soils inoculated with the chlorobenzoate degrader. As the chlorobenzoate-
utilizing inoculants were not able to metabolize PCBs, it would
appear that genetic exchange between the chlorobenzoate-degrading inoculants and the indigenous biphenyl utilizers might
have occurred.

The occurrence of genetic exchange in soil was first demonstrated 20 years ago [196]. The subject has been reviewed by
Stotzky [197] and Slater [198]. Pertsova et al. [194] provided
evidence for genetic exchange between indigenous bacteria and
an inoculant of *P. aeruginosa* that contained a plasmid coding for
the benzoate catabolic pathway. Neither the culture nor any of
the indigenous soil bacteria were able to utilize 3-chlorobenzoate.
However, 3-chlorobenzoate was completely metabolized in the inoculated soil, but not in the uninoculated soil. The diversity of
pseudomonad isolates, obtained from spread plates with 3-chloro-
benzoate as the sole carbon source, suggested that genetic exchange between the inoculant and indigenous pseudomonads was
not restricted to a narrow species diversity. Genetic exchange
among potential members of a complementary catabolic pathway
could be very significant, because strong selection pressure
would obviate a low recombination frequency. Moreover, there
would be little competition for the xenobiotic substrate because
the indigenous bacteria could not utilize it.

C. Aerobic Biodegradation of PCBs
   in Aquatic Environments

Biodegradation of chlorinated biphenyls in aquatic environments is
slower than in soil, because of the lower concentration of nutrients

and microorganisms. However, because of the low water solubility of PCBs, the concentrations in aquatic environments are lower. The first comparative study of biodegradation rates in sewater was conducted by Reichardt et al. [199], who found that the relative rates of biodegradation were 1.0, 0.43, 0.33, and 0.27 for biphenyl 2-chloro-, 4-chloro-, and 3-chlorobiphenyl, respectively. They estimated a turnover time of 1 year for 0.1 ppb or less for the most refractory congener, which was comparable to results from pure cultures [34,200] in which much higher concentrations of cells (10^9 to 10^{10}/ml) and chlorobiphenyls (100 to 1000 ppm) were used. Thus, the rate of PCB degradation could be characterized as a second-order process [201] that is dependent on both substrate and microbial concentration.

Evidence for the use of the meta-fission pathway in nature was provided by the isolation of the catechol, 4,4'-dichloro-2,3-dihydroxybiphenyl, that was produced from 4,4'-dichlorobiphenyl in a mixed culture from activated sludge [202]. Biodegradation of all three monochlorobiphenyls in river water was shown to progress through the corresponding monochlorobenzoates [203]. The non-chlorinated ring was preferentially metabolized, and about half of the theoretical $^{14}CO_2$ was recovered in the corresponding mono-chlorobenzoates, whereas no accumulation of benzoate was observed with biphenyl. The authors' conclusion that growth of the chlorobenzoate metabolizers was significantly slower than growth of biphenyl utilizers [203] is consistent with the observations discussed in the previous section about chlorobenzoates being a rate-limiting step in the biodegradation of chlorobiphenyls.

Recent evidence indicates that genetic transfer from a 3-chlorobenzoate—degrading inoculum to indigenous bacteria occurs readily in lake water microcosms and that selection pressure of the added substrate is important in maintaining bacteria that carry the conjugative plasmid (pBRC60) encoding 3-chlorobenzoate degradation [204,205]. Plasmid transfer in microcosms amended with 3-chlorobenzoate and 3-chlorobiphenyl resulted in the selection of three phenotypic clusters of chlorobenzoate degraders, only one of which was related to the original donor. After the addition of 3-chlorobenzoate, the numbers of 3-chlorobenzoate utilizers increased from a nondetectable level to 10^6/g sediments. The addition of 3-chlorobiphenyl also selected for higher numbers of bacteria bearing the plasmid than in the unamended sediment control.

D. Anaerobic Dehalogenation of PCBs

Because aerobic transformation of PCBs appears to be progressively limited with increased chlorination, reductive dehalogenation of highly chlorinated congeners to lesser chlorinated ones would be very significant in the environment. However, earlier studies with samples of marine mud [40] and silage [206] did not show any transformation

of PCBs. Carey and Harvey [40] concluded that "These anaerobic
environments may then serve as long term sinks for PCBs." However, Fries [206] noted that the rate of anaerobic transformation
of DDT in silage was much less than that observed in other anaerobic environments and suggested that silage may not be a suitable
environment. More recent studies with Hudson River sediments
[207–210] revealed a significantly lower fraction of highly chlorinated congeners and a corresponding enrichment of less chlorinated congeners in deeper sediments than in more aerobic surface
sediments receiving more recent inputs of PCBs. The anomaly between older and more recent studies may be explained by at least
two factors: (1) sulfate (marine muds) and acetate (silage) inhibit
the dehalogenation process [210–212]; (2) uncontaminated microcosms take from 12 to 16 weeks [210] or 21 to 29 weeks [213] before dehalogenation begins.

Definitive evidence for the reductive dehalogenation of higher
to lower chlorinated congeners was established by Quensen et al.
[214] in laboratory experiments that showed considerable dehalogenation of Aroclor 1242 at high concentrations (700 ppm). The
distribution of mono-, di-, tri-, tetra-, penta-, and hexachlorobiphenyls had changed from 0, 9, 49, 36, 5, and 1% at the beginning to 67, 21, 8, 3, 1 and 0%, respectively, after 16 weeks. The
gas chromatographic data of Nies and Vogel [211] clearly show
(Figure 10) that the decrease in the more chlorinated congeners,
represented by the longer retention times, was accompanied by
an increase in less chlorinated congeners, represented by shorter
retention times. The lack of change among the congeners between
the highly and less chlorinated ones is the net result of production
and metabolism.

The anaerobic conversion of PCBs to less chlorinated congeners
makes the interpretation of environmental analyses and sources of
PCBs difficult. Brown et al. [208] observed that dechlorination
of Aroclor 1260 in Silver Lake sediments would give a congener
mixture easily mistaken for Aroclor 1254. However, the manufacturing plant that was the sole source of contamination used Aroclor
1260 exclusively.

Reductive dehalogenation results primarily in the removal of meta
and para chlorines [213–217]. The transformation of para-chloro-
isomers is important environmentally because of the reduction in
the more toxic coplanar toxic congeners, as discussed in Section II.
Preferential differences between meta and para dehalogenation have
been observed and appear to be related to differences in microbial
communities. Quensen et al. [215] found differences in dechlorination patterns between sediments obtained from the Hudson River
and Silver Lake, which were previously contaminated with Aroclor
1242 and 1260, respectively. Although transformation of Aroclor
1242, 1248, and 1254 was similar between the two sediments, with

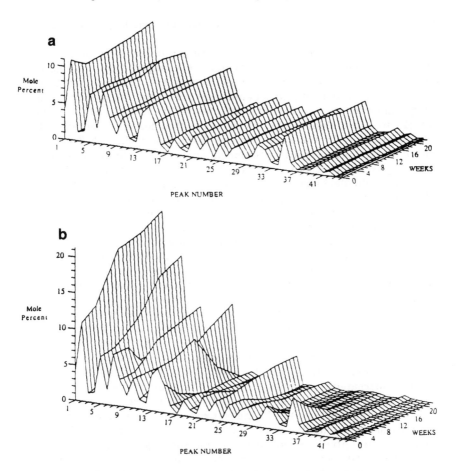

Figure 10 Gas chromatographic data for anaerobic sediments showing reductive dehalogenation of Aroclor 1242 (b) compared to sterilized control (a). More highly chlorinated congeners (higher peak numbers) show reduction with time to less highly chlorinated congeners (lower peak numbers). From Ref. 211 with permission.

maximal rates of 0.3, 0.3, and 0.2 mg-atoms Cl/g sediment/week, respectively, Silver Lake sediments showed a greater preference for meta dehalogenation, whereas Hudson River sediments preferentially removed para chlorines. However, the Hudson River sediments were only 20% as effective as the Silver Lake sediments in dechlorination of Aroclor 1260. The authors concluded that there were different PCB-dechlorinating microorganisms at the different sites that had characteristic specificities for dehalogenation of PCBs.

Van Dort and Bedard [213] studied the transformation of 2,3,5,6-tetrachlorobiphenyl in an unacclimated methanogenic pond sediment and observed a sequential reduction during the first stage (21 to 29 weeks), in which 92% was dehalogenated, with 79% of this total being converted to 2,3,6-trichlorobiphenyl by loss of meta chlorine and 21% to 2,5-dichlorobiphenyl by sequential losses of ortho and meta chlorines. In the second stage (after 28 weeks), 2,3,6-trichlorobiphenyl was rapidly dechlorinated to 2,6-dichlorobiphenyl. They suggested that the first population had a long acclimation time and could dechlorinate 2,3,5,6-tetrachloro- and 2,3,5-trichlorobiphenyls but not 2,3,6-trichlorobiphenyl. The second population, in contrast, had a shorter acclimation time and readily dechlorinated 2,3,6-trichlorobiphenyl to 2,6-dichlorobiphenyl. However, the second population could not dehalogenate 2,3,5,6-tetrachlorobiphenyl (the starting substrate) or 2,3,5-trichlorobiphenyl (the transient product from the first population). This study was unique and encouraging in that it demonstrated reductive ortho dechlorination of PCB congeners in this sediment.

A systematic approach to verify that different bacterial populations are involved in preferential dehalogenations of PCBs was undertaken by Ye et al. [216]. One series from Hudson River sediments was treated with heat or ethanol, which would select for spore formers (e.g., *Clostridium*), whereas the other series was left untreated. Samples were analyzed for Aroclor 1242 and methane after several 1% serial transfers (with heat-ethanol treatments following transfer). Methane production was absent in treated samples at all times but not in untreated samples. Treated samples preferentially removed meta chlorines, yielding congeners with 2,4-substitution, while untreated cultures removed virtually all meta and para chlorines from Aroclor 1242. Thus, the para chlorines were removed by the consortium that was heat and alcohol sensitive (i.e., containing non-spore formers).

Different results with respect to anaerobic dehalogenation of PCBs have been observed with Hudson River sediments. Although Rhee et al. [218] found that all PCB congeners from mono- to pentachlorobiphenyls were transformed, they found no accumulation of less chlorinated congeners. Moreover, the best rate of transformation of PCBs was observed with the addition of biphenyl to the sediments, which is what occurs under aerobic conditions [191,192]. Chen et al. [219] found that the rate of transformation of PCBs was greater under anaerobic conditions with the less chlorinated congeners. When sediments were incubated with ^{14}C-labeled monochlorobiphenyls, $^{14}CO_2$ was detected, but methane was not. The lack of methanogenic conditions in both studies [218,219] suggests the establishment of a different microbial population than others [208—217] have observed during reductive dehalogenation of PCBs.

The mechanism of reductive dehalogenation is unclear. In nutrient-limited sediment samples, energy sources have been shown to accelerate the process [210,211,217]. A deuterium isotope study of the anaerobic dehalogenation of 2,3,4,5,6-pentachlorobiphenyl to 2,3,4,6-tetrachlorobiphenyl showed that the proton source came from water and not from H_2 or protons evolved during metabolism [212]. The authors proposed a two-step (one electron each step) mechanism with a carbanion intermediate. The reductive dehalogenation of chlorobenzoates by *Desulfomonile tiedje* also involves the proton of water, but only at the 3-position that contained the chlorine [220]. Inasmuch as this observation would rule out partial hydrogenation at other ring positions, Mohn and Tiedje [113] suggested a direct electron addition and nucleophilic attack involving transition metal—containing coenzymes, a hydride ion, or NADH. Abramowicz [221] proposed the participation of vitamin B_{12}, a known hydride transfer agent, in a single-step reaction.

E. Anaerobic-Aerobic Biodegradation of DDT and PCBs

Transformation of DDT to DDD was observed in the gut of mice by Barker and Morrison [222] and subsequently shown to be brought about by a strain of *Proteus vulgaris* isolated from the mouse gut [223]. Reductive dehalogenation of DDT is widespread in the microbial world and has since been reported in yeast [224], fungi [225—227], protozoa [228], and algae [229]. The process was first observed in soil by Guenzi and Beard [230,231].

The pathway for anaerobic transformation of DDD involves the removal of both aliphatic chlorines by dehydrodehalogenations to produce DDA (2,2'-bis[p-chlorophenyl]acetic acid) [232]. The first dehydrodehalogenation of DDD produces DDMU (1-chloro-2,2-bis[p-chlorophenyl]ethylene), which is reduced to the corresponding ethane. A second dehydrodehalogenation, followed by two hydrations, gives the corresponding alcohol (2,2-bis[p-chlorophenyl]-ethanol) and DDA. Decarboxylation of DDA with subsequent hydrations and reductions leads to 4,4'-dichlorodiphenylmethane (DDM), 4,4'-dichlorobenzhydrol (DBH), and 4,4'-dichlorobenzophenone (DCB) in sequential order [233]. Two of these metabolites, DDM and DBH, have been shown to undergo oxidation and ring fission by the meta-fission pathway in *Hydrogenomonas* sp., as mentioned earlier [25,26] and shown in Figure 11.

The aerobic fate of DCB is uncertain, as it was found to be recalcitrant to attack by *Hydrogenomonas* sp. Although Plapp et al. [234] postulated that 4-chlorobenzoate was formed by fission of DCB, the evidence, based entirely on R_f comparisons by paper

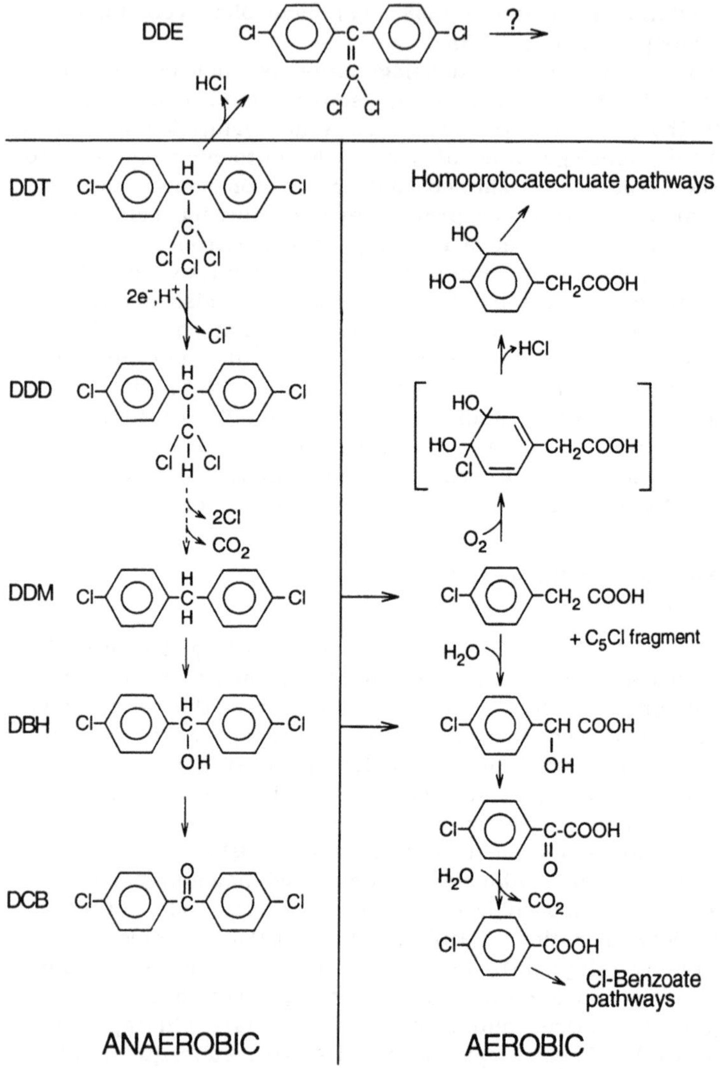

Figure 11   Anaerobic and aerobic biodegradation of DDT.   The steps
between DDD and DDM are omitted for brevity to show the important juncture of the two pathways at DDM and DBH.   Production of
DDE is independent of aeration, and its fate is unknown.   From
Refs. 29–31, 230, 231, 233–236.   Abbreviations: DDT, 1,1,1-tri-
chloro-bis(*p*-chlorophenyl)ethane; DDD, 1,1-dichloro-bis(*p*-chloro-
phenyl)ethane; DDE, 1,1-dichloro-bis(*p*-chlorophenyl)ethylene; DDM,
4,4'-dichlorodiphenyl methane; DBH, 4,4'-dichlorodiphenyl benzhy-
drol; DCB, 4,4'-dichlorodiphenyl benzophenone.

chromatography, must be considered equivocal. Nevertheless, the
possibility of a Bayer-Villiger oxidation through a diphenyl ester
cannot be excluded by analogy with the formation of 4-chloro-
phenylacetate by a monooxygenase-catalyzed esterification of 4-
chloroacetophenone [68].

Pfaender and Alexander [235] found that resting cells of the
obligately aerobic bacterium *Hydrogenomonas* also reduced DDT,
when incubated anaerobically, by the same pathway shown by
Wedemeyer [232,233]. Thus, a single aerobic organism was capable under alternative anaerobic and aerobic conditions of ef-
fecting substantial degradation of DDT. A pathway for the complete degradation of DDT was established upon isolation of an
Arthrobacter sp. capable of utilizing p-chlorophenylacetic acid
as a sole carbon source. However, rapid turnover of the substrate precluded establishment of a biodegradation pathway. The
metabolism of p-chlorphenylacetic acid (Figure 11) is based on
the pathway suggested by Markus et al. [236] in *Pseudomonas*
sp. CBS3 for the formation or protocatechuic acid through a
3,4-dioxygenase attack. However, p-chlorophenylglyoxal [237]
has been identified as a degradation product of DDM, and phen-
ylgloxal and benzoate were reported in trace quantities from the
degradation of diphenylmethane [29]. Thus, an alternative pathway via 4-chlorobenzoate is also proposed in Figure 11. Pfaender
and Alexander [235] were unable to demonstrate mineralization of
[^{14}C]DDT in the presence of *Hydrogenomonas* sp. and *Arthrobacter* sp. or in sewage cultures, despite extensive transformation of DDT, which they attributed to the inability to maintain
optimal conditions for both anaerobic and aerobic reaction sequences.

Synchronous anaerobic and aerobic degradation of DDT by an
immobilized mixed culture system was demonstrated by Beunink and
Rehm [238]. *Alcaligenes* sp. isolated by enrichment on DPM cometabolized DDM and also released chloride under aerobic conditions. *Enterobacter cloacae*, maintained by growth on lactose, reduced DDT to DDD and other metabolites only below an E_H of −200
mV in pure culture. DDT was not metabolized aerobically by either
strain, except for the production of trace quantities of DDE from
E. cloacae. In a chemostat with immobilized cells (Ca alginate),
there was little dependence on E_H, and equal amounts of DDD and
DDE were formed by *E. cloacae*. However, in synchronous anaerobic-aerobic degradation by both cultures, DDE was not produced,
and the degradation pathway proceeded via reductive dehalogenation of DDD to DDM by *E. cloacae*. The latter product was then
further degraded by *Alcaligenes* sp. Although both oxidative and
reductive processes could occur simultaneously at low aeration rates,
diffusion was a problem in the transport of DDT to the anaerobic
center of the immobilized cells.

When Quensen et al. [214] demonstrated the anaerobic dehalogenation of PCBs, they suggested that if the process was followed
by an aerobic incubation, complete biodegradation of PCBs would
be feasible. Fathepure and Vogel [239] demonstrated rapid and
complete biodegradation of hexachlorobenzene, tetrachloroethylene,
and chloroform in a tandem reactor column, with the first one containing an aerobic biofilm and the second an aerobic biofilm. In
the anaerobic film, 99, 80, and 32% of the respective compounds
were degraded to di- and trichlorinated products before introduction into the aerobic column. The final amounts transformed into
CO_2 and nonvolatile intermediates were 94, 96, and 83%, respectively.

Sequential anaerobic and aerobic biotransformation of Aroclor
1242 (300 ppm) was studied in 100-ml serum bottles containing 30 g
of Hudson River sediment to which methanol (6700 ppm) had been
added [240]. After 20 weeks, trichloro- and tetrachlorobiphenyls
were reduced to one-half and one-fifth, respectively, of their original composition; pentachloro- and hexachlorobiphenyls were virtually absent; and dichloro- and monochlorobiphenyls increased to
about four and two times their original composition, respectively.
When the bottles were inoculated with a biphenyl-utilizing bacterium, designated as strain S3, and aerated for 96 h, further
reduction of monochloro- and dichloro- but not trichloro- or tetrachlorobiphenyls occurred. Strain S3, an indigenous bacterium, was
noted to be less active than PCB-cometabolizing strains described
by others [37, 41, 43, 44, 193]. However, the authors' objective
was to determine if total biodegradation of PCBs could be achieved
by indigenous bacteria in the sediment under sequential aerobic
and anaerobic incubations.

F. Bioremediation of PCBs in Soil

Bioremediation can be defined as the use of microorganisms to promote the biodegradation of toxic chemicals in the environment. This
can be done in situ, by working with the contaminated material in
place, or by removing it for treatment in closed reactors. In situ
bioremediation, if applicable in all cases, would be a very cost-effective method. For example, the treatment of contaminated top-
soils by "landfarming" involves practices similar to growing crops,
namely the addition of N and P, tilling the soil, and irrigation. In
situ bioremediation is most applicable to environments that are contained, as high densities of microorganisms can be achieved to effect metabolism of the contaminant. Bioremediation of aquatic systems without containment is impractical because temperature, pH,
nutrients, and other environmental factors which affect the numbers and activity of microorganisms cannot be controlled. The
cleanup of the *Exxon-Valdez* oil spill is an example in which oil

on the beaches was removed by bioremediation [241]. Because of
containment, N and P could be added and retained in the sand, and
the temperature could be maintained near optimal levels by solar
radiation. In contrast, nothing could be done to raise the temperature of the ocean or to ensure tha N and P would be maintained
at levels commensurate with optimal microbial activity.

The major problem in demonstrating biodegradation in situ is the
inability to conduct a thorough mass balance of substrates, products, biomass, and CO_2. Thus, disappearance of the contaminant
is, in many cases, the sole criterion used to demonstrate efficacy
of bioremediation. The ubiquitous enzyme "disappearase" has been
discussed by Unterman [243], who noted that 90% of all the PCBs
that had disappeared from a sterile soil over a 19-day incubation
could be recovered by analyzing all of the physical components of
the reactor, as well as coalesced droplets in the bottom. Because
it is not possible to do mass balances or to make comparisons between sterilized and inoculated soils in the field, it is necessary
to make these comparisons in the laboratory, to negate the null
hypothesis that disappearance of a contaminant is strictly the result of volatilization, adsorption, or some other nonbiological process.

Using an internal standard that is less biodegradable than the
target chemical is another way to ascertain that disappearance is
a result of biodegradation. Phytane, a branched-chain hydrocarbon that is known to be less biodegradable than the other components of crude oil, was used as an internal standard in assessing
the efficacy of bioremediation of the Exxon-Valdez oil spill [241].
This practice also obviates the problem of spatial variability, as
the internal standard and the components of the mixture have the
same proportional composition prior to the advent of biodegradation.

In contrast to the metabolism of petroleum hydrocarbons, which
is a rapid process effected by the indigenous microorganisms, complete metabolism (i.e., biodegradation) of PCBs and DDT is normally a slow process for several reasons. First, the process is
brought about by several different bacteria, many of which have
not been isolated or characterized. Second, it is uncertain whether
all the requisite microorganisms are distributed ubiquitously in soil
or sediments. Third, the survival of inoculants, except with engineered strains [98], normally requires the addition of biphenyl or
some related exogenous growth substrate. Fourth, PCBs and DDT
are not uniformly distributed, and their contact with and availability to microorganisms will be more limited in the environment than
in reactors or laboratory batch systems. Fifth, the aerobic pathway for PCBs is probably limited to chlorinated congeners containing less than six chlorines (i.e., $\leq$pentachlorobiphenyls). Sixth,
there is no known aerobic pathway for bacterial degradation of DDT.

Bioremediation has been used successfully with soils contaminated with pentachlorophenol, as reviewed by Häggblom [242], but it is not a commercially viable strategy for soils contaminated with PCBs or DDT. Several field trials on aerobic bioremediation of PCBs have been conducted by the General Electric Company [47,70,221, 242]. Unterman et al. [244] used two PCB congeners, 2,4,6,2',4'-pentachloro- and 2,4,5,2',4',5'-hexachlorobiphenyl, not known to be transformed by any aerobic PCB-cometabolizing strains, as internal standards for assessing the efficacy of bioremediation in soil contaminated with PCBs. 2,4,6,2',4'-Pentachlorobiphenyl has also been shown to be recalcitrant in activated sludge [245]. A 25% disappearance of PCBs occurred in the top 1 cm after 19 weeks, but significantly less was observed below the surface. Rototilling improved the amount of PCB transformed (about 19%) throughout a 15-cm depth.

One of the major problems encountered with natural PCB-contaminated soils, in contrast to pristine soils spiked with PCBs for scientific experiments, is the presence of other industrial oils, such as chlorobenzenes, mineral oil, di-(2-ethylhexyl)phthalates (DEHPs), kerosene, and diesel fuel No. 2 [47]. These materials could have beneficial effects if they maintained populations of PCB-metabolizing bacteria. However, adverse effects can occur, particularly with DEHP, which greatly impeded the transformation of PCBs not by inhibiting microbial metabolism but by sequestering PCBs and making them unavailable for metabolism [47].

The diversity of PCB congeners, when coupled with the diversity of bacteria that metabolize them, indicates that inoculation with a single microorganism would be an inferior approach to inoculation with a consortium of microorganisms. The differences in bacterial strains with respect to 2,3- or 3,4-dioxygenase attack prompted Unterman et al. [244] to utilize combinations of two physiological types of bacteria (e.g., *Pseudomonas* sp. LB400 and *Corynebacterium* sp. MB1), which gave the best results in a field study. Viney and Bewley [246] took this procedure one step further by using combinations of these isolates with *Phanerochaete chrysosporium*. The best results were noted when all cultures were mixed together.

Anid et al. [240] studied the sequential anaerobic-aerobic biotransformation of Aroclor 1242 in sealed vials of Hudson River sediment. Anaerobic dehalogenation of Aroclor 1242 was also conducted at the same time in a microcosm (1.8 × 0.6 × 0.6 m) containing 0.8 ton of Hudson River sediment. Water was recirculated above the sediments, and a mineral medium plus methanol was periodically injected into eight locations in the sediment. After 20 weeks, there was considerable transformation of the higher to the lower chlorinated congeners. The authors noted that the rates of dehalogenation in the continuous-flow microcosm were lower than in the batch incubations, which they attributed to the faster onset of methanogenesis in the batch incubations. Although the change from anaerobic to aerobic conditions was done only in the batch incubations, the authors suggested that it would be possible to do the same in the river microcosm by the addition of hydrogen peroxide.

VIII. CONCLUDING REMARKS

At this point, it is relevant to think of catabolic pathways of recalcitrant compounds, such as PCBs and DDT, not in terms of single organisms but rather in terms of their contribution to the total gene pool that is scattered throughout nature. The real challenge is to identify and isolate the members of catabolic consortia that together can effect total destruction of the target compound. One strategy for the aerobic biodegradation of chlorinated biphenyls is to recombine the genes from two or more bacteria into hybrid strains able to utilize the contaminant as a growth substrate. In most cases in which this does not appear possible, the alternative strategy is to maximize the activity of the consortium by adjusting the environmental parameters or by adding beneficial growth factors, or both.

When all of the studies of anaerobic and aerobic metabolism of PCBs, DDT, and their metabolites are considered, there is clearly sufficient microbial diversity and catabolic capacity for the biodegradation of all of these compounds, including the highly chlorinated congeners of Aroclor 1260. The major focus for moving biodegradation research on PCBs and DDT from the laboratory to the field will inevitably involve the application of sequential anaerobic and aerobic processes to large volumes of soils or sediments, or possibly the growth of aerobic and anaerobic bacteria in close proximity to one another.

REFERENCES

1. Hutzinger, O., and W. Veerkamp. 1991. *In* T. Leisinger, R. Hütter, A. M. Cook, and J. Nüesch (eds.). Microbial degradation of xenobiotics and recalcitrant compounds, p. 3–45. Academic Press, London.
2. Carson, R. 1962. Silent Spring, Houghton Mifflin, Boston.
3. Jensen, S. 1972. The PCB story. Ambio 1:123–131.
4. Metcalf, R. L., and T. R. Fukuto. 1968. The comparative toxicity of DDT and analogues to susceptible and resistant houseflies and mosquitos. Bull. WHO 38:633–647.
5. Miller, S. 1983. The PCB imbroglio. Environ. Sci. Tehnol. 17:11a–14a.

6. National Academy of Sciences. 1979. Polychlorinated biphen-
    yls. Washington, D.C.
7. Riseborough, R. W., P. Reiche, D. B. Peakall, S. G. Hersman,
    and M. N. Kirven. 1968. Polychlorinated biphenyls in the
    global ecosystem. Nature (London) 220:1098–1102.
8. Hutzinger, O., S. Safe, and V. Zitko. 1974. The chemistry
    of PCBs. CRC Press, Cleveland.
9. Tucker, E. S., V. W. Saeger, and O. Hicks. 1975. Acti-
    vated sludge primary degradation of polychlorinated biphenyls.
    Bull. Environ. Contam. Toxicol. 14:705–712.
10. Pal, D., J. B. Weber, and M. R. Overcash. 1980. Fate of
    polychlorinated biphenyls (PCBs) in soil-plant systems. Resi-
    due Rev. 74:45–98.
11. Bergh, A. K., and R. S. People. 1977. PCB distribution in
    sewage wastes and their environmental and community effects,
    p. 4–6. Proc. 1977 Natl. Conf. on Treatment and Disposal of
    Industrial Wastewaters and Residues, Houston.
12. Fujiwara, K. 1975. Environmental and food contamination with
    PCB's in Japan. Sci. Total Environ. 4:219.
13. Tanabe, S., N. Kannan, An. Subramanian, S. Watanabe, M.
    Picer, and N. Picer. 1991. Long-term trends of DDTs and
    PCBs in sediment samples collected from the eastern Adriatic
    coastal waters. Bull. Environ. Contam. Toxicol. 47:864–873.
14. Williams, L. L., J. P. Glesy, N. Degalan, D. A. Verbrugge,
    D. E. Tillitt, G. T. Ankley, and R. L. Welch. 1992. Predic-
    tion of concentrations of 2,3,7,8-tetrachlorodibenzo-*para*-dioxin
    equivalents from total concentrations of polychlorinated biphen-
    yls in fish fillets. Environ. Sci. Technol. 26:1151–1159.
15. Safe, S. 1984. Polychlorinated biphenyls (PCBs) and poly-
    brominated biphenyls (PBBs): biochemistry, toxicology, mech-
    anisms of action. CRC Crit. Rev. Toxicol. 13:319–395.
16. Safe, S., S. Bandiera, T. Sawyer, L. Robertson, L. Safe,
    A. Parkinson, P. E. Thomas, D. E. Ryan, L. M. Reik, W.
    Levin, M. Denomme, and T. Fujita. 1985. PCBs: structure-
    function relationships and mechanism of action. Environ. Health
    Perspect. 60:47–56.
17. Safe, S., L. Safe, and M. Mullin. 1985. Polychlorinated bi-
    phenyls: congener specific analysis of a commercial mixture
    and a human milk-extract. J. Agric. Food Chem. 33:24–29.
18. Sassa, S., O. Sugita, A. Kappas, N. Ohnuma, S. Imajo, T.
    Okumura, and T. Noguchi. 1986. Chloro-substituted sites
    and probability of coplanarity in polychlorinated biphenyls in
    determining uroporphyrin formation in cultured liver cells.
    Biochem. J. 240:622–623.
19. Dewailly, E., J. P. Weber, S. Gingras, and C. Laliberte. 1991.
    Coplanar PCBs in human milk in the province of Quebec,

Canada: are they more toxic than dioxin for breast fed infants? Bull. Environ. Contam. Toxicol. 47:491–498.

20. Tanabe, S., N. Kannan, An. Subramanian, S. Watanabe, and R. Tatsukawa. 1987. Highly toxic coplanar PCBs: occurrence, source, persistency and toxic implications to wildlife and humans. Environ. Pollut. 47:147–163.

21. Kannan, N., S. Tanabe, M. Ono, and R. Tatsukawa. 1989. Critical evaluation of polychlorinated biphenyl toxicity in terrestrial and marine mammals: increasing impact of non-ortho and mono-ortho coplanar polychlorinated biphenyls from land to ocean. Arch. Environ. Contam. Toxicol. 18: 850–857.

22. Vandenbergh, P. A., R. H. Olsen, and J. F. Colaruotolo. 1981. Isolation and genetic characterization of bacteria that degrade chloroaromatic compounds. Appl. Environ. Microbiol. 42:737–739.

23. Horvath, R. S. 1972. Microbial cometabolism and the degradation of organic compounds in nature. Bacteriol. Rev. 36: 146–155.

24. Dagley, S. 1984. Introduction, p. 1–11. *In* D. T. Gibson (ed.), Microbial degradation of organic compounds. Marcel Dekker, New York.

25. Focht, D. D., and M. Alexander. 1970. DDT metabolites and analogs: ring fission by *Hydrogenomonas.* Science 170: 91–92.

26. Focht, D. D., and M. Alexander. 1971. Aerobic cometabolism of DDT analogues by *Hydrogenomonas* sp. J. Agric. Food Chem. 19:20–22.

27. Dagley, S., P. J. Chapman, D. T. Gibson, and J. M. Wood. 1964. Degradation of the benzene nucleus by bacteria. Nature (London) 202:775–778.

28. Gibson, D. T., and V. Subramanian. 1984. Microbial degradation of aromatic hydrocarbons, p. 181–252. *In* D. T. Gibson (ed.), Microbial degradation of organic compounds. Marcell Dekker, New York.

29. Focht, D. D., and M. Alexander. 1970. Bacterial degradation of diphenylmethane, a DDT model substrate. Appl. Microbiol. 20:608–611.

30. Focht, D. D., and H. Joseph. 1974. Degradation of 1,1-diphenylethylene by mixed cultures. Can. J. Microbiol. 20: 631–635.

31. Francis, A. J., R. J. Spanggord, G. I. Ouchi, R. Bramhall, and N. Bohonos. 1976. Metabolism of DDT analogues by a *Pseudomonas* sp. Appl. Environ. Microbiol. 32:213–216.

32. Lunt, D., and W. C. Evans. 1970. The microbial metabolism of biphenyl. Biochem. J. 118:54.

33. Catelani, D., S. Sorlini, and V. Trecanni. 1971. The metabolism of biphenyl by *Pseudomonas putida*. Experientia 27:1173–1174.

34. Ahmed, M., and D. D. Focht. 1973. Degradation of polychlorinated biphenyls by two species of *Achromobacter*. Can. J. Microbiol. 19:47–52.

35. Furukawa, K., K. Tonomura, and A. Kamibayashi. 1978. Effect of chlorine substitution of the biodegradability of polychlorinated biphenyls. Appl. Environ. Microbiol. 35:223–227.

36. Furukawa, K., K. Tonomura, and A. Kamibayashi. 1979. Effect of chlorine substitution on the bacterial metabolism of various polychlorinated biphenyls. Appl. Environ. Microbiol. 38:301–310.

37. Kohler, H.-P. E., D. Kohler-Staub, and D. D. Focht. 1988. Cometabolism of PCBs: enhanced transformation of Aroclor 1254 by growing bacterial cells. Appl. Environ. Microbiol. 54:1940–1945.

38. Kilpi, S., K. Himberg, K. Yrhälä, and V. Backström. 1988. The degradation of biphenyl and chlorobiphenyls by mixed bacterial cultures. FEMS Microbiol. Ecol. 53:19–26.

39. Fava, F., S. Zappoli, L. Marchett, and L. Morselli. 1991. Biodegradation of chlorinated biphenyls (Fenclor 42) in batch cultures with mixed and pure aerobic cultures. Chemosphere 22:3–14.

40. Carey, A. E., and G. R. Harvey. 1978. Metabolism of polychlorinated biphenyls by marine bacteria. Bull. Environ. Contam. Toxicol. 20:527–534.

41. Bedard, D. L., R. E. Wagner, M. J. Brennan, M. L. Haberl, and J. F. Brown, Jr. 1987. Extensive degradation of Aroclors and environmentally transformed polychlorinated biphenyls by *Alcaligenes eutrophus* H850. Appl. Environ. Microbiol. 53:1094–1102.

42. Bedard, D. L., M. L. Haberl, R. J. May, and M. J. Brennan. 1987. Evidence for novel mechanisms of polychlorinated biphenyl metabolism in *Alcaligenes eutrophus* H850. Appl. Environ. Microbiol. 53:1103–1112.

43. Bopp, L. H. 1986. Degradation of highly chlorinated PCBs by *Pseudomonas* strain LB400. J. Ind. Microbiol. 1:23–29.

44. Furukawa, K., N. Tomizuka, and A. Kamibayashi. 1983. Metabolic breakdown of Kaneclors (polychlorobiphenyls) and their products by *Acinetobacter* sp. Appl. Environ. Microbiol. 46:140–145.

45. Bedard, D. L., R. Unterman, L. H. Bopp, M. J. Brennan, M. L. Haberl, and C. Johnson. 1986. Rapid assay for screening and characterizing microorganisms for the ability to degrade polychlorinated biphenyls. Appl. Environ. Microbiol. 51:761–768.

46. Yates, J. R., and F. J. Mondello. 1989. Sequence similarities in the genes encoding polychlorinated biphenyl degradation by *Pseudomonas* strain LB400 and *Alcaligenes eutrophus* H850. J. Bacteriol. 171:1733–1735.
47. Bedard, D. L. 1990. Bacterial Transformation of polychlorinated biphenyls, p. 369–388. *In* D. Kamely, A, Chakrabarty, and G. Omenn (eds.), Biotechnology and biodegradation. Portfolio Publishing Co., Woodlands, Texas and Gulf Publishing Co., Houston.
48. Springer, W., and H. G. Rast. 1988. Biologischer Abbau mehrfach halogenierter mono- und polyzyklischer Aromaten. Wasser Abwasser 129:70–75.
49. Yagi, O., and R. Sudo. 1980. Degradation of polychlorinated biphenyls by microorganisms. J. Water Pollut. Control Fed. 52:1035–1043.
50. Higson, F. K., and D. D. Focht. 1989. Bacterial metabolism of hydroxylated biphenyls. Appl. Environ. Microbiol. 55:946–952.
51. Kohler, H.-P. E., D. Kohler-Staub, and D. D. Focht. 1988. Degradation of 2-hydroxybiphenyl and 2,2'-dihydroxybiphenyl by *Pseudomonas* sp. strain HBP1. Appl. Environ. Microbiol. 54:2683–2688.
52. Omori, T., K. Sugimura, H. Ishigooka, and Y. Minoda. 1986. Purification and some properties of a 2-hydroxy-6-oxo-6-phenyl-2,4-dienoic acid hydrolyzing enzyme from *Pseudomonas cruciviae* S93B1 involved in the degradation of biphenyl. Agric. Biol. Chem. 50:1513–1518.
53. Catelani, D., A. Colombi, C. Sorlini, and V. Treccani. 1973. Metabolism of biphenyl. 2-Hydroxy-6-oxo-6-phenylhex-2,4-dienoate: the meta-cleavage product from 2,3-dihydroxybiphenyl by *Pseudomonas putida*. Biochem. J. 134:1063–1066.
54. Ahmad, D., M. Sylvestre, and R. Massé. 1991. Bioconversion of 2-hydroxy-6-oxo-6-(4'-chlorophenyl)hexa-2,4-dienoic acid, the meta-cleavage product of 4-chlorobiphenyl. J. Gen. Microbiol. 137:1375–1385.
55. Massé, R., F. Messier, C. Ayotte, M. F. Levesque, and M. Sylvestre. 1989. A comprehensive gas chromatographic/mass spectrometric analysis of 4-chlorobiphenyl bacterial degradation products. Biomed. Environ. Mass Spectrom. 18:27–47.
56. Omori, T., H. Ishigooka, and Y. Minoda. 1988. A new metabolic pathway for meta ring-fission compounds. Agric. Biol. Chem. 52:503–509.
57. Dagley, S., W. C. Evans, and D. W. Ribbons. 1960. New Pathways in the oxidative metabolism of aromatic compounds by micro-organisms. Nature (London) 188:560–566.
58. Takase, I., T. Omori, and Y. Minoda. 1986. Microbial degradation products from biphenyl-related compounds. Agric. Biol. Chem. 50:681–686.

59. Schmidt, S., R.-M. Wittich, D. Erdmann, H. Wilkes, W.
     Francke, and P. Fortnagel. 1992. Biodegradation of di-
     phenyl ether and its monohalogenated derivatives in *Sphingo-
     monas* sp. strain SS3. Appl. Environ. Microbiol. 58:2744–
     2750.
60. Gibson, D. T., R. L. Roberts, M. C. Wells, and V. M. Kobal.
     1973. Oxidation of biphenyl by a *Beijerinckia* sp. Biochem.
     Biophys. Res. Commun. 50:211–219.
61. Wittich, R. M., H. Wilkes, V. Sinnwell, W. Francke, and P.
     Fortnagel. 1992. Metabolism of dibenzo-*para*-dioxin by
     Sphingomonas sp. strain-RW1. Appl. Environ. Microbiol.
     58:1005–1010.
62. Engesser, K. H., V. Strubel, K. Christoglou, P. Fischer,
     and H. G. Rast. 1989. Dioxygenolytic cleavage of aryl
     ether bonds—1,10-dihydro-1,10-dihydroxyfluoren-9-one, a
     novel arene dihydrodiol as evidence for angular dioxygena-
     tion of dibenzofuran. FEMS Microbiol. 65:205–209.
63. Fortnagel, P., H. Harms, R. M. Wittich, S. Krohn, H. Meyer,
     V. Sinnwell, H. Wilkes, and W. Francke. 1990. Metabolism
     of dibenzofuran by *Pseudomonas* sp. strain HH69 and the
     mixed culture HH27. Appl. Environ. Microbiol. 56:1148–
     1156.
64. Strubel, V., K. H. Engesser, P. Fischer, and H.-J. Knack-
     muss. 1991. 3-(2-Hydroxyphenyl)catechol, a substrate for
     proximal meta ring cleavage in dibenzofuran degradation by
     Brevibacterium sp. strain DPO-1361. J. Bacteriol. 173:116–
     122.
65. Baxter, R. M., and S. A. Sutherland. 1984. Biochemical
     and photochemical processes in the degradation of chlorinated
     biphenyls. Environ. Sci. Technol. 18:608–610.
66. Barton, M. R., and R. L. Crawford. 1988. Novel biotrans-
     formations of 4-chlorobiphenyl by a *Pseudomonas* sp. Appl.
     Environ. Microbiol. 54:594–595.
67. Cripps, R. E. 1975. The microbial metabolism of acetophenone.
     Biochem. J. 152:233–241.
68. Higson, F. K., and D. D. Focht. 1990. The bacterial degra-
     dation of ring-chlorinated acetophenones. Appl. Environ. Mi-
     crobiol. 56:3678–3685.
69. Higson, F. K. 1991. Degradation of xenobiotics by white rot
     fungi. Rev. Environ. Contam. Toxicol. 122:111–152.
70. Unterman, R., F. J. Mondello, M. J. Brenna, R. E. Brooks,
     D. P. Mobley, J. B. McDermott, and C. C. Schwartz. 1987.
     Bacterial treatment of PCB-contaminated soils: prospects for
     the application of recombinant DNA technology, p. 259–264.
     In Proc. 2nd Int. Conf. New Front. Hazard. Waste Manage.
     U.S. Environmental Protection Agency, Res. Dev. [Rep.]
     EPA/600/9-87/018F.

71. Burdsall, H. H., and W. E. Eslyn. 1974. A new *Phanero-chaete* with a *chrysosporium* imperfect state. Mycotaxon 1: 123–133.

72. Bumpus, J. A., M. Tien, D. Wright, and S. D. Aust. 1985. Oxidation of persistent pollutants by a white rot fungus. Science 228:1434–1436.

73. Fernando, T., S. D. Aust, and J. A. Bumpus. 1989. Effects of culture parameters on DDT [1,1,1-trichloro-2,2-bis(4-chlorophenyl)ethane] biodegradation by *Phanerochaete chrysosporium*. Chemosphere 19:1387–1398.

74. Bumpus, J. A., and S. D. Aust. 1987. Biodegradation of DDT [1,1,1-trichloro-2,2-bis(4-chlorophenyl)ethane] by the white rot fungus *Phanerochaete chrysosporium*. Appl. Environ. Microbiol. 53:2001–2008.

75. Eaton, D. C. 1985. Mineralization of polychlorinated biphenyls by *Phanerochaete chrysosporium*: a ligninolytic fungus. Enzyme Microb. Technol. 7:194–196.

76. Koenigs, J. W. 1974. Production of hydrogen peroxide by wood-rotting fungi in wood and its correlation with weight loss, depolymerization and pH changes. Arch. Microbiol. 99:129–145.

77. Walling, C. 1975. Fenton's reagent revisited. Accounts Chem. Res. 8:125–131.

78. Bumpus, J. A., and S. D. Aust. 1987. Biodegradation of chlorinated organic compounds by *Phanerochaete chrysosporium*, a wood-rotting fungus, p. 340–390. *In* J. H. Exner (ed.), Solving hazardous waste problems. Learning from dioxins. ACS Symp. Ser. 338. American Chemical Society, Washington, D.C.

79. Katayama, Y., T. Nishida, N. Morohoshi, and K. Kuroda. 1989. The metabolism of biphenyl structures in lignin by the wood-rotting fungus *Coriolus versicolor*. FEMS Microbiol. 61:307–314.

80. Axcell, B. C., and P. J. Gearey. 1975. Purification and some properties of a soluble benzene-oxidizing system from a strain of *Pseudomonas*. Biochem. J. 146:173–183.

81. Gearey, P. J., F. Saboowalla, D. Patil, and R. Cammack. 1984. An investigation of the iron-sulphur proteins of benzene dioxygenase from *Pseudomonas putida* by electron-spin-resonance spectroscopy. Biochem. J. 217:667–673.

82. Yeh, W.-K., D. T. Gibson, and T.-N. Liu. 1977. Toluene dioxygenase: a multicomponent enzyme system. Biochem. Biophys. Res. Commun. 78:401–410.

83. Subramanian, V., T.-N. Liu, W.-K. Yeh, and D. T. Gibson. 1979. Toluene dioxygenase: purification of an iron-sulfur protein by affinity chromatography. Biochem. Biophys. Res. Commun. 91:1131–1139.

84. Subramanian, V., T.-N. Liu, W.-K. Yeh, M. Narro, and D. T. Gibson. 1981. Purification and properties of NADH-ferredoxin$_{TOL}$ reductase, a component of toluene dioxygenase from *Pseudomonas putida*. J. Biol. Chem. 256:2723–2730.
85. Subramanian, V., T.-N. Liu, W.-K. Yeh, C. M. Serdar, L. P. Wackett, and D. T. Gibson. 1985. Purification and properties of ferredoxin$_{TOL}$, a component of toluene dioxygenase from *Pseudomonas putida* F1. J. Biol. Chem. 260:2355–2363.
86. Ensley, B. D., and D. T. Gibson. 1983. Naphthalene dioxygenase: purification and properties of a terminal oxygenase component. J. Bacteriol. 149:948–954.
87. Ensley, B. D., D. T. Gibson, and A. L. Laborde. 1982. Oxidation of naphthalene by a multicomponent enzyme system from *Pseudomonas* sp. strain NCIB9816. J. Bacteriol. 149:948–954.
88. Haigler, B. E., and D. T. Gibson. 1990. Purification and properties of ferredoxin$_{NAP}$, a component of naphthalene dioxygenase from *Pseudomonas* sp. strain NCIB 9816. J. Bacteriol. 172:465–468.
89. Haigler, B. E., and D. T. Gibson. 1990. Purification and properties of NADH-ferredoxin$_{NAP}$ reductase, a component of naphthalene dioxygenase from *Pseudomonas* sp. strain NCIB 9816. J. Bacteriol. 172:457–464.
90. Sariaslani, F. S. 1989. Microbial enzymes for oxidation of organic molecules. Crit. Rev. Biotechnol. 9:171–257.
91. Zylstra, G. J., and D. T. Gibson. 1989. Toluene degradation by *Pseudomonas putida* F1. Nucleotide sequence of the todC1C2BADE genes and their expression in *Escherichia coli*. J. Biol. Chem. 264:14940–14946.
92. Baker, M. E. 1990. Sequence similarity between *Pseudomonas* dihydrodiol dehydrogenase, part of the gene cluster that metabolizes PCBs, and dihydrodiol dehydrogenases involved in metabolism of ribitol and glucitol and synthesis of antibiotics and 17 β-oestradiol, testosterone and corticosterone. Biochem. J. 267:839–841.
93. Hayaishi, O. 1966. Crystalline oxygenases of pseudomonads. Bacteriol. Rev. 30:720–731.
94. Furukawa, K., and N. Arimura. 1987. Purification and properties of 2,3-dihydroxybiphenyl dioxygenases from polychlorinated biphenyl–degrading *Pseudomonas pseudoalcaligenes* and *Pseudomonas aeruginosa* carrying the cloned bphC gene. J. Bacteriol. 169:924–927.
95. Kuhm, A. E., A. Stolz, and H.-J. Knackmuss. 1991. Metabolism of naphthalene by the biphenyl-degrading bacterium *Pseudomonas paucimobilis* Q1. Biodegradation 1:115–120.
96. Taira, K., N. Hayase, N. Arimura, S. Yamashita, T. Miyazaki, and K. Furukawa. 1988. Cloning and nucleotide sequence of

the 2,3-dihydroxybiphenyl dioxygenase gene from the PCB-degrading strain of *Pseudomonas paucimobilis* Q1. Biochemistry 27:3990–3996.

97. Higson, F. K., and D. D. Focht. 1992. Utilization of 3-chloro 2-methylbenzoic acid by *Pseudomonas cepacia* MB2 through the meta fission pathway. Appl. Environ. Microbiol. 58:2501–2504.

98. Mondello, F. J. 1989. Cloning and expression in *Escherichia coli* of *Pseudomonas* strain LB400 genes encoding polychlorinated biphenyl degradation. J. Bacteriol. 171:1725–1732.

99. Khan, A., R. Tewari, and S. Walia. 1988. Molecular cloning of 3-phenylcatechol dioxygenase involved in the catabolic pathway of chlorinated biphenyl from *Pseudomonas putida* and its expression in *Escherichia coli*. Appl. Environ. Microbiol. 54:2664–2671.

100. Khan, A., and S. Walia. 1989. Cloning of bacterial genes specifying degradation of 4-chlorobiphenyl from *Pseudomonas putida* OU83. Appl. Environ. Microbiol. 55:798–805.

101. Khan, A. A., and S. K. Walia. 1990. Identification and localization of 3-phenylcatechol dioxygenase and 2-hydroxy-6-oxo-6-phenylhexa-2,4-dienoate hydrolase genes of *Pseudomonas putida* and expression in *Escherichia coli*. Appl. Environ. Microbiol. 56:956–962.

102. Khan, A. A., and Walia, S. K. 1991. Expression, localization, and functional analysis of polychlorinated biphenyl degradation genes *cbp*ABCD of *Pseudomonas putida*. Appl. Environ. Microbiol. 57:1325–1332.

103. Ahmad, D., R. Massé, and M. Sylvestre. 1990. Cloning and expression of genes involved in 4-chlorobiphenyl transformation by *Pseudomonas testosteroni*: homology to polychlorobiphenyl-degrading genes in other bacteria. Gene 86:53–61.

104. Hansen, J. B., and R. H. Olsen. 1978. Isolation of large bacterial plasmids and characterization of the P1 incompatibility group plasmids pMG1 and pMG5. J. Bacteriol. 135:227–238.

105. Furukawa, K., and T. Miyazaki. 1986. Cloning of a gene cluster encoding biphenyl and chlorobiphenyl degradation in *Pseudomonas pseudoalcaligenes*. J. Bacteriol. 166:392–398.

106. Furukawa, K., N. Hayase, K. Taira, and N. Tomizuka. 1989. Molecular relationship of chromosomal genes encoding biphenyl/polychlorinated biphenyl catabolism: some soil bacteria possess a highly conserved *bph* operon. J. Bacteriol. 171:5467–5472.

107. Walia, S., A. Khan, and N. Rosenthal. 1990. Construction and applications of DNA probes for detection of polychlorinated

biphenyl—degrading genotypes in toxic organic—contaminated soil environments. Appl. Environ. Microbiol. 56:254–259.

108. Furukawa, K., J. R. Simon, and A. M. Chakrabarty. 1983. Common induction and regulation of biphenyl, xylene/toluene, and salicylate catabolism in *Pseudomonas paucimobilis*. J. Bacteriol. 154:1356–1362.

109. Nakai, C., H. Kagamiyama, M. Nozaki, T. Nakazawa, S. Nouye, Y. Ebina, and A. Nakazawa. 1983. Complete nucleotide sequence of the metapyrocatchase gene on the TOL plasmid of *Pseudomonas putida* mt-2. J. Biol. Chem. 258: 2923–2928.

110. Minton, N. P., T. Atkinson, C. J. Bruton, and R. F. Sheerwood. 1984. The complete nucleotide sequence of the *Pseudomonas* gene coding for carboxypeptidase G1. Gene 31:31–38.

111. Khan, A. A., and S. K. Walia. 1992. Use of a genetically engineered *Escherichia coli* strain to produce 1,2-dihydroxy-4'-chlorobiphenyl. Appl. Environ. Microbiol. 58:1388–1391.

112. Reineke, W., and H.-J. Knackmuss. 1988. Microbial degradation of haloaromatics. Annu. Rev. Microbiol. 42:263–287.

113. Mohn, W. W., and J. M. Tiedje. 1992. Microbial reductive dehalogenation. Microbiol. Rev. 56:482–507.

114. van den Tweel, W. J. J., J. B. Kok, and J. A. M. de Bont. 1987. Reductive dechlorination of 2,4-dichlorobenzoate to 4-chlorobenzoate and hydrolytic dehalogenation of 4-chloro-, 4-bromo-, and 4-iodobenzoate by *Alcaligenes denitrificans* NTB-1. Appl. Environ. Microbiol. 53:810–815.

115. Hickey, W. J., and D. D. Focht. 1990. Degradation of mono-, di-, and trihalogenated benzoates by *Pseudomonas aeruginosa* JB2. Appl. Environ. Microbiol. 56:3842–3850.

116. Marks, T. S., R. Wait, A. R. W. Smith, and A. V. Quirk. 1984. The origin of the oxygen incorporated during the dehalogenation/hydroxylation of 4-chlorobenzoate by an *Arthrobacter* sp. Biochem. Biophys. Res. Commun. 124:669–674.

117. Higson, F. K., and D. D. Focht. 1990. Degradation of 2-bromobenzoic acid by a strain of *Pseudomonas aeruginosa*. Appl. Environ. Microbiol. 56:1615–1619.

118. Müller, R., J. Thiele, U. Klages, and F. Lingens. 1984. Incorporation of [^{18}O]water into 4-hydroxybenzoic acid in the reaction of 4-chlorobenzoate dehalogenase from *Pseudomonas* sp. CBS3. Biochem. Biophys. Res. Commun. 124:178–182.

119. Adriaens, P., H. P. Kohler, D. Kohler-Staub, and D. D. Focht. 1989. Bacterial dehalogenation of chlorobenzoates and coculture biodegradation of 4,4'-dichlorobiphenyl. Appl. Environ. Microbiol. 55:887–892.

120. Groenewegen, P. E. J., W. J. J. van den Tweel, and J. A. M. deBont. 1992. Anaerobic bioformation of 4-hydroxybenzoate from 4-chlorobenzoate by the coryneform bacterium NTB-1. Appl. Microbiol. Biotechnol. 36:541–547.

121. Elsner, A., F. Löffler, K. Miyashita, R. Müller, and F. Lingens. 1991. Resolution of 4-chlorobenzoate dehalogenase from *Pseudomonas* sp. strain CBS3 into 3 components. Appl. Environ. Microbiol. 57:324–326.

122. Copley, S. D., and G. P. Crooks. 1992. Enzymic dehalogenation of 4-chlorobenzoyl coenzyme-A in *Acinetobacter* sp strain 4-CB1. Appl. Environ. Microbiol. 58:1385–1387.

123. Löffler, R., R. Müller, and F. Lingens. 1991. Dehalogenation of 4-chlorobenzoate by 4-chlorobenzoate dehalogenase from *Pseudomonas* sp CBS3—an ATP-coenzyme-A dependent reaction. Biochem. Biophys. Res. Commun. 176: 1106–1111.

124. Groenewegen, P. E. J., A. J. M. Dreissen, W. N. Konings, and J. A. M. deBont. 1990. Energy-dependent uptake of 4-chlorobenzoate in the coryneform bacterium NTB-1. J. Bacteriol. 172:419–423.

125. Hartmann, J., W. Reineke, and H.-J. Knackmuss. 1979. Metabolism of 3-chloro-, 4-chloro-, and 3,5-dichlorobenzoate by a pseudomonad. Appl. Environ. Microbiol. 37:421–428.

126. Hernandez, B. S., F. K. Higson, R. Kondrat, and D. D. Focht. 1991. Metabolism of and inhibition by chlorobenzoates in *Pseudomonas putida* P111. Appl. Environ. Microbiol. 57:3361–3366.

127. Sleeper, B. P. 1951. The bacterial oxidation of aromatic compounds. V. Metabolism of benzoic acids labelled with C^{14}. J. Bacteriol. 62:657–662.

128. Engesser, K. H., and P. Schulte. 1989. Degradation of 2-bromobenzoate, 2-chlorobenzoate and 2-fluorobenzoate by *Pseudomonas putida* CLB-250. FEMS Microbiol. Lett. 60: 143–148.

129. Fetzner, S., R. Müller, and F. Lingens. 1989. Degradation of 2-chlorobenzoate by *Pseudomonas cepacia* 2CBS. Biol. Chem. Hoppe-Seyler 370:1173–1182.

130. Sylvestre, M., K. Mailhiot, D. Ahmad, and R. Massé. 1989. Isolation and preliminary characterization of a 2-chlorobenzoate degrading *Pseudomonas*. Can. J. Microbiol. 35:439–443.

131. Zaitsev, G. M., and Y. N. Karasevich. 1985. Preparatory metabolism of 4-chlorobenzoic acids and 2,4-dichlorobenzoic acids in *Corynebacterium sepedonicum*. Mikrobiologiya 54: 356–359.

132. Fetzner, S., R. Müller, and F. Lingens. 1989. A novel metabolite in the microbial degradation of 2-chlorobenzoate. Biochem. Biophys. Res. Commun. 161:700–705.

133. Bartles, I., H.-J. Knackmuss, and W. Reineke. 1984. Suicide inactivation of catechol-2,3-dioxygenase from *Pseudomonas putida* mt-2 by 3-halocatechols. Appl. Environ. Microbiol. 47:500–505.

134. Reineke, W., and H.-J. Knackmuss. 1984. Microbial metabolism of haloaromatics: isolation and properties of a chlorobenzene-degrading bacterium. Appl. Environ. Microbiol. 47:395–402.

135. Dorn, E., and H.-J. Knackmuss. 1978. Chemical structure and biodegradability of halogenated aromatic compounds. Two catechol 1,2-dioxygenases from a 3-chlorobenzoate-grown pseudomonad. Biochem. J. 174:73–84.

136. Dorn, E., and H.-J. Knackmuss. 1978. Chemical structure and biodegradability of halogenated compounds. Substituent effects on 1,2-dioxygenation of catechol. Biochem. J. 174:85–94.

137. Schmidt, E., and H.-J. Knackmuss. 1980. Chemical structure and biodegradability of halogenated aromatic compounds. Conversion of chlorinated muconic acids into maleylacetic acid. Biochem. J. 192:339–347.

138. Stanier, R. Y., and L. N. Ornston. 1973. The β-keto-adipate pathway. Adv. Microbiol. Phys. 9:89–151.

139. Reineke, W. 1984. Microbial degradation of halogenated aromatic compounds, p. 319–360. *In* D. T. Gibson (ed.), Microbial degradation of organic compounds. Marcel Dekker, New York.

140. Tiedje, J. M., and M. Alexander. 1969. Enzymatic cleavage of the ether bond of 2,4-dichlorophenoxyacetate. J. Agric. Food Chem. 17:1080–1084.

141. Tiedje, J. M., J. M. Duxbury, M. Alexander, and J. E. Dawson. 1969. 2,4-D metabolism: pathway of degradation of chlorocatechols by *Arthrobacter* sp. J. Agric. Food Chem. 17:1021–1026.

142. Duxbury, J. M., J. M. Tiedje, M. Alexander, and J. E. Dawson. 1970. 2,4-D metabolism: enzymatic conversion of chloromaleylacetic acid to succinic acid. J. Agric. Food Chem. 18:1989–201.

143. Chapman, P. J. 1979. Degradation mechanisms, p. 28–66. *In* A. W. Bourquin and P. H. Pritchard (eds.), Microbial degradation of pollutants in marine environments. EPA-600/9-79-012, U.S. Environmental Protection Agency, Gulf Breeze, Fla.

144. Miguez, C. B., C. W. Greer, and J. M. Ingram. 1990. Degradation of monochlorobenzoic and dichlorobenzoic acid

isomers by two natural isolates of *Alcaligenes denitrificans*. Arch. Microbiol. 154:139–143.

145. Timmis, K. N., F. Rojo, and J. L. Ramos. 1988. Prospects for laboratory engineering of bacteria to degrade pollutants, p. 61–79. *In* G. S. Omenn (ed.), Environmental biotechnology: reducing risks from environmental chemicals through biotechnology. Plenum Publishing, New York.

146. Adriaens, P., and D. D. Focht. 1991. Cometabolism of 3,4-dichlorobenzoate by *Acinetobacter* sp. strain 4CB1. Appl. Environ. Microbiol. 57:173–179.

147. Adriaens, P., and D. D. Focht. 1991. Evidence for inhibitory substrate interactions of 3,4-dichlorobenzoate by *Acinetobacter* sp. strain 4CB1. FEMS Microbiol. Ecol. 85:293–300.

148. Gerritse, J., and J. C. Gottschal. 1992. Mineralization of the herbicide 2,3,6-trichlorobenzoic acid by a co-culture of anaerobic and aerobic bacteria. FEMS Microbiol. Ecol. 101:89–98.

149. Ghosal, D., I. S. You, D. K. Chatterjee, and A. M. Chakrabarty. 1985. Genes specifying degradation of 3-chlorobenzoic acid in plasmids pAC27 and pJP4. Proc. Natl. Acad. Sci. USA 82:1638–1642.

150. Chatterjee, D. K., S. T. Kellog, D. R. Watkins, and A. M. Chakrabarty. 1981. Plasmid specifying total degradation of 3-chlorobenzoate by a modified ortho-pathway. J. Bacteriol. 146:639–646.

151. Don, R. H., and J. M. Pemberton. 1981. Properties of six pesticide degradation plasmids isolated from *Alcaligenes paradoxus* and *Alcaligenes eutrophus*. J. Bacteriol. 145:681–686.

152. Don, R. H., A. J. Weightman, H.-J. Knackmuss, and K. M. Timmis. 1985. Transposon mutagenesis and cloning analysis of the pathways for degradation of 2,4-dichlorophenoxyacetic acid and 3-chlorobenzoate in *Alcaligenes eutrophus* JMP134(pJP4). J. Bacteriol. 161:85–90.

153. Frantz, B., K.-L. Ngai, D. K. Chatterjee, L. N. Ornston, and A. M. Chakrabarty. 1987. Nucleotide sequence and expression of *clcD*, a plasmid borne dienelacetone hydrolase gene from *Pseudomonas* sp. B13. J. Bacteriol. 169:704–709.

154. Ngai, K.-L., M. Scholmann, H.-J. Knackmuss, and L. N. Ornston. 1987. Dienelactone hydrolase from *Pseudomonas* sp. B13. J. Bacteriol. 169:699–703.

155. Weisshaar, M. P., F. C. H. Franklin, and W. Reineke. 1987. Molecular cloning and expression of the 3-chlorobenzoate–degrading genes from *Pseudomonas* sp. strain B13. J. Bacteriol. 169:394–402.

156. Weightman, A. J., R. H. Don, P. R. Lehrbach, and K. N. Timmis. 1984. The identification and cloning of genes coding haloaromatic catabolic enzymes and the construction of hybrid

pathways for substrate mineralization, p. 47–80. *In* G. S.
Omenn and A. Hollaender (eds.), Genetic control of environmental pollutants. Plenum Publishing, New York.

157. Timmis, K. N., P. R. Lehrbach, S. Harayama, R. H. Don,
N. Mermod, S. Bas, R. Leppik, A. J. Weightman, W. Reineke,
and H.-J. Knackmuss. 1985. Analysis and manipulation of
plasmid-coded pathways of the catabolism of aromatic compounds by soil bacteria, p. 719–739. *In* D. R. Helinski,
S. N. Cohen, D. B. Clewell, D. A. Jackson, and A. Hollaender (eds.), Plasmids in bacteria. Plenum Publishing,
New York.

158. Yen, K. M., and I. C. Gunsalus. 1982. Plasmid gene organization: naphthalene/salicylate oxidation. Proc. Natl.
Acad. Sci. USA 79:874–878.

159. Chatterjee, D. K., and A. M. Chakrabarty. 1984. Restriction mapping of chlorobenzoate degradative plasmid and
molecular cloning of the degradative genes. Gene 27:173–
181.

160. Wyndham, R. C., and N. A. Straus. 1988. Chlorobenzoate
catabolism and interactions between *Alcaligenes* and *Pseudomonas* species from Bloody Run Creek. Arch. Microbiol. 150:
230–236.

161. Ghosal, D., I.-S. You, D. K. Chatterjee, and A. M. Chakrabarty. 1985. Plasmids in the degradation of chlorinated
aromatic compounds, p. 667–686. *In* D. R. Helinski, S. N.
Cohen, D. B. Clewell, D. A. Jackson, and A. Hollaender
(eds.), Plasmids in bacteria. Plenum Publishing, New York.

162. Frantz, B., and A. M. Chakrabarty. 1987. Organization
and nucleotide sequence determination of a gene cluster involved in 3-chlorocatechol degradation. Proc. Natl. Acad.
Sci. USA 84:4460–4464.

163. Neidle, E. L., M. K. Shapiro, and L. N. Ornston. 1987.
Cloning and expression in *Escherichia coli* of *Acinetobacter
calcoaceticus* genes for benzoate degradation. J. Bacteriol.
169:5496–5503.

164. Furukawa, K., and A. M. Chakrabarty. 1982. Involvement
of plasmids in total degradation of chlorinated biphenyls.
Appl. Environ. Microbiol. 44:619–626.

165. Shields, M. S., S. W. Hooper, and G. S. Sayler. 1985.
Plasmid-mediated mineralization of 4-chlorobiphenyl. J.
Bacteriol. 163:882–889.

166. Parsons, J. R., D. T. H. M. Sijm, A. Van Laar, and O.
Hutzinger. 1988. Biodegradation of chlorinated biophenyls
and benzoic acids by a *Pseudomonas* strain. Appl. Microbiol.
Biotechnol. 29:81–84.

167. Bedard, D. L., and M. L. Haberl. 1990. Influence of chlorine substitution pattern on the degradation of polychlorinated

biphenyls by eight bacterial strains. Microb. Ecol. 20:87–102.

168. Focht, D. D. 1988. Performance of biodegradative microorganisms in soil: xenobiotic chemicals as unexploited metabolic niches, p. 15–29. *In* G. S. Omenn (ed.), Environmental biotechnology. Plenum Publishing, New York.

169. Reineke, W., and H.-J. Knackmuss. 1979. Construction of haloaromatic utilizing bacteria. Nature (London) 277:385–386.

170. Williams, P. A., and K. J. Murray. 1974. Metabolism of benzoate and the methylbenzoates by *Pseudomonas putida* (arvilla) MT-2: evidence for the existence of a TOL plasmid. J. Bacteriol. 120:416–423.

171. Chatterjee, D. K., and A. M. Chakrabarty. 1982. Genetic rearrangements in plasmids specifying total degradation of chlorinated benzoic acids. Mol. Gen. Genet. 188:279–285.

172. Hartmann, J., K. Engelberts, B. Nordhaus, E. Schmidt, and W. Reineke. 1989. Degradation of 2-chlorobenzoate by in vivo constructed hybrid pseudmonads. FEMS Microbiol. Lett. 61:17–21.

173. Schwein, U., and E. Schmidt. 1982. Improved degradation of monochlorophenols by a constructed strain. Appl. Environ. Microbiol. 44:33–39.

174. Kröckel, L., and D. D. Focht. 1987. Construction of chlorobenzene-utilizing recombinants by progenitive manifestation of a rare event. Appl. Environ. Microbiol. 53:2470–2475.

175. Mokross, H., E. Schmidt, and W. Reineke. 1990. Degradation of 3-chlorobiphenyl by in vivo constructed hybrid pseudomonads. FEMS Lett. 71:179–185.

176. Adams, R. H., C.-M. Huang, F. K. Higson, V. Brenner, and D. D. Focht. 1992. Construction of a 3-chlorobiphenyl–utilizing recombinant from an intergeneric mating. Appl. Environ. Microbiol. 58:647–654.

177. Adriaens, P., C.-M. Huang, and D. D. Focht. 1991. Biodegradation of PCBs by aerobic microorganisms, p. 311–326. *In* R. A. Baker (ed.), Organic substances and sediments in water, Vol. 1. Lewis Publishers, Chelsea, MI.

178. Havel, J., and W. Reineke. 1991. Total degradation of various chlorobiphenyls by cocultures and in vivo constructed hybrid pseudomonads. FEMS Microbiol. Lett. 78:163–170.

179. Sondossi, M., M. Sylvestre, and D. Ahmad. 1992. Effects of chlorobenzoate transformation on the *Pseudomonas testosteroni* biphenyl and chlorobiphenyl degradation pathway. Appl. Environ. Microbiol. 58:485–495.

180. Sylvestre, M., R. Massé, C. Ayotte, F. Messier, and J. Fauteux. 1985. Total biodegradation of 4-chlorobiphenyl

(4-CB) by a two-membered bacterial culture. Appl. Microbiol. Biotechnol. 21:192–195.

181. Haller, H. D., and R. K. Finn. 1979. Biodegradation of 3-chlorobenzoate and formation of black color in the presence and absence of benzoate. Eur. J. Appl. Microbiol. Biotechnol. 8:191–205.

182. Perry, J. J. 1984. Microbial metabolism of cyclic alkanes, p. 61–97. *In* R. M. Atlas (ed.), Petroleum microbiology. Macmillan, New York.

183. Beam, H. W., and J. J. Perry. 1973. Co-metabolism as a factor in microbial degradation of cycloparaffinic hydrocarbons. Arch. Microbiol. 91:87–90.

184. Beam, H. W., and J. J. Perry. 1974. Microbial degradation of cycloparaffinic hydrocarbons via co-metabolism and commensalism. J. Gen. Microbiol. 82:163–169.

185. deClerk, H., and A. C. van der Linden. 1974. Bacterial oxidation of cyclohexane: participation of a co-oxidation reaction. Antonie van Leeuwenhoek 40:7–15.

186. Adriaens, P., and D. D. Focht. 1990. Continuous coculture degradation of selected polychlorinated biphenyl congeners by *Acinetobcter* spp. in an aerobic reactor system. Environ. Sci. Technol. 24:1042–1049.

187. Fava, F., and L. Marchetti. 1991. Degradation and mineralization of 3-chlorobiphenyl by a mixed aerobic bacterial culture. Appl. Microbiol. Biotechnol. 36:240–245.

188. Pettigrew, C. A., A. Breen, C. Corcoran, and G. S. Sayler. 1990. Chlorinated biphenyl mineralization by individual populations and consortia of freshwater bacteria. Appl. Environ. Microbiol. 56:2036–2045.

189. Hooper, S. W., T. C. Dockendorff, and G. S. Sayler. 1989. Characteristics and restriction analysis of the 4-chlorobiphenyl catabolic plasmid pSS50. Appl. Environ. Microbiol. 55:1286–1288.

190. Layton, A. C., J. Sanseverino, W. Wallace, C. Corcoran, and G. S. Sayler. 1992. Evidence for 4-chlorobenzoic acid dehalogenation mediated by plasmids related to pSS50. Appl. Environ. Microbiol. 58:399–402.

191. Brunner, W., F. H. Sutherland, and D. D. Focht. 1985. Enhanced biodegradation of polychlorinated biphenyls in soil by analog enrichment. J. Environ. Qual. 14:324–328.

192. Focht, D. D., and W. Brunner. 1985. Kinetics of biphenyl and polychlorinated biphenyl metabolism in soil. Appl. Environ. Microbiol. 50:1058–1063.

193. Furukawa, K., F. Matsumura, and K. Tonomura. 1978. *Alcaligenes* and *Acinetobacter* strains capable of degrading polychlorinated biphenyls. Agric. Biol. Chem. 42:543–548.

194. Pertsova, R. N., F. Kunc, and L. A. Golavleta. 1984. Degradation of 3-chlorobenzoate in soil by pseudomonads carrying biodegradative plasmids. Folia Microbil. 29:242–247.

195. Focht, D. D., and D. Shelton. 1987. Growth kinetics of *Pseudomonas alcaligenes* C-O relative to inoculation and 3-chlorobenzoate metabolism in soil. Appl. Environ. Microbiol. 58:1846–1849.

196. Weinberg, S. R., and G. Stotzky. 1972. Conjugation and genetic recombination of *Escherichia coli* in soil. Soil Biol. Biochem. 4:171–180.

197. Stotzky, G. 1989. Gene transfer among bacteria in soil, p. 165–222. *In* S. B. Levy and R. V. Miller (eds.), Gene transfer in the environment. McGraw-Hill, New York.

198. Slater, J. H. 1985. Gene transfer in microbial communities, p. 89–98. *In* H. O. Halvorson, D. Pramer, and M. Rogul (eds.), Engineered organisms in the environment: scientific issues. American Society for Microbiology, Washington, D.C.

199. Reichardt, P. B., B. L. Chadwick, M. A. Cole, B. R. Robertson, and D. K. Button. 1981. Kinetic study of the biodegradation of biphenyl and its monochlorinated analogues by a mixed marine microbial community. Environ. Sci. Technol. 15:75–79.

200. Wong, P. T. S., and K. L. E. Kaiser. 1975. Bacterial degradation of polychlorinated biphenyls. II. Rate studies. Bull. Environ. Contamin. Toxicol. 13:249–255.

201. Paris, D. F., W. C. Steen, G. L. Baughman, and J. T. Barnett, Jr. 1981. Second-order model to predict microbial degradation of organic compounds in natural water. Appl. Environ. Microbiol. 41:603–609.

202. Tulp, M. T. M., R. Schmitz, and O. Hutzinger. 1980. The bacterial metabolism of 4,4'-dichlorobiphenyl, and its supression by alternative carbon sources. Chemosphere 1:103–108.

203. Bailery, R. E., S. J. Gonsior, and W. L. Rhinehart. 1983. Biodegradation of the monochlorobiphenyls and biphenyl in river water. Environ. Sci. Technol. 17:617–621.

204. Fulthorpe, R. R., and R. C. Wyndham. 1981. Transfer and expression of the catabolic plasmic pBRC60 in wild bacterial recipients in a freshwater ecosystem. Appl. Environ. Microbiol. 57:1546–1553.

205. Fulthorpe, R. R., and R. C. Wyndham. 1992. Involvement of a chlorobenzoate-catabolic transposon, Tn5271, in community adaptation to chlorobiphenyl, chloroaniline, and 2,4-dichlorophenoxyacetic acid in a freshwater ecosystem. Appl. Environ. Microbiol. 58:314–325.

206. Fries, F. G. 1971. Degradation of chlorinated hydrocarbons under anaerobic conditions. Adv. Chem. Ser. 111:256–270.
207. Brown, J. F. Jr., R. E. Wagner, D. L. Bedard, M. J. Brennan, J. C. Carnahan, R. J. May, and T. Toffle. 1984. PCB transformations in upper Hudson sediments. Northeast Environ. Sci. 3:167–179.
208. Brown, J. F., Jr., R. E. Wagner, H. Feng, D. L. Bedard, M. J. Brennan, J. C. Carnahan, and R. J. May. 1987. Environmental dechlorination of PCBs. Environ. Toxicol. Chem. 6:579–593.
209. Brown, J. F., D. L. Bedard, M. J. Brennan, J. C. Carnahan, H. Feng, and R. E. Wagner. 1989. Polychlorinated biphenyl declorination in aquatic sediments. Science 236:709–712.
210. Morris, P. J., W. W. Mohn, J. F. Quenson III, J. M. Tiedje, and S. A. Boyd. 1992. Establishment of polychlorinated biphenyl-degrading enrichment culture with predominantly meta dechlorination. Appl. Environ. Microbiol. 58:3088–3094.
211. Nies, L., and T. M. Vogel. 1990. Effects of organic substrates on dechlorination of Aroclor 1242 in anaerobic sediment. Appl. Environ. Microbiol. 56:2612–2617.
212. Nies, L., and T. M. Vogel. 1991. Identification of the proton source for the microbial reductive dechlorination of 2,3,4,5,6-pentachlorobiphenyl. Appl. Environ. Microbiol. 57:2771–2774.
213. Van Dort, H. M., and D. L. Bedard. 1991. Reductive ortho dechlorination and meta dechlorination of a polychlorinated biphenyl congener by anaerobic microorganisms. Appl. Environ. Microbiol. 57:1576–1578.
214. Quensen, J. F. III, J. M. Tiedje, and S. A. Boyd. 1988. Reductive dechlorination of polychlorinated biphenyls by anerobic microorganisms from sediments. Science 242:752–754.
215. Quensen, J. F., S. A. Boyd, and J. M. Tiedje. 1990. Dechlorination of 4 commercial polychlorinated biphenyl mixtures (Aroclors) by anaerobic microorganisms from sediments. Appl. Environ. Microbiol. 56:2360–2369.
216. Ye, D. Y., J. F. Quensen, J. M. Tiedje, and S. A. Boyd. 1992. Anaerobic dechlorination of polychlorobiphenyls (Aroclor 1242) by pasteurized and ethanol-treated microorganisms from sediments. Appl. Environ. Microbiol. 58:1110–1114.
217. Assaf-Anid, N., L. Nies, and T. M. Vogel. 1992. Reductive dechlorination of a polychlorinated biphenyl congener and hexachlorobenzene by vitamin-B12. Appl. Environ. Microbiol. 58:1057–1060.
218. Rhee, G. Y., B. Bush, M. P. Brown, M. Kane, and L. Shane. 1989. Anaerobic biodegradation of polychlorinated biphenyls in Hudson River sediments and dredged sediments in clay encapsulation. Water Res. 23:957–964.

219. Chen, M., C. S. Hong, B. Bush, and G. Y. Rhee. 1988. Anaerobic biodegradation of polychlorinated biphenyls by bacteria from Hudson River sediments. Ecotoxicol. Environ. Saf. 16:95–105.

220. Griffith, G. D., J. R. Cole, J. F. Quensen, and J. M. Tiedje. 1992. Specific deuteration of dichlorobenzoate during reductive dehalogenation by *Desulfomonile tiedjei* in D_2O. Appl. Environ. Microbiol. 58:409–411.

221. Abramowicz, D. A. 1990. Aerobic and anaerobic biodegradation of PCBs: a review. Crit. Rev. Biotechnol. 10:241–249.

222. Barker, P. S., and F. O. Morrison. 1964. Breakdown of DDT to DDD in mouse tissue. Can. J. Zool. 42:324–325.

223. Barker, P. S., F. O. Morrison, and R. S. Whitaker. 1965. Conversion of DDT to DDD by *Proteus vulgaris*, a bacterium isolated from the intestinal flora of a mouse. Nature (London) 205:621–622.

224. Kalman, B. J., and A. K. Andrews. 1963. Reductive dehydrochlorination of DDT to DDD by yeast. Science 141: 1050–1051.

225. Anderson, J. P. E., E. P. Lichtenstein, and W. F. Wittingham. 1970. Effect of *Mucor alternans* on the persistance of DDT in culture and in soil. J. Econ. Entomol. 11:63: 1595–1599.

226. Matsumura, F., K. C. Patil, and G. M. Boush. 1971. DDT metabolized by microorganisms from Lake Michigan. Nature (London) 230:325–326.

227. Matsumura, F., and G. M. Boush. 1968. Degradation of insecticides by a soil fungus, *Trichoderma viride*. J. Econ. Entomol. 61:610–612.

228. Kuches, A. I., and D. C. Church. 1971. ^{14}C-DDT metabolism by rumen bacteria and protozoa in vitro. J. Dairy Sci. 54:540–543.

229. Patil, K. C., F. Matsumura, and G. M. Boush. 1972. Metabolic transformation of DDT, dieldrin, aldrin and endrin by marine microorganisms. Environ. Sci. Technol. 6:629–632.

230. Guenzi, W. D., and W. E. Beard. 1967. Anaerobic biodegradation of DDT to DDD in soil. Science 156:1116–1117.

231. Guenzi, W. D., and W. E. Beard. 1968. Anaerobic conversion of DDT to DDD and aerobic stability of DDT in soil. Proc. Soil Sci. Soc. Am. 32:522–524.

232. Wedemeyer, G. 1967. Dechlorination of 1,1,1-trichloro-2,2-bis(p-chlorophenyl)ethane by *Aerobacter aerogenes*. Appl. Environ. Microbiol. 15:569–574.

233. Wedemeyer, G. 1967. Biodegradation of dichlorodiphenyltrichloroethane: intermediates in dichlorodiphenylacetic acid metabolism by *Aerobacter aerogenes*. Appl. Microbiol. 15: 1494–1495.

234. Plapp, F. W., Jr., G. A. Chapman, and J. W. Morgan. 1965. DDT resistance in *Culex tarsalis* coquillett: cross resistance to related compounds and metabolic fate of a C^{14}-labelled DDT analog. J. Econ. Entomol. 58:1064–1069.

235. Pfaender, F. K., and M. Alexander. 1972. Extensive microbial degradation of DDT in vitro and DDT metabolism by natural communities. J. Agric. Food Chem. 20:842–846.

236. Markus, A., U. Klages, S. Krauss, and F. Lingens. 1984. Oxidation and dehalogenation by a component enzyme system from *Pseudomonas* sp. strain CBS3. J. Bacteriol. 160:618–621.

237. Subba-Rao, and M. Alexander. 1977. Cometabolism of 1,1,1-trichloro-2,2-bis(*p*-chlorophenyl)ethane (DDT) by *Pseudomonas putida.*

238. Beunink, J., and H. J. Rehm. 1988. Synchronous anaerobic and aerobic degradation of DDT by an immobilized mixed culture system. Appl. Microbiol. Biotechnol. 29:72–80.

239. Fathepure, B. Z., and T. M. Vogel. 1991. Complete degradation of polychlorinated hydrocarbons by a 2-stage biofilm reactor. Appl. Environ. Microbiol. 57:3418–3422.

240. Anid, P. J., L. Nies, and T. M. Vogel. 1991. Sequential anaerobic-aerobic biodegradation of PCBs in the river model, p. 428–436. *In* R. E. Hinchee and R. F. Olfenbuttel (eds.), On site bioreclamation: processes for xenobiotic and hydrocarbon treatment. Butterworth-Heinemann, Boston.

241. Pritchard, P. H., and C. F. Costa. 1991. EPA's Alaska oil spill bioremediation project. Environ. Sci. Technol. 25:372–379.

242. Häggblom, M. M. 1992. Microbial breakdown of halogenated aromatic pesticides and related products. FEMS Microbiol. Rev. 63:277–300.

243. Unterman, R. 1991. What is the K_m for disappearase? p. 159–162. *In* G. S. Sayler, R. Fox, and J. W. Blackburn (eds.), Environmental biotechnology for waste treatment. Plenum Publishing, New York.

244 Unterman, R., D. L. Bedard, M. J. Brennan, L. H. Bopp, F. J. Mondello, R. E. Brooks, D. P. Mobley, J. B. McDermott, C. C. Schwartz, and D. K. Dietrich. 1988. Biological approaches for polychlorinated biphenyl degradation, p. 253–269. *In* G. S. Omenn (ed.), Environmental biotechnology: reducing risks from environmental chemicals through biotechnology. Plenum Publishing, New York.

245. Herbst, E., I. Sheunert, W. Klein, and F. Korte. 1977. Contributions to ecological chemistry. CXXXVIII. Fate of PCBs-^{14}C in sewage treatment: laboratory experiments with activated sludge. Chemosphere 6:725–730.

246. Viney, I., and R. J. F. Bewley. 1990. Preliminary studies on the development of a microbiological treatment for poly-chlorinated biphenyls. Arch. Environ. Contam. Toxicol. 19: 789–796.
247. Focht, D. D. 1987. Ecological and evolutionary considerations on the metabolism of xenobiotic chemicals in soil, p. 157–168. *In* Future developments in soil science research. American Society of Agronomy, Madison, WI.
248. Sander, P., R. M. Wittich, P. Fortnagel, H. Wilkes, and W. Francke. 1991. Degradation of 1,2,4-trichloro- and 1,2,4,5-tetrachlorobenzene by *Pseudomonas* strains. Appl. Environ. Microbiol. 57:1430–1440.

Index